Acclaim for Liza Mundy's

EVERYTHING CONCEIVABLE

"Mundy embed[s] the facts within the human stories. . . . This remarkable work provokes a spectrum of emotions ranging from alarm to wonder."
                                                    —*San Francisco Chronicle*

"Mundy writes accessibly about science, but the human dimension is at least as prominent in her work. . . . Mundy bring[s] to bear impressive reporting skills and a sharp analytical mind. With empathy and wit, she illuminates the ironies and absurdities, the tragedies and dilemmas, but also the joys, of assisted reproduction."
                                                    —*The Nation*

"Engaging. . . . A panoramic and unsettling view. . . . Mundy digs deep into [her subjects'] lives."
                                                    —*Chicago Tribune*

"Mundy tells [these] stories with insight and sensitivity. . . . These are tales of miracles, but miracles with a dark side. They throw up quandaries our ancestors never dreamed of, and answers that raise yet more questions."
                                                    —*The Economist*

"Mundy is not a doom-and-gloom sensationalist. . . . Her book offers an important chronicle not only of the existing technology, but also of the unanswered questions about the short- and long-term implications."
                                                    —*Slate*

"Mundy does an excellent job of rendering a lot of very technical information accessible. . . . What makes the book compelling are the people she writes about. . . . Such details humanize the ethical questions."
                                                    —*Austin-American Statesman*

LIZA MUNDY

EVERYTHING CONCEIVABLE

Liza Mundy received her A.B. degree from Princeton University and an M.A. at the University of Virginia. She is a feature writer at *The Washington Post Magazine*, and her work was selected by Oliver Sacks for inclusion in *The Best American Science Writing 2003*. She has won awards from the Sunday Magazine Editors Association, among others. She lives in Arlington, Virginia, with her husband and two children.

EVERYTHING CONCEIVABLE

HOW ASSISTED REPRODUCTION IS CHANGING OUR WORLD

LIZA MUNDY

ANCHOR BOOKS
A DIVISION OF RANDOM HOUSE, INC.
NEW YORK

To my parents
And to Mark, Anna, and Robin
With love and gratitude

FIRST ANCHOR BOOKS EDITION, MAY 2008

Copyright © 2007, 2008 by Liza Mundy

All rights reserved. Published in the United States by Anchor Books, a division of Random House, Inc., New York, and in Canada by Random House of Canada Limited, Toronto. Originally published in hardcover in slightly different form in the United States by Alfred A. Knopf, a division of Random House, Inc., New York, in 2007.

Anchor Books and colophon are registered trademarks of Random House, Inc.

The Library of Congress has cataloged the Knopf edition as follows:
Mundy, Liza, [date].
Everything conceivable : how assisted reproduction is changing our
world / Liza Mundy.—1st ed.
p. cm.
Includes bibliographical references (p.).
1. Human reproductive technology—Social aspects—United States.   2. Human
reproductive technology—Economic aspects—United States.   I. Title.
RG133.5.M86 2007
362.196'69206—dc22
2006051432

Anchor ISBN: 978-1-4000-9537-7

Author photograph © Claudio Vazquez

www.anchorbooks.com

Printed in the United States of America
10   9   8   7   6   5   4   3   2   1

AUTHOR'S NOTE

Of the hundreds of individuals interviewed for this book, a small number wished to preserve their anonymity because of the private nature of their reproductive decision-making. In most cases, this was achieved by omitting a name. In a few instances where people's narratives are presented at length, names have been changed. The names of the Ramirez family members are pseudonyms. In this case, the parents were happy to discuss childbearing issues openly, but because of the nature of the mother's police work, she preferred that her name and those of her children not be used. Melanie, in chapter two, and Jane and Emma, in chapter eleven, are pseudonyms, for reasons explained in the text. Every other name, and all events described in the book, are real.

CONTENTS

AN UNEXPECTED
DEVELOPMENT

1

One day in early May of 2005, the Reverend Beth Parab—the personable, sandy-haired young associate rector of the Episcopal Church of St. Matthew in San Mateo, California—sat down at her desk and opened her e-mail. As usual, the morning's mail brought messages from parishioners with various questions and suggestions regarding church business, and Reverend Beth proceeded to reply to them, scrolling down the list until she came to a message that was not usual at all.

That message came from Laura Ramirez, a local police officer whose children Reverend Beth was scheduled to baptize on May 22, after the regular Sunday morning service. The children were triplet boys. Their names were Preston, Edward, and Hunter Ramirez, and they would be exactly six months old on their baptism day. The twenty-second happened to be Trinity Sunday—the Sunday after Pentecost—and in thinking about the prayer she would write for the baptism, Reverend Beth had already started toying with the happy juxtaposition of "trinity" and "triplets." This baptism would include six godparents—one pair for each child—and of course the parents, Laura and her husband, Hector Ramirez, as well as family members and friends. Which was fine; there was plenty of room in the nave of St. Matthew's, a Gothic church built in the 1860s and expanded over the years to comprise several graceful stone outbuildings, a leafy courtyard, and a splendid Nativity window overlooking the baptismal font.

Now Laura Ramirez was writing to ask whether Reverend Beth could find a way to incorporate another participant in the baptism ceremony. That would be the Ramirez boys' egg donor, a young woman who had in a sense engendered the occasion.

The egg donor, whose name was Kendra Vanderipe, lived in a suburb of Denver, Colorado. Not so long ago she had been a stranger to the Ramirez family. Laura Ramirez had located her over the Internet, with the help of a commercial, Web-based agency that procures egg donors for infertile couples. The Ramirez boys had been conceived by in vitro fertilization, the scientific procedure in which sperm and egg are brought together in a petri dish, and the resulting human embryos are grown for several days in the lab, and then transferred into the womb of the mother. In this case, the procedure had been carried out using the eggs of Kendra, the donor. The embryos had been gestated, and delivered, by Laura, their mother. After the birth of the boys a warm relationship had developed, unexpectedly, between the two women. At Laura's invitation Kendra had agreed to attend the baptism. It would be the first time she had seen the children conceived with her help. Laura wondered if Reverend Beth could find a way, in the service, to thank Kendra for all she had given her.

Reverend Beth had never written an egg donor into a baptismal ceremony. She didn't know any clergy member who had. But she was glad to try. She did not know Laura and Hector well, but everything about the Ramirezes suggested that they were a committed and deeply likable couple. One of the first contacts Reverend Beth had had with them was when Laura (an energetic multitasker who does a lot of e-mailing from her squad car) had e-mailed, several months back, to ask whether the church knew of any families who might welcome some secondhand newborn clothing, of which the Ramirezes, with three growing infants, had even more than most first-time parents. Reverend Beth also knew that Laura and Hector had tried hard to have children naturally. Only after a devastating series of miscarriages had the couple turned to fertility medicine, and a willing stranger, for help in doing what they could not do by themselves.

Recently married, Reverend Beth, who was in her early thirties, was looking forward to starting a family of her own. She felt acute empathy for anybody who wanted children and could not have them easily. Just three years out of seminary, she had already counseled a surprising number: not just women but men, and not just people who were actively trying but older couples who had confided to her their permanent sense of loss at never having had the child they longed for. Reverend Beth knew that suffering is part of human existence. She got the situation, theologically. Even so, infertility had always seemed to her an especially hard form of suffering to inflict. The fact that there are people in the world who can have children easily and do not want them, and people who want children ardently but cannot have them, seemed to her a kind of terrible cosmic joke, "one of these things I'm just going to have words with God about when I get there." Unlike some Christian leaders, she had no knee-jerk antipathy toward technologies that can help people have the families that will give their lives pleasure and meaning. It did not seem to her unnatural, or wrong, to make a family with the help of a third party. It did not seem to her unnatural, or wrong, to be the third party who helped.

But what about the baptism itself? How, logistically, to incorporate an egg donor into a ceremony that dates back more than two millennia, to the days when St. John the Baptist baptized Christ himself in the waters of the river Jordan? It's true that baptism symbolizes the creation of a new and deliberately constructed family, a family based on religious faith rather than on blood or genetic connection. True, too, that something similar was going on here. Love and parental commitment were understood in the Ramirez family as more important than any direct genetic bond. And yet genetics clearly mattered. So what exactly did an egg donor stand for? Where should an egg donor stand? The egg donor was not the triplets' mother. She wasn't a birth mother, as in a conventional adoption. She was a new entity and at the same time a very old one: a progenitor in the purest, most stripped-down sense of the word. It was Kendra Vanderipe's genetic history that was preserved in the chromosomes of Preston, Edward, and Hunter Ramirez, her unique history that came together with the unique genetic line of Hector Ramirez, a man unknown to her at the time of the boys' conception, and barely known to her now. Two strangers, two gene lines that had evolved over two human migratory paths; now intertwined in the famous double helix to form a unique gene line—three unique gene lines—that would continue into the future. Part of Kendra Vanderipe lived in the bodies of the Ramirez children, and would live in their children's children, and so on, perhaps, until the end of time.

There were no precedents to help Reverend Beth in writing a prayer for this ceremony. St. Augustine had not confronted the role of the egg donor in the formation of the human family, nor had St. Paul, nor Archbishop Tutu, nor any other of the great Christian preachers to whom, in other instances, a newly minted priest might look for inspiration. Homiletics class had little to offer. Never before in history has it been possible for a woman to give birth to an infant who is genetically unrelated to her. Never before has it been possible for a woman to be the genetic parent of living children to whom she has not given birth. Never—it seemed safe to assume—had two such women stood together in the nave of St. Matthew's to celebrate the rather resounding results of their joint efforts. Authorially and theologically, Reverend Beth was on her own.

Reverend Beth hit Reply. In her response to Laura Ramirez, she said that she would be glad to incorporate an egg donor into the baptism ceremony. Then—caught up, like millions of others, in a wholly unprecedented chapter of human history—the Reverend Beth Parab sat at her desk and started to write.

2

Several years ago, as a reporter for the *Washington Post,* I was asked by a socially aware editor to write an article on infertility as experienced by the poor: people who would like to have a child; physically cannot, for any number of reasons; and cannot afford the high-tech solutions available to the better-off and/or well insured. It seemed a rich and promising topic, new and yet inevitable. I had always been drawn to the topic of families and children, and had written about reproductive issues and the conflict they inspire in American culture. Much has been said during these conflicts about the provision of abortion services and the lack of care for the poor. But what about help *bearing* children? Were there many poor people who needed it?

Researching that article, I discovered to my surprise that the less money you have, the more likely you are to have difficulty conceiving. Much infertility has always been caused by infections that can damage reproductive passageways; the lower your tax bracket, the less likely you are to have received the fairly simple medical treatment that can stave off these consequences. Reproductive rights groups do little to help people in this situation, yet neither do those who oppose reproductive rights groups and what they stand for. Interviewing for that article, I encountered women who were trying to belly-dance their way to a baby, because an abdominal workout was all the treatment they could afford; fertility doctors quietly discounting procedures for determined, impecunious patients; blue-collar workers who had sold belongings and maxed out credit cards to pay for a single round of IVF; women who mailed leftover fertility drugs to reduce the cost of treatment for an acquaintance they had met on a chat room.

After the article ran, reader reaction was deep and deeply split. A number of people sent checks, or even a few dollars, to assist the couples named. Others sent scathing letters to the effect that a person who couldn't afford fertility treatment didn't deserve a child.

The poor, concluded some of the letter writers, bred quite enough, thank you, and certainly didn't need help.

This was my introduction to the fact that the spectacle of someone trying to have a child can be even more inflammatory than the spectacle of someone trying not to have one.

At the same time I was realizing how common the same struggle was becoming among my own co-workers and friends. An editor I worked with had been diagnosed with infertility, as had another close colleague, who, without telling anyone in our office, had undergone three cycles of failed IVF. So, I realized, thinking about it, had an editor I worked with earlier: a man, who had confided his and his wife's arduous experience with treatment. So had a male writer I knew. After the article ran, I had lunch with a lovely, vibrant reporter for another news organization who was pregnant with IVF twins. She had sent a generous check to one of the couples named in my article, saying only that she had some idea what they were going through. During our lunch this woman talked about how taboo infertility is in her own African American community; how it angered her when people, white or black, assumed that she herself was infertile because she had "waited" too long to conceive.

Waited! As if the arrival of a family is something you can snap your fingers and make happen! As if you can choose when you meet the man you want to spend your life with—the man who wants to spend his life with you. As if you can choose the moment when the man in your life decides *he's* ready for the family you have been wanting all along. As if your boss is going to be out there cheering you on: Oh yeah, go girl, have that baby! "Life isn't a restaurant where things are served in courses," my lunch partner reflected, with feeling. "Life is more like a deli, where they're throwing plates out and you just have to catch them when they fly out of the kitchen." Over lunch, this woman mentioned that she and her husband were now facing the question of what to do with their frozen embryos. It was far harder than she had expected. At the outset they had assumed they would donate any surplus embryos to another infertile couple, but now that she was pregnant with actual babies, it seemed unthinkable to give those children's potential siblings away. It was much, much harder to know how to think about early human life and morally what was the right thing to do with it.

At this point it became clear that she and I, who came of age at approximately the same time, were talking and thinking about issues unanticipated in the late 1970s and early 1980s, when both of us were in college and coming into social and political consciousness. Back then, forming an opinion on reproductive rights meant forming an opinion on abortion. Back then, how you felt about "choice" meant how you felt about a woman's right to choose the outcome of a pregnancy she didn't feel she could bring to term. Abruptly, the scope of these terms had gotten larger, the terrain more confusing and unmapped. What does "choice" mean, given the range of medical and scientific procedures that are rather suddenly possible? If you support reproductive freedom, do you support everything now offered in the reproductive marketplace? Thawing frozen embryos and letting them lapse? Donating eggs to science? Selecting the sex of your baby? Conceiving triplets, and "reducing" the pregnancy to twins?

If you oppose abortion, does this mean you oppose a science that has given life to millions of children, while keeping hundreds of thousands of human embryos in cold storage?

These are not questions that just affect individuals in private, anguished conversations, though they are affecting lots of individuals I know. In past few years the nation as a whole has begun to engage in a prolonged debate about the nature and disposition of the lab-made human embryo, a wondrous, problematic entity that until recently did not exist. One of President George W. Bush's first domestic policy decisions after he took office in January 2001 was whether to provide federal funding for new stem-cell research using some of this country's half-million surplus IVF embryos, with the hope that such research might benefit people suffering from afflictions including Parkinson's disease and juvenile diabetes. Bush chose not to approve the funding. In 2006 the issue re-emerged when the Republican-controlled U.S. Congress voted in favor of funding this research; in response, Bush issued his first presidential veto, saying that "each of these human embryos is a unique human life." Clearly, even as pro-choice sensibilities are being affected by new technologies, pro-lifers are undergoing a wrenching realignment. Utah's Senator Orrin Hatch, a pro-life Republican, is in favor of stem-cell research using IVF embryos. So is former first lady Nancy Reagan.

Talking to so many people who had been touched by the newest reproductive technologies, I felt there was a real and even urgent need for a book exploring the impact that reproductive science is having on our ideas about human life, the changes that reproductive science is imposing on families, medicine, culture, schools, the women's movement, the human race itself. Clearly, that impact is being felt everywhere. In the greater Washington, D.C., area, where I live, there are at least seventeen fertility clinics, all of them intensely competitive. Their area of most rapid expansion is not the central city but outlying areas like Loudoun County, Virginia, and Hagerstown, Maryland, where middle-class Americans are pouring into clinics that are striving to make treatment accessible, offering easy payment plans, financing schemes, and money-back guarantees. It isn't just the poor and the well-heeled who need this technology: it's all kinds of people, everywhere. In Johnson City, Tennessee, a doctor named Sam Thatcher has brought the cost of a complete round of IVF down to $7,000, a reasonably affordable price for someone who's determined. Janice Grimes, a nurse who has written books for children conceived with donor egg and donor sperm, works in a clinic in Iowa, where she sees parents struggling to explain to children the complex truth of their origins. Wendy Kramer, who founded an Internet group whereby children of sperm donors can try to make contact with progenitors as well as half siblings, lives in Colorado. People everywhere want to have children; people everywhere are dealing with the consequences of the technology that enables them to do so.

As I got further into the reporting, the dilemmas were more complex, the people facing them more diverse, than I had anticipated. I interviewed married couples, single mothers, lesbian partners, gay fathers. I interviewed parents of twins, parents of triplets, parents who had twins and triplets both. I interviewed egg donors who had contributed the genetic material to make a baby possible for another person. I interviewed surrogates or, as they're now called, "gestational carriers": women who gestate a child for another person, a child who is sometimes biologically related to the surrogate herself, but often not related to her, biologically, at all. Very quickly the categories converged. Very quickly I was interviewing single mothers of twins; straight women who had decided, in the absence of marriageable men, to "co-parent" together;

women who had used egg donation and gone on to donate surplus embryos, becoming birth mothers—or maybe adoption agencies—for potential children to whom they were not genetically related.

As with abortion technologies, assisted reproduction is being used by people with a diversity of outlooks, backgrounds, and moral codes. In most cases, assisted reproduction is being used by women and men who want a baby, can't have one without help, and are willing to surmount the highest barriers to attain what for many people comes effortlessly. In a few cases, the technology is being used by, and enabling, people who may not fully understand the time and commitment required to raise children. Then again, who does, starting out? All of these people, at some point, find themselves facing wrenching decisions for which nothing—*nothing*—has prepared them.

The cliché used to describe fertility medicine is the "Wild West." The impression many have is that the field is populated by rogue scientists and multimillionaire doctors willing to stuff pretty much anything into a woman's expensively prepped womb. There is some truth to this. IVF is a high-paying medical specialty, one in which doctors can invoke "reproductive liberty" to justify performing any procedure a patient wants and can pay for, and to make a lot of money standing on this ideological high ground. The field attracts many deeply dedicated doctors and gifted scientists, but their skill and commitment are sometimes eclipsed by a small group of opportunists, charlatans, embryo thieves, self-professed baby cloners, and the unforgivably reckless. It is usually possible to tell the difference between the wild-eyed and the dedicated, though not always: this is a field where the borders of "normal" are fluid, and a willingness to try extreme solutions, and keep your fingers crossed about the outcome, seems to be part of the job requirement. Some of the most prestigious practitioners have, from time to time, engaged in practices that led to their experiments being curtailed by federal authorities, in rare bursts of governmental oversight.

Many doctors and scientists were kind enough to give prolonged interviews for this book, in most cases because they find their work important, and hard, and misunderstood. Many patients also shared their stories, in the hope that their narratives might educate others about what it feels like to be the only one at a baby shower who can't have a baby. For many, I think, it was a relief to talk. Fertility treatment is excruciatingly private, and often isolating, and when it's collective it tends to be competitive. One woman spoke of taking the hospital elevator to early-morning blood checks and ultrasound appointments, riding up with women who would dash out as soon as the doors opened, clattering down the hall in Ferragamo heels, determined to be the first in line for the ultrasound. Roughly one-third of IVF procedures work and two-thirds don't, and nobody forgets that, looking around the waiting room.

"You sit there and go, 'It's going to work for one of the three of us,'" said this woman. "What are the odds? Is it not going to be me, or is it not going to be you?"

Abortion clinics can be unhappy places, but there is a sense of shared unhappiness. People are there for the same reason and the outcome in each case is going to be the same. In fertility clinic waiting rooms, there is no solidarity; people tend to build invisible walls around themselves, difficult as that may be in close quarters. At one point I was sitting in the waiting room of a clinic in suburban Philadelphia, where a couple was huddled together filling out one of many introductory forms. "I don't know; do we?" whispered wife to husband, as they faced a number of issues they would have to reach marital consensus on. Elsewhere, women were sitting by themselves, looking withdrawn. Elsewhere, men were sitting while their partners were off undergoing procedures; these men tended to wear the unnerved expression men adopt in lingerie stores and other places that bring them into close contact with the mysteries of the female anatomy. At one point a woman brought in a strawberry blond toddler, announcing to the delighted staff that treatment had worked for her and she was back to try for a sibling. This visit was unusual: many clinics forbid babies in the waiting room, so as not to make the contrast between those who have children and those who want them quite so stark and merciless.

At this clinic, a man was sitting by himself, wearing jeans and a button-down shirt, not looking guarded or uncomfortable at all. He was sitting casually with his legs crossed, speaking into a cell phone, wearing that distant look people get when the person they're talking to is at the other end of a wireless connection. "I'm in a fertility clinic!" he was saying, loudly. "A fertility clinic! Windows XP! Don't get the full version! I can get you the student version! About $50. I can get it for you this afternoon, after this appointment."

Listening to him—there was no choice—I reflected that assisted reproduction has inserted itself into the loop of technology we use every day. The cell phone, the latest Windows software, the fact of being in a fertility clinic: to this man they were different technologies, no more, no less, nothing to hide, all ways to efficiently get the thing you want, be it better computing power, or a quick conversation, or a child. Assisted reproduction is like so many technologies in that it makes certain situations possible that never were possible before and it suffers from unforeseen glitches; it sometimes delivers the desired outcome faster, and in greater number, than a person can handle. It solves problems, and creates them. This is the way we have babies now, many of us, and it's not going to get easier, or simpler, and the world—the species—is never going to be the same.

EVERYTHING CONCEIVABLE

CHAPTER ONE

THE NEW REPRODUCTIVE LANDSCAPE

"Eye Hoop They All Have Babies"

Every industrial convention has its own eccentric flavor, and the 2005 gathering of the American Society for Reproductive Medicine was no exception. That year the annual meeting of American fertility doctors was held in conjunction with the annual meeting of Canadian fertility doctors; the massive conference, which took place in Montreal over five days in October, was attended by emissaries from North America as well as from England, France, Europe, Japan, China, Africa, India, Asia, Israel: anywhere that humans live and wish, as humans usually do, to be fruitful and multiply. So numerous were the babymakers that airport immigration was bogged down and the city's downtown was transformed; the hospitality rooms of the Fairmont Queen Elizabeth were booked for events like "Cocktails with the Middle East Fertility Society." Converging on the downtown convention center, reproductive endocrinologists, embryologists, andrologists, urologists, therapists, and psychologists attended courses in packed seminar rooms. But the real action was in the cavernous exhibition hall, where an array of twenty-first century conception technology was on display, rivaling anything unveiled by the military-industrial complex.

At the entrance to the hall, unavoidable to all who entered, was a booth maintained by Scandinavian Cryobank, a subsidiary of Cryos, one of the world's largest sperm banks. As one might expect, Scandinavian Cryobank specializes in Scandinavian sperm donors: specifically Danish donors enrolled in graduate programs at "major Scandinavian universities," men so mentally and physically superior that they passed "some of the most exacting genetic testing in the industry." Deliberately recalling another era when northern European men inflicted their genes on women of other nations, sales staff were distributing wry little buttons announcing "Congratulations! It's a Viking!" underneath which was a photo of a very blond, very sturdy-looking baby. A banner advertisement noted that the company caters to gay and straight, black and white, male and female. Under the happy we-are-the-world tableau of patients, it added that it serves patients "as energetically as our ancestors once grabbed countries."

Not far away, one of the other principal players in the realm of international genetic redistribution, Los Angeles–based California Cryobank, was advertising its sperm bank by means of an indoor hockey game. It was not clear what hockey was supposed to symbolize. Maybe it was an homage to Canada. Maybe it was supposed to underscore the importance, in this crowd, of being deft and competent enough to shoot a small, frenetically moving object into a stationary target. No matter: setting down the espressos and Belgian chocolates that were being freely dispensed, the medical men and women lined up to whack away at the puck, cheering whenever a colleague, you know, scored.

Nearby, Cryogenic Laboratories was hoping to edge out this competition by offering a service called Lifetime Photos. For a price, clients can obtain photos of a sperm donor, from infancy to adulthood, and thereby see how their child's own appearance might unfold if they select that donor's genetic product to conceive their baby.

The conference was dominated and underwritten by the pharmaceutical industry. Standing everywhere were cheerful representatives from Ferring Pharmaceuticals, Organon USA, Serono Inc., Wyeth Pharmaceuticals, and others, who together do an estimated $3 billion a year business selling the drugs and medical devices that are an integral part of childbearing through assisted reproduction technology (ART). By now, ART comprises a spectrum of procedures of varying levels of sophistication. They include the fertility drugs that control and stimulate ovaries to produce more eggs; artificial insemination, or the injection of washed and treated sperm directly into a woman's cervix or uterus; in vitro fertilization, the more high-tech laboratory procedure in which sperm and egg are removed from the body and brought together in a culture dish; and a host of speedily developing related technologies such as genetic testing of embryos.

There were booths operated by the companies that make products to facilitate these procedures, sometimes all of them at once: media (Life Global: The ART Media Company!) for culturing embryos; flexible catheters for removing eggs and transferring embryos into uteruses; and long, terrifying surgical scissors for—one didn't want to think what. There were companies that make specialized petri dishes (test-tube babies are never made in test tubes); incubators for keeping developing embryos warm; freezers for keeping frozen embryos cold. There were software programs with names like BabySentry, for keeping track of the contents of all those dishes and incubators and avoiding that most dreaded of laboratory mishaps: the wrong embryo going into, oops, the wrong uterus.

There were microscopes with joysticks controlling hollow needles that enable lab technicians to suck a single cell out of a three-day-old, eight-cell human embryo. That cell can then be fixed onto a slide and sent off to a lab so that its chromosomes might be tested for any one of almost a thousand genetic diseases. After the testing is done, embryos that carry a genetic disease can be discarded and only unaffected embryos used, with the hope that these will grow into healthy children. "Cystic Fibrosis Testing: There is a difference!" said the advertisement for one of the labs that weeds out defective embryos. "RMA Genetics: Technology for New Beginnings, Offering Power through Knowledge!" said another.

Nearby was a booth run by the Genetics and IVF Institute, a Fairfax, Virginia–based fertility clinic that was distributing pink or blue M&Ms, scooped into urine specimen cups, as a way of advertising a patented sperm-sorting technique called Microsort®, which offers parents a way to select the sex of their baby.

The hall was an enormous rectangle. The biggest and most profitable entities were located prominently at the front, where they lured passersby with everything from sperm-shaped pens to ice cream pellets (a favorite way to advertise any technology involving cryopreservation). But equally interesting were the smaller outfits located toward the back of the hall, jostling to attract browsers to their bunting-covered folding tables, and often not prosperous enough to be offering freebies. There were support groups for women with endometriosis and polycystic ovarian syndrome. There were general advocacy groups for the infertile. There were cutting-edge groups dedicated to helping women find ways to delay childbearing and still bear children. One of these is Fertile Hope, run by a cancer survivor named Lindsay Nohr Beck, whose mission is to help cancer patients preserve their fertility during treatment. One of Beck's mentors is a businesswoman named Christy Jones, a former dot-commer who now runs a for-profit company called Extend Fertility, which offers career women the chance to freeze their eggs with the hope of becoming pregnant later, when relationships and/or work schedules permit.

Since egg freezing is in its infancy, however, what the modern woman often needs to conceive—if things have been left too long—are the eggs of a younger woman. Snuggled against the back wall were egg-donation agencies, none of them as large or gleaming as the front-of-the-room sperm banks, since it is not—yet—possible to stockpile human eggs in the mass-market, quasi-industrial way in which human sperm can be stored and shipped. Egg-donation agencies are a sort of cross between a real estate brokerage and a dating service: for a fee, they connect infertile patients with live, real-time egg donors, and manage what is, legally, a property transfer. Egg donation is an invasive, time-consuming medical procedure, requiring physical risk on the donor's part. Which is not to say you can't build up a decent inventory: all of the banks were offering databases of winsome yet wholesome, sexy yet motherly young women, with profiles that detailed their height, weight, SAT scores, and lifetime goals. You could see how hard the agencies had to work to recruit them. One booth belonged to Global ART, an international outfit with a branch in Richmond, Virginia, that procures egg donors from Romania. Circumventing those aspects of reproductive technology (like egg freezing) that do not work reliably yet, and taking advantage of those (like sperm freezing) that do, Global ART rather ingeniously conducts transactions by shipping a prospective father's frozen sperm to the lab in Bucharest, where it is thawed and used to fertilize the eggs of a Romanian donor. The resulting human embryos—half-American, half-Romanian—are then frozen and shipped back to the United States, where they are thawed and transferred into the prospective American mother, all for much, much cheaper than can be done with a U.S. donor, in part because Romanian egg donors are paid so much less than U.S. donors are. And you don't even need a passport for the embryos!

Also there was an L.A.-based agency, Fertility Futures International, which does a brisk trade in providing egg donors to gay men, another rapidly growing customer base. Surrogacy agencies were also there, catering to straight and gay alike.

There were also, of course, lawyers. Not so long ago, running a "family-building" legal practice meant handling adoptions, foreign and domestic. Increasingly, attorneys are called upon to negotiate scenarios that involve a transfer of sperm or egg—part of the babymaking process—rather than the entire baby. "Half adoptions" you could call them: adoption of half the child's genetic makeup.

And then were the companies that have evolved to deal with the problematic presence of the frozen embryo. Though it's still pretty hard to freeze and successfully thaw human eggs, it is strangely easy to freeze and thaw human embryos. Embryos don't get freezer burn. Unlike, say, hamburgers, human embryos can be frozen, and thawed, and frozen, and thawed again, and used. There are about a half-million frozen embryos in storage in the United States alone. These embryos present terrible moral difficulties for patients, and for doctors, who for fear of lawsuits are reluctant to destroy or thaw frozen embryos, even when patients divorce or move or disappear or otherwise fail to pay "storage fees." Enter ReproTech: standing by one display was a man named Russell Bierbaum, who operates a company that for a fee will take over a practice's frozen embryos, and also is willing, collection-agency style, to track down delinquent patients and persuade them to make what has come to be known as the "disposition decision."

"There are ways of getting people to respond," said the affable Bierbaum, who declined, for proprietary reasons, to reveal how he locates patients and encourages them to decide what to do with their frozen embryos. He did not seem to recognize the menacing significance of any phrase beginning "There are ways." Keeping things upbeat, Bierbaum would say only that "the Internet is a wonderful tool for finding people." Also nearby was the National Embryo Donation Center, one of a number of brokerages that help one couple "donate" surplus human embryos to another. Really good quality embryo batches are sometimes passed among three or four families before they get all used up, or born, or both.

Standing in yet another cubicle was—could it be true?—Professor Robert Edwards, *the* Bob Edwards, the British scientist who with his partner, the gynecological surgeon Patrick Steptoe, enabled the birth of the first IVF child in Oldham, England, in 1978. The very man who set this elaborate reproductive machinery into motion. Edwards was wearing a tan suit jacket, pale gray slacks that did not match the coat, and beige slip-on shoes. He was grayer but otherwise little changed from the photos that show him and Steptoe celebrating the birth of the infant

Louise Joy Brown almost three decades ago. There was the same voluminous, side-parted haircut, the same big rectangular glasses, the same stout and genial look, more like a satisfied fly fisherman, or a Rotarian, than the scientific visionary he is.

Robert Edwards, who is probably the most knowledgeable embryologist in the world, now edits a Web publication called *Reproductive BioMedicine Online,* a British-based journal that publishes scientific papers and essays on the many ethical issues raised by the field he helped create. He was standing in the *RBM Online* cubicle for the purpose of saying hello to a long line of visitors, and, when possible, to sign them up as subscribers. Edwards was also, it emerged, brooding. I stood in line with the vague hope of asking whether back in 1978 he had had any idea of the array of services and situations that would arise from his work. I knew the answer in part: Edwards has a reputation for having been remarkably prescient. He had an early fascination with genetics and is widely credited with having foreseen that science someday would be able not only to produce embryos but to diagnose their genetic makeup before placing them in the womb.

Still, it would be interesting to hear what the man himself had to say.

As it turned out, the man had a lot to say and not much time to say it: Edwards, who was raised in the north of England, speaks in a wonderfully non-establishment, workingman's burr. He had been standing in the *RBM Online* cubicle for two days and needed to leave to catch a plane. An assistant was meaningfully clasping a rolling suitcase. Nevertheless, almost before I had finished my question Edwards began by commenting on a speech given by a prominent stem-cell scientist. "Did you hear the talk this morning?" he wanted to know, smoldering over an assertion that embryonic stem-cell research—one of the most promising, and controversial, realms of modern medicine—was an unforeseen consequence of IVF.

Unforeseen? Edwards wanted to correct the record here. Well before Louise Brown was perking along in her dish, he had indeed envisioned that the cells of the human embryo might be coaxed into making a medical therapy. And so many other things! Babies, period! Millions of babies! "Four percent of the babies in Finland are from IVF!" pointed out Edwards with a kind of defensive glee. It seemed that he, Bob Edwards, had seen coming much of what surrounded us, and found it, for the most part, good. Not just babies but delighted parents, of all stripes and varieties and ages. "Eye hoop they all have babies!" Edwards called out as he was being pulled away by his assistant, leaving behind a line of disappointed pilgrims who had hoped to shake his hand. "What coood be better than a baby?"

"Cancer Patients Aren't as Motivated as Infertility Patients"

What indeed? Through the displays wandered doctors, male and female, young and old, many of whom find it hard to believe that Steptoe and Edwards never received a Nobel Prize for what they did. What they did, after all, was conceive human life—*human life*—outside the womb. What they did was create a situation in which millions of human beings would be born who otherwise never would have existed. What they did was find the first effective treatment for infertility, an ancient affliction as old as humankind itself, and for most of history one of the most dreaded and untreatable; if you don't believe that, why are fertility totems found among the earliest human artifacts? According to more than one doctor, what Steptoe and Edwards accomplished in 1978 was one of the medical breakthroughs of the twentieth century, ranking with the discovery of penicillin and Christiaan Barnard's first human heart transplant.

Observers differ on whose achievement was greater: that of Steptoe, a deft and innovative surgeon who figured out how to use a laparoscope to retrieve a newly ovulated egg—a single cell, invisible to the naked eye—from an anesthetized female patient; or that of Edwards, the Julia Child of the petri dish, a laboratory genius who spent a lifetime trying to understand how something that develops inside the body could thrive and grow outside if it. Edwards managed to take one of those eggs, mix it with sperm, achieve fertilization, and immerse the resulting embryo in a propitious and painstakingly devised culture mix, coaxing it to live and grow until it could be transferred into the uterus of a prospective mother named Lesley Brown. Or—let them not be forgotten—at least eighty other women who lay down on Steptoe's examining table while the team was attempting to perfect IVF in the 1970s, women willing to undergo any amount of gynecological probing, any form of surgical discomfort, in pursuit not of fame or renown but of something infinitely more precious: a child of their own.

"I think it's interesting that he had the balls to operate on [so many] people in a row and keep going," said one doctor at the conference, Robert Nachtigall, saluting not only the tenacity of Steptoe, who died in 1988, but also the fortitude of the women who permitted their abdomens to be opened. Then again, Nachtigall reflected, "We're talking infertility here. There is no stronger motivation. Even cancer patients aren't as motivated as infertility patients."

To many, the fact that Steptoe and Edwards were not rewarded by the Nobel committee illustrates how little respect is given to the infertile—and to the men and women who treat them. Which is odd, considering that the inability to reproduce is a fundamental affliction, maybe *the* fundamental affliction, affecting one out of seven adults in the United States and elsewhere. To take just one example of how little regarded are this group of patients: when commissions convened under U. S. President Jimmy Carter and, later, Ronald Reagan struggled to decide whether the federal government should fund research involving human embryos, one official who opposed funding explained that infertility is not a disease and does not warrant taxpayer-funded research.

If infertility is not a disease, what is it? Reproductive biologists like to joke that "the goal of the human body is to reproduce; everything else is details." While this is clearly reductive, it is also clearly true. To watch a child enter puberty is to marvel at the painstaking, years-long, fully transforming preparation of the human body for its one essential task: that of creating another human body. And if the body cannot reproduce, why is that not a legitimate affliction? What worse affliction, really, is there? Not just for the individual, but for the species?

Certainly, the patients know what a profound affliction unwanted childlessness is. These days, those patients are a rapidly expanding group. In the early days, IVF was envisioned as therapy for one kind of patient: young married women whose fertility was intact but for the fact that their Fallopian tubes—the narrow passages where sperm meets egg and fertilization occurs—were missing or blocked. Today, the group has expanded to include, first and foremost, men. Men are responsible for half of all infertility cases, and men are now the *majority* of patients at the Center for Reproductive Medicine and Infertility, the Manhattan clinic run by Cornell University's Weill Medical College and New York–Presbyterian Hospital, arguably the most prestigious and one of the most expensive fertility clinics in the United States. Among women, there are now patients who suffer from endometriosis, fibroids, ovulatory problems, and missing uteruses, as well as women caught up in one of the most significant reproductive developments in the history of the world; the massive shift of childbearing to later ages, what one scientist,

Patricia Hunt, refers to as "an amazing shift in our species and the way we handle reproduction."

The growth in the patient base is also being driven by those whose problem is not so much infertility as the lack of a willing partner, or the presence of a willing partner who happens to be of the same sex. Single mothers, lesbians, and gay men are among the fastest-growing groups of assisted reproductive technology patients. Another group are patients who suffer from, or carry a gene for, a genetic disease such as cystic fibrosis, Huntington's disease, hemophilia, Tay-Sachs disease, or even a propensity for certain adult-onset cancers, and who want to use IVF, combined with genetic testing, to create children who are unafflicted.

The doctors themselves see all of these patient profiles. They see homeless people off the street. They see women with crippling auto-immune diseases. They see people suffering from psychiatric problems that render them unable to have sexual intercourse, and who want to use IVF to do an end run around that little problem. They see ordinary married couples. They see men with an extra female chromosome, which prevents them from making viable sperm. They see blue-collar workers whose infertility may or may not be caused by exposure to toxic substances. They see plumbers, schoolteachers, lawyers. They see immigrant women who, having migrated to the United States believing it will be a better place to raise a family, find themselves unable to have that family, putting them at risk of being rejected by the men they love and live with. They see women with cancer who before they die want to use IVF to conceive a sibling for an existing child. They see spinal-cord injuries: young men, often, who as a result of car accidents or gunshot wounds or hang-gliding mishaps are now paralyzed, unable to perform many of the ordinary functions of daily life, but who can—thanks to a veterinarian who invented a machine that can compel a paralyzed man to ejaculate—father their own biological children.

So broad is the patient base, and so eager is the field to accommodate them, that assisted reproduction has gone from being an oddball, fringe technology to being perhaps the most socially influential reproductive technology of the twenty-first century. As Robert Edwards pointed out, in many European countries—where IVF is often regulated by government and funded by national health services—as many as 4 percent of children born are the result of high-tech lab procedures. In 2006, the International Committee for Monitoring Assisted Reproductive Technologies announced that 3 million children have been born, worldwide, from IVF. Millions more have been born thanks to related treatments such as sperm donation, artificial insemination, and fertility drugs. And their numbers are rapidly escalating: In the United States, according to the U.S. Centers for Disease Control and Prevention, about 130,000 rounds or "cycles" of IVF were conducted in 2004 (the most recent year for which figures are available), twice as many as in 1996. Close to 50,000 children were born from IVF in the United States in 2004, a 128 percent increase from 1996. At least 30,000 more were born from donor sperm. About 12 percent of women—7.3 million in 2002—find themselves unable to conceive or bring to term the children they want. Infertility is equally or more common among men, whose reproductive experiences tend to be less closely tracked, but as deeply felt.

Every American adult now has either undergone fertility treatment or knows someone who has. Sign in to Yahoo's directory of chat groups; search on "reproductive." You'll find—or I found, the day I searched—16 groups for discussing abortion, 63 on birth control, and 644 related to infertility, its treatments and consequences. Hence the profusion of ovulation kits, early pregnancy tests, at-home fertility tests whereby a man can gauge the swimming capacity of his sperm. Niche publications like *Conceive Magazine* and *Fertility Today* are being marketed to a generation that when it wants a baby wants that baby now. There are of course infertility blogs.

Like many transformative technologies—the automobile, the cotton gin, the Internet—a technology that enables men and women to have children without having sex has had consequences both expected and unforeseen. Notwithstanding the array of apparatuses marketed at conferences, the fact is that assisted reproduction at the beginning of the twenty-first century is breathtakingly sophisticated and unbelievably inadequate: certain aspects have rocketed forward while others have lagged so far as to be primitive. Both because of what it can do and what it can't, assisted reproduction is affecting human evolution in contradictory ways, driving us forward and backward at the same time. It is challenging the women's movement, which is being forced to deal with the ambivalent feelings of a generation who, having come of age at a time when women were being urged to put childbearing aside in favor of a career, now find that the only way they can have a child is to use the genetic material of another woman. It is creating new schisms on the Right. It is confounding the clergy. Reproductive technology is confusing, and confused. It both affirms and denies the importance of genetic inheritance. It creates life, and puts it on hold. Parenting has been divided up; it has been compartmentalized; it has been outsourced. At the same time it has been made deliciously possible.

Reproductive technology is mirroring social change, but it also enables and drives that change, in ways that will affect every single citizen, and probably already have.

A Small City of Twins Born Every Year

To track the impact, a good place to look is a report issued by the National Center for Health Statistics, which provides an annual survey of U.S. childbearing patterns. Virtually every trend reflected in the National Vital Statistics Report issued on September 8, 2005—which analyzed birth patterns in 2003, the most recent year for which statistics are available—either explained, or was driven by, the growing need for assisted reproduction technology.

The report notes the striking rise in maternal age that has occurred, a shift that is being experienced in other first-world countries, where every year women are having their first child at an older age than ever before. In 2003, the birth rate for U.S. women between the ages of 20 and 24 declined to 102.6 per 1,000 women, the lowest rate on record for this age group. Births to women from 25 to 29 have dropped 4 percent since 1990. Noting that women in their twenties "historically account for the largest share of all births," the report points out that the birth rate for twentysomething women has steadily declined over the last three decades, falling from 65 percent of all births in 1980 to 52 percent in 2003.

Meanwhile, the number of births to women between 30 and 34 rose to its highest rate since 1964. Births to women between 35 and 39 reached a record high in 2003: the number of births to women in their late thirties has risen by 47 percent. The birth rate for women 40 to 44 is up 58 percent since 1990, and the number of births to women aged 50 to 54 years—ages when it is extremely unlikely a woman will conceive naturally—has risen by an average of 14 percent every year since 1997, to 323 births in 2003. Prior to that year, the federal government did not specifically track births to fiftysomething women, because they were too rare. In 2003, for the first time, the number of babies born to women over 40 topped 100,000. The mean age of first childbirth has risen steadily, also reaching an all-time high—25.1 years—in 2003. In short, women are beginning their families much later than ever before, with the result that sometimes they are unable to have babies at all, or can have them only with help.

"The increase in birth rates for women 35 years of age and over during the last 20 years has been linked, at least in part, to the use of fertility-enhancing therapies," says the report. "The proportion of childless women aged 35–44 reporting impaired fecundity who sought fertility treatment rose considerably from 1982 to 1995."

The report also details one of the consequences of treatment: multiple births. Once upon a time, the birth of three babies simultaneously was a rarity, a miracle, a freak occurrence. Twins were novel enough that they became creatures of legend and myth: Romulus and Remus, Castor and Pollux. Thanks to fertility medicine, the rate of multiple births in this country has skyrocketed. Between 1980 and 1998, the triplets rate surged, from 37 to almost 200 per 100,000 births, or one in 500. In 2003 there were more triplets born in the United States—more than 8,000—than in any previous year.

Even more striking has been the rise in twin births, which continues to shape family dynamics, popular culture, classroom composition, and children's health. Who doesn't have twins these days? Julia Roberts has twins. Joan Lunden has twins. Cheryl Tiegs has twins. Geena Davis has twins. Lance Armstrong has twins. George W. and Laura Bush have twins. The disgraced lobbyist Jack Abramoff has twins. The exposed CIA agent Valerie Plame has twins. Kelli Conlin, the lesbian mother who heads the National Abortion Rights Action League in New York, has twins. You can't do a good Nexis search for "twins," because there are too many hits to be manageable. There are groups with such names as Twins World and Twins Network. There is a *Twins Magazine*. Coney Island has an annual Twins Day; the annual Twins festival in Twinsburg, Ohio, gets bigger every year. There is a twin talent search; a twin reality show. Since 1980, the number of twins has climbed by 75 percent, from about 70,000 in 1980 to close to 130,000 twins born in 2003, a small city of twins born every year.

And then there is the convergence of both trends. The older the mother, the more likely she is to give birth to more than one child at the same time. In part, this is because as women age, they ovulate more erratically and are more likely to conceive twins naturally. But the trend is also being fueled by technology. "The rise in twins and triplet/+ birth rates has been most pronounced among older mothers, and especially those aged 40 years and over," says the National Vital Statistics Report. "The number of singletons born to women aged 45–49 years has risen nearly threefold between 1990 and 2003 (from 1599 to 4371 live births), but the number of multiples has climbed even more dramatically. Between 1990 and 2003 the number of twins born [alive] to women in this age group surged from only 39 to 1045 and the number of triplet/+ births from 0 to 106."

"In 2003," the report notes, "one out of 18 births to women aged 35 years or older was a multiple delivery, an outcome associated with fertility treatment."

Think about that last statistic. If you are a women conceiving after 35, there is a 1-in-18 chance that you will give birth to twins, or more.

It is hard to overstate the impact that later pregnancies, multiple births, and the convergence of both are having—on women, children, schools, sports teams, health statistics. The older the mother, the greater the chance of miscarriage, chromosomal problems resulting in afflictions such as Down syndrome, and serious complications of pregnancy, such as preeclampsia and gestational diabetes. The consequences of multiple births are even more striking. The meticulous course of human evolution designed women's bodies to carry one child at a time. One baby has lots of room for brain development. One baby is less likely to kill the mother while being born. One baby will have the full attention of his or her mother and father for at least nine months, cementing the child's attachment to the parents, and vice versa.

Thanks to fertility drugs, the ovary now can be induced to produce far more than one egg at a pop. Twenty eggs regularly result from the drug regimens that precede IVF. From that, ten or fifteen or even twenty human embryos are caused to develop. Doctors transfer two, or three, or five to the uterus, not knowing how many will take. Sometimes they all do. In the United States, more than a third of IVF births involve multiples, which means that more than half of all IVF children are born as part of a set.

It's true—as doctors inevitably argue—that many patients want more than one baby. A paper presented at the 2005 conference shows that more than half of all IVF patients want twins, and it's hard to talk them (especially the women) out of it. In part, this is because fertility treatment is so expensive and stressful that many patients prefer to get it over in one fell swoop. In part it's because patients and doctors want to increase the odds of success. In part it's because of what James Gleick diagnoses in his book *Faster*: our widespread expectation that events, including childbirth, should happen as soon as we are ready. We want the things we want to be here, now. We are impatient. This is true of elevators and Internet connections; it's true, too, of families. Like a high-speed broadband connection, twins are a way to get the thing you want in half the time.

But multiple births are dangerous both to mother and to children. Doctors refer to the increase in twinning in the United States as an "epidemic." Children born as a set are more likely to be born prematurely; premature babies are more likely to suffer from cerebral palsy, developmental delay, serious respiratory ailments, and other nerve and brain disorders. Premature babies are also more likely to die. Among twins, infant mortality rates are four to five times greater than that of singletons; for triplets it is much higher. In the Washington, D.C., metropolitan area, Inova Fairfax Hospital, which delivers some 11,000 babies a year, has an annual memorial service for the infants who died there without ever going home. October 15, in case you didn't know it, is National Pregnancy and Infant Loss Day.

In hospitals, there has been a wholesale expansion in neonatal intensive care units, and in high-risk maternity wards. There women can be seen who are spending months on enforced bed rest. Here is what a woman on bed rest, pregnant with multiples, looks like: she looks like a Volkswagen Beetle, all belly, often forced to lie with her head lower than her body, and to lift herself out of the bed—assuming she can do so—with the help of a trapeze hanging from the ceiling. Many hospitals have a wing for these patients, who are trying to accomplish one thing: keep those babies in the oven for as long as possible. Every day in the womb, the calculation goes, saves two days in the neonatal intensive care unit.

Although many of these pregnancies turn out well, or reasonably well, in the United States several major indicators of child well-being are moving in the wrong direction, and multiple births are a primary reason. Between 2003 and 2004, the rate of premature birth increased by 2 percent. Now, one out of every eight babies in the United States is born prematurely, as a major 2006 report from the Institute of Medicine noted with alarm. The preterm delivery rate has risen 16 percent since 1990, and more than 30 percent since 1981. "The rise in the incidence of plural births, which are much more likely than singletons to be born preterm, had an important influence on the overall preterm birth rate over the past two decades," says the Vital Statistics report, noting that births to teenage mothers, formerly a major factor in prematurity, have declined, meaning that these trends are being driven not by younger women but by older ones.

The rates of low birth weight (LBW) among infants have also increased. Low birth weight (birth weight of less than 2,500 grams or 5 lb, 8 oz) is also associated with problems later in life. In 2003, children born with low birth weight accounted for 7.9 percent of all births, the highest level since 1970. The report noted: "Recent trends in LBW are influenced by the strong growth in the multiple birth rate: twins and higher order multiples are much more likely to be born LBW than singletons; 58.2 percent of all plural births were born LBW in 2003." Small wonder, then, that in 2002 the U.S. infant mortality rate also rose, reversing four decades of progress and galvanizing such organizations as the March of Dimes to raise awareness of the causes, and consequences, of premature birth. A sort of cognitive dissonance is at work among pregnant women and their partners. Women are less likely than ever to smoke or drink alcohol during pregnancy; yet a twins pregnancy hardly draws a second glance, and in some cases is actively courted.

The rise in multiple births is one way in which IVF is arguably setting us back, creating babies who are—some of them—at a disadvantage. This is an important point, often ignored. When bioethicists worry about assisted reproduction and its impact, they worry that the technology may be used to make some babies better. They worry about picky white-collar parents who patronize outfits like Scandinavian Cryobank, agonizing over sperm and egg donor profiles, comparing IQ scores, bent on ensuring that their baby is in every controllable way above average. Bioethicists worry, too, about hardcore eugenics: genetic technologies that may someday allow parents to tinker with the genes in the culture dish, adding a soupçon of charisma there, a touch of tenderness here, altering height and hair color to fashion a superbaby. That day is not here yet, but when it arrives, many experts are prepared to comment. A number of authors have thoughtfully explored the ethics of enhanced children, in books like *The Perfect Baby* (by bioethicist Glenn McGee), *Redesigning Humans* (by the futurist Gregory Stock), and *Our Posthuman Future* (by the philosopher Francis Fukuyama). The shared concern, often, is that people may use genetic engineering, together with assisted reproduction, to realize a desire for top-quality offspring. The shared question, often, is how to think about the new eugenic opportunities being afforded parents.

Certainly, eugenics is something to worry about. Or it may be someday. But the truth is more complex, and in some ways less pretty. The truth is that it's not yet possible to genetically enhance offspring. The genes in the culture dish cannot yet be improved upon. Real designer babies are a long way away. The truth is that many fertility patients are deeply grateful for what they get, and what they get sometimes are children who are at a disadvantage, not unfairly enhanced. When this happens, the vast majority are only too happy to do what they need to do to care for those children, even if it means spending days at the pediatrician's office to ward off the respiratory illnesses to which premature babies are susceptible; taking three babies to three pediatrician appointments at three different times; coping as infections sweep through households; living for eight months with a child in quarantine; driving children not to soccer games but to weekly visits with hearing and vision specialists; balancing a family in which two children are fine and the third is a little different; or even just trying to find a way to lovingly raise two or three siblings who happen to be the same age.

Being born at a disadvantage is not only a problem for many babies in multiple births. It is sometimes true of ART-conceived singleton babies, in part because of the advances made in the treatment of men. In 1992, reproductive technology underwent a quiet revolution with the advent of a technique that enables a single sperm to be injected into an egg, ensuring that almost any sperm has the ability to reproduce. Now almost any man who can afford treatment can be a biological father. Yet scientists are realizing that some male infertility exists for a reason: as a safeguard for the species, to prevent genetic problems from being passed on. When men produce very little sperm, sometimes it's because there are heritable chromosomal problems with the man. Thanks to the new techniques, infertility itself is being passed from father to son. Heritable infertility sounds like a contradiction in terms, but now, thanks to science, it isn't. Other genetic problems are being transmitted as well, and possibly magnified. Evolution is being thwarted. Problems are being admitted into the gene line that Nature would have headed off at the pass. This new technique, intracytoplasmic sperm injection (ICSI), is so popular that in some clinics it's done for all patients, male and female. It has never been regulated by the U.S. government. There is no standard for counseling patients for whom it is done.

There is a growing suspicion that some female infertility may also be genetic; that some infertile women are passing problems to their offspring, in ways that subtly affect infant health. In the United States, the first IVF baby was conceived in Norfolk, Virginia, with the help of the husband-and-wife medical team of Howard and Georgeanna Jones. When Howard Jones appeared to announce the birth of Elizabeth Carr, he had two press releases in his pocket, one to read if the child was born normal, the other to read if she turned out deformed. To everyone's relief, she was healthy. Most IVF babies are. This, too, is an important point. But some, puzzlingly, are not: studies have now proved that even singleton IVF babies are more likely to be born prematurely, and with lower birth weight, than children conceived naturally. Other studies have suggested that IVF children are more likely to be born with certain birth defects, such as deformations of the bowels, urogenital malformations, and retinoblastoma, a form of eye cancer. At this point the fertility community has grudgingly come to accept that some IVF babies, even singletons, are different from their naturally conceived peers.

"IVF children are different from children who are conceived from parents who have no fertility problems," says Arthur Leader, professor of Obstetrics, Gynecology and Medicine and former chief of reproductive medicine at the University of Ottawa, former president of the Canadian Fertility and Andrology Society, and himself the parent of a much-loved IVF child. "People who suffer from infertility, or subfertility, appear to be genetically different than people who have no trouble having the families they want. The outcomes for their pregnancies are different. We don't understand what it is, but it's clear that they are different."

"You Can't Say, 'Oh, in Seventeenth Century France They Did This'"

The Vital Statistics Report notes another emerging trend: more babies than ever before are being born into families with no father. Notably, 2003 saw the highest proportion of births to single mothers that has ever been reported in the history of the United States. "Childbearing by unmarried women rose steeply in 2003," the report notes. An unprecedented 34.6 percent of births that year were to single mothers. They amounted to 1.4 million babies, an increase of 4 percent from the year before, which itself had seen an all-time high.

In part, the rise in births to single mothers is driven by unmarried women getting pregnant through sexual intercourse. But it also is being driven by women using sperm donation to conceive a child they intend to raise alone. A national group founded in 1981, Single Mothers by Choice, now has a worldwide membership, somewhat to the astonishment of its founder, Jane Mattes. Among other things, the phenomenon is inviting a change in the way society thinks about children of single parents. Up to now, statistics have shown that children of single mothers are more likely to live in poverty and do less well later in life. Many women using sperm donation are white-collar and financially secure, and they are creating a new generation of offspring who are fatherless, but not poor.

Some feminists celebrate this rise in women-headed households; they argue that "maverick moms" make better mothers, of boys in particular, and that a family without a father is a family where a new type of male can be incubated and raised. Others—conservatives, but not only conservatives—take, of course, the opposite view. In recent legislative sessions in Virginia and Indiana, bills were introduced that would ban artificial insemination or IVF for unmarried women. These measures so far have failed, but they reflect a certain public level of unease. How much *does* a father matter? If a social conservative thinks single motherhood to be problematic, should the liberal position be that there is no problem? What about single mothers of twins? Can a maverick mom serve the needs of two infants simultaneously?

Of course, most women using donated sperm are not trying to ignite a cultural movement, or raise a brave new boy, or anything so high-minded or ideological. They simply want a child. For many women, the birth of a sperm-donor child is the result of a reluctant awareness that single motherhood is the only path to motherhood that's open. The single mother movement is one that's being driven, at least in part, by men. Unready men. One British fertility doctor, Gillian Lockwood, calls this "bio-panic": a state of mind induced by "commitment-phobic" boyfriends who, interested in conceiving someday but chronically unready to do so, eventually compel their partners—or, rather, at a certain point, ex-partners—to turn to sperm donors.

The widespread use of donated sperm is also changing and challenging our definition of parenthood. Among other things, it raises the questions that affect adoptive families: How much do genetic progenitors matter? Does a child who grows up father-free—or any child conceived through donated sperm or egg—have the right, or the need, to know the identity of the donor who helped bring him or her into being? For decades, the widespread but secretive use of donor sperm to circumvent male infertility resulted in hundreds of thousands of children who grew up as ignorant of their origins—or half their origins, anyway—as adopted children once were. Now, a movement of donor offspring, like adopted children before them, are seeking to find out more about their roots. Recently, a boy plugged his genetic information into an Internet service that traces generations using the Y chromosome, which is passed down unchanged from father to son. The boy located his sperm-donor dad, even though the man had not registered on that service. But *is* a donor a dad? If not, what is he?

For children who are adopted, it is widely accepted that secrecy about their origins is unfair and destructive. So what about children of donor gametes? Do they need to know the truth? Do they need to know the donor? On a number of Internet chat rooms there is an anguished debate going on about the impact on children of growing up without knowledge of their biological origins.

Yet, even as these concerns are being raised about sperm donation, anonymous egg donation continues to grow exponentially, largely as a result of the number of women trying to conceive later, and failing. According to the U.S. Centers for Disease Control and Prevention, in 2003, 12 percent of all IVF procedures involved the use of donor eggs. That means that some 15,000 rounds of IVF were performed for mothers who, if they succeeded, would give birth to children unrelated to them. Each year, the rate increases dramatically: The country's largest fertility clinic, Shady Grove Fertility in Rockville, Maryland, did about 800 donor egg procedures between 2000 and 2004. It was scheduled to do 500 in 2005 alone. Many of these children will remain unaware of their origins, because more than half of parents who conceive using egg donation never tell their children the truth of how they were conceived.

Scientists who study "paternal discrepancy" have found that at any given time in any population, a certain percentage of children—as high as one in twenty-five—are misled about the identity of their genetic father (and a certain percentage of men mistakenly believe their children are theirs). Egg donation has now given humanity the phenomenon of maternal discrepancy. Now, even the fact that your mother gave birth to you doesn't mean she is your genetic mother. For the first time it is possible for a woman to conceal the fact that she is not related to the child she gave birth to.

"Let's face it: donor gametes is an experiment," says Robert Nachtigall, a San Francisco doctor who has practiced fertility medicine for more than thirty years, and has done sociological studies on donor-conceived

families. "There's nothing in our programming—nothing in our accul-
turation—to prepare us for this. It's not possible. You can't have a child
come out of your body that is not genetically yours at any other time until
the late twentieth century. Thousands and thousands of years of programming doesn't do you any good. This is a huge social experiment
mediated by technology. Who the hell knows how it's going to turn
out? It's meant to fill the needs of infertile people in the late twentieth
century, many of whom became infertile for their own social, cultural
reasons, part of another experiment: the baby boom experiment of post-
poning parenthood. That created a demand that can only be met by
another experiment. We've got this experiment-within-an-experiment
type of thing. That's why on the one hand it's so fascinating, and on the
other hand, there's no easy answer for it because there's no precedent.
You can't say, 'Oh, in seventeenth century France they did this.'"

What assisted reproductive technology does is replace one uncertainty—will I ever conceive naturally?—with other uncertainties. Will
my child reject me if she finds out I am not her genetic mother? If I tell
her the truth, how will that affect her? Should I form a relationship with
the sperm or egg donor? Should I thank the donor? Should I get to know
other people who have conceived children using the same donor? What
if I'm straight and they're gay? What if I'm gay and they're straight?
What if we don't like each other?

What would raising twins be like?

What will we do with our surplus embryos?

In all these respects, assisted reproduction is having a social impact
as profound as the widespread availability of the birth control pill in the
1960s, and the passage of *Roe v. Wade*, legalizing abortion in the United
States, in 1973. Both of these technologies altered society in ways both
immediate and long-lasting. They affected millions of lives, and changed
the broader culture. Few people fully anticipated the impact they would
have, not even those involved in the movements that produced them.
Even as *Roe* was being argued in the U.S. Supreme Court, some in the
reproductive rights community didn't realize that a favorable ruling
would permit the instant, nationwide operation of abortion clinics. All at
once it became much easier for women to forestall or end unwanted
pregnancy; to enter graduate school and the workforce in a critical mass.
These technologies provided the rallying cry for feminism in the 1970s
and '80s; abortion rights, along with issues like equal pay and opportunity, was the crucible in which the consciousness of a generation was

forged. These technologies also provoked, of course, a lasting social backlash, fueling culture wars that continue to polarize the country today. Now, the same sort of impact is being created by newer technologies that permit people to create children rather than avoid them.

In the twenty-first century the radical thing may not be to end a pregnancy, but to begin one.

CHAPTER TWO

WOMEN AND THE DILEMMAS OF MODERN MOTHERHOOD

"Your Numbers Look Good"

Lunchtime in downtown Washington. Spezie, a crowded restaurant specializing in upscale Italian for the downtown power diner. The woman across from me: a friend of a friend. I'll call her Melanie. She prefers not to be identified by her real name, because all of this is still tentative and ongoing. She is wide-eyed, freckled, uncosmeticized, likable, humorous, smart, self-deprecating, slender, married. She looks thirty-five but is forty-two. In much of her story she typifies the modern fertility patient, who when she begins treatment swears, "I will never do IVF."

Melanie grew up in a working-class town in the mid-Atlantic region. Summers, as a teenager, she held down two jobs. Having worked hard to get into a top-flight university, having done well there and pursued her MBA afterward, Melanie pursued what is, for her, extremely satisfying public service work in the nonprofit sector. Melanie is number two in her office, and has been number two for several years. Being number two is in some ways harder than being number one; being number two means taking care of the administrative scutwork number one doesn't bother with; it means running the office when number one is away; it means striving to be good enough to be, someday, number one herself. Among other things, being number two has meant that there has not been a natural opportunity to have a baby. Nevertheless, she has always wanted children; so three years ago, when she was thirty-nine, she and her husband decided they'd better get started.

It was, she thinks now, about three years too late.

She did get pregnant naturally. At least, that's what the blood test confirmed. The pregnancy turned out to be a blighted ovum, something she had never heard of but would soon became all too familiar with. "Blighted ovum" means that the sperm fertilizes the egg, but the embryo never develops, and instead halts as an empty sac, a not-fetus, a missed opportunity. A blighted ovum is a form of miscarriage. Sometimes it is voided naturally by the body, and sometimes, as in Melanie's case, it is scraped out of the uterus by a procedure known as dilation and curettage. After that unpleasant experience—knowing that even this unsuccessful pregnancy had taken months to achieve—her gynecologist proposed that she take clomiphene citrate, a low-level, orally administered fertility drug, commonly known as Clomid, that would stimulate her ovaries to make sure they were still expelling a monthly egg. This is known as the Clomid challenge; it's a diagnostic test of the female reproductive system, like something a mechanic would perform on an engine.

A challenge! Excellent! Melanie took the Clomid.

"Your numbers look good," her gynecologist told her afterward, which was sort of like saying "Here is a yummy liver bit" to a well-trained dog. It was the kind of comment to which she has been programmed to respond.

Numbers looking good is how Melanie has lived her life. Her numbers have always looked good, and she has worked hard to make those numbers look better. If her numbers looked good, then clearly treatment would eventually work. Melanie decided to ratchet things up a notch. On her doctor's recommendation she visited a private fertility clinic, where she tried intrauterine insemination (IUI), often regarded as the natural next step after Clomid. IUI involves injecting a concentrated dose of sperm directly into a woman's uterus, to increase the odds of fertilization. IUI usually also involves taking a stronger drug, aimed at persuading the woman's body to produce more than a single egg each month. A drug like this must be injected. Melanie found herself sitting in her bathroom every night, palms sweating, flicking the vial to dispel air bubbles, sure that she would be the one patient to die of an embolism. Night after night she made the injection. Morning after morning she woke up with bruises on her thigh. She was hooked now. She was engaged. She was trying.

"It was my makeup," she says now. "There hasn't been anything I've wanted that I haven't gotten. It's awful but it's true."

Problem was, the shots weren't working. Or maybe they were. It was hard to know. She would go into the doctor's office for the ultrasound tests that would tell whether her Graafian follicles, the fluid-filled cysts where eggs grow and mature, were enlarging in response to the drug. But the nurse was reticent and Melanie couldn't make out what was happening on the ultrasound screen. The night came when she was supposed to take the "trigger" shot: another drug, one that causes the eggs to be released. But now—finally—the doctor was on the phone and, without saying what was wrong, told her just to take her trigger shot and "have sex with your husband." It wasn't worth coming in for the insemination procedure. Apparently what had looked like a follicle wasn't. Or something.

Melanie decided she hated that doctor. So she went to a hospital practice, where women queue up when the door opens at seven. There is little privacy in a hospital. That part was bad. But she happened to get a terrific young female doctor who fully understood what she was going through. They decided it was time for full-blown IVF, which months earlier Melanie had considered unthinkable, but now seemed like a natural transition. Again Melanie would take the shots to stimulate her ovaries. This time, the eggs would be surgically removed before the hormones could cause her own ovaries to ovulate, and fertilized with her husband's sperm through IVF. Following this regimen Melanie produced thirteen eggs, which was good, and about half of those eggs were fertilized, which was also promising. The best two embryos were transferred, and a blood test confirmed she was pregnant. One morning she joined the women lining up for the big eight-week ultrasound. She sat there in her paper gown while other elated women emerged saying things like "It's twins!" to their beaming partners. She went in and found out that what was growing inside her was . . . a blighted ovum.

It was the day before Thanksgiving. Melanie and her husband were supposed to spend the holiday with her in-laws, who lived in another state. Her husband—figuring somebody had to go—went. She stayed in their apartment, curled up in a little ball. Her blood tests were continuing to come back normal. "I'm just old," she says now. "If we had started when I was thirty-six or thirty-seven, I'm sure we would have a child. There's nothing aside from age that's wrong with me."

How Women Got Here: A Brief History of Female Infertility

And yet "age" is a diagnosis few foresaw thirty years ago, when reproductive technology experienced a quantum leap forward. In fact, when one of the first IVF clinics in the United States opened, there was an age limit for female patients. In the early 1980s, no woman older than thirty-five was accepted into the Jones Institute for Reproductive Medicine. IVF was seen as a technology for the younger woman, recently married, who wanted a family and because of specific physiological problems couldn't have one naturally.

"We limited our cases at first to those women who had had their Fallopian tubes removed," explains Howard Jones, elegant and white-haired, who today happens to be wearing a fuchsia shirt that a British colleague literally gave him off his back at a meeting when Jones admired it. Sitting in his capacious office in Norfolk, Virginia, Jones, now in his nineties and semiretired, describes the first group of patients treated in the Jones clinic, which is affiliated with Eastern Virginia Medical School. Back then, he says, the clinic wanted to maximize the chances of pregnancy, and also wanted to be sure this new technique, IVF, was responsible for any children that were conceived. If a woman got pregnant who had no Fallopian tubes at all—no natural passageway between ovaries and uterus, no way for an egg to slip through so that fertilization could occur naturally—then success could be due only to science. And if IVF worked for tubeless women, presumably it would work for other young women who suffered from tubal blockages. These were considered the target group for a new therapy that some doctors still viewed as freakish, irrelevant, a kind of unimportant scientific parlor trick.

And there were many such women. Among women, infertility has always existed. Contrary to popular belief, female infertility is not a new condition that has been inflicted as punishment on career women who have dallied overlong in fulfilling their reproductive duty. Historically, and today, female infertility often results from tubal scarring, the result of pelvic infections that can be due to any number of causes, including sexually transmitted diseases but also childbirth itself. In fact, childbirth is a frequent cause of infertility in third-world countries, where, ironically, both fertility rates *and* infertility rates tend to be highest. Other common causes of infertility are fibroids, or large uterine growths; endometriosis, a painful condition where the menstrual lining grows outside the uterus; and hormonal imbalances, such as polycystic ovarian syndrome (PCOS), that interfere with ovulation. All these conditions have waxed and waned over time. In this country there is less infection now than formerly, thanks to condoms and antibiotics, but there are more ovulatory problems, owing in part to the nation's obesity epidemic as well as other, contradictory trends: there is a form of PCOS—you could call it the Upper East Side version—that comes from being too thin.

In addition, as Margaret Marsh and Wanda Ronner outline in *The Empty Cradle: Infertility in America from Colonial Times to the Present,* there has always been infertility imposed on women by outside forces. In the nineteenth century, double standards for sexual behavior caused many virgin brides to suffer pelvic pain not long after their wedding night; visiting their doctor, they learned that they had contracted gonorrhea, and with it infertility, from their new husbands. In the twentieth century, fads for reproductive surgery prompted the gratuitous removal of organs. One contemporary doctor, David Keefe, remembers training with a gynecologist who, every time he removed a woman's uterus during a hysterectomy, held it up and said, with relish: "Another brick for the villa." Keefe himself was deeply affected by a young woman he knew growing up, who was diagnosed with a cancer called vaginal clear-cell adenocarcinoma. During pregnancy the girl's mother had been given diethylstilbestrol, or DES, a drug widely prescribed in the 1950s and '60s by doctors who believed, wrongly, that it would prevent miscarriage. Instead, it had devastating effects on female fetuses. Some DES daughters, like the young woman Keefe knew, would die of cancer; others would grow up to find their own reproductive organs malformed. A coalition of scientists, the Collaborative on Health and the Environment, organized by a women's health group at Stanford University medical school, is attempting to discover whether other forms of female infertility are the result of the chemical revolution—whether plastics and other chemical compounds are affecting human reproduction in subtle ways. The U.S. government has also launched a major project, Life Study, that will track the fertility of a thousand men and women, to evaluate the reproductive effects of the tens of thousands of chemicals that are now a routine part of our sleeping, eating, and working lives.

Let it not be forgotten either that some have always unfairly blamed women for creating their own barrenness, through book learning and ambition. Excessive education "is accountable for much of the sterility and physical degeneracy of American womanhood," tut-tutted doctor and social commentator Horace Bigelow in 1883; he was appalled by the trend of women attending colleges in substantial numbers, which, he believed, led to "a want of reverence for her special vocation," motherhood. His concerns were shared by Harvard physician Edward H. Clarke, who wrote in 1873 that female "sterility" was caused by education of women, which diverted energy from the womb to the brain, leaving women with "monstrous brains and puny bodies."

And naturally there have always been folk willing to peddle a solution. "A baby in every bottle" is what Lydia Pinkham, purveyor of a nineteenth century "vegetable compound," promised women who drank her fertility tonic. A poster from the time shows a chubby baby sitting in a bottle, and a woman guzzling tonic while a man—her husband? her doctor?—looks fiercely on. For many years, as Marsh and Ronner also document, the field of fertility cures was populated both by quacks and by genuine healers: one early surgeon, suspecting women were infertile because their cervical openings were too small, created a vogue for surgical enlargements of the vagina, done with a scalpel, often without anesthetic. During the twentieth century, some real strides began to be made: surgery got better and in some cases may even have helped. Equally important, doctors discovered the reproductive hormones. Chief among these pioneers was John Rock, whose midcentury experiments with estrogen and progesterone would famously lead to the birth control pill, but also—with less notice—to advances in fertility treatment. Women now could be given drugs that in some cases would regulate their ovulation and result in much-wanted pregnancies. As early as 1942, Rock achieved the first reported attempt at IVF (it is not clear whether fertilization occurred), but never transferred an embryo. An article in Look prompted hundreds of women to write to Rock begging to be guinea pigs in his research, including one who wrote that life was "empty and useless without children."

Rock would later comment that he found infertile women who wanted babies far more willing to subject themselves to experiments than fertile ones who wanted contraception. And of course, Edwards and Steptoe found no end of eager subjects. When Louise Brown was born, Robert Edwards remarked that IVF signaled "the end of the beginning": the end of gory interventions and the beginning of something like effective treatment for millions of women worldwide.

The End of the Beginning

In the United States, the first IVF child was born in 1981 thanks to the work of two Joneses: Howard and his wife, Georgeanna, who both had distinguished careers caring for the infertile. Howard Jones was a gynecological surgeon who had collaborated with Robert Edwards during a stay Edwards made in the United States. Equally eminent, Georgeanna Jones was a reproductive endocrinologist (an ob-gyn who specializes in treating infertility) who had much success with gonadotropins, the hormonal drugs that stimulate ovulation. The Joneses themselves had three children and an extremely close marriage. In 1977—forced to retire from Johns Hopkins University, which had a mandatory retirement age of sixty-five—they were persuaded to take over the department of obstetrics and gynecology at Eastern Virginia Medical School. They happened to move there on the day Louise Brown was born in England. During the hullabaloo, a television reporter asked Howard Jones whether a test-tube pregnancy could be achieved in the States. Jones replied that it could, easily. Asked what it would take, he replied, "Money."

The next day, the Joneses received a call from a former patient who had had a child thanks to a drug regimen of Dr. Georgeanna Jones's devising. The grateful woman wanted to fund the first U.S. IVF clinic. She offered $5,000 to purchase a microscope and an incubator. Other benefactors followed, and within a few months, the Joneses had $25,000, which, amazingly, was enough to set up an IVF lab. Women, it should be noted, have always driven infertility research, not just with their bodies but, when they had some, with their money. Edwards and Steptoe also owed part of their success to the fact that a California woman stepped in to underwrite their controversial IVF project, after other, more conventional funders bowed out.

Money also came from unlikely sources: When the Norfolk clinic was being established, it was bitterly opposed by pro-life groups, who—leery of any work involving human embryos—wanted to stop IVF before it got started. A Virginia pro-life organization tried to prevent the medical school from obtaining the "certificate of need" required by the state to set up any new clinic. Having failed at that, pro-life groups continued to send letters and organize protests. When the Norfolk *Virginian-Pilot* ran an editorial erroneously saying that the Joneses were forcing patients to get prenatal testing and to agree in advance to abort any abnormal pregnancy, Georgeanna Jones sued the paper's parent company for $5.5 million, and plowed the money from the settlement back into the clinic. Still, the pro-lifers exacted some concessions: in the early days, the clinic promised that all embryos would be used, and that none would be disposed of.

In those days, the Jones clinic attracted staffers who would go on to become the architects and in some cases the superstars of the field, along with colleagues in countries such as Australia, England, and France. Among the first generation was Lucinda Veeck (later Lucinda Veeck Gosden), a young scientist who helped advance the field of embryology by, among other things, producing a picture atlas of human embryos. Another was Zev Rosenwaks, a reproductive endocrinologist who, like Veeck Gosden, would later move to the Manhattan clinic affiliated with Cornell University and New York–Presbyterian, where he trained many in the next generation of doctors. In the United States, IVF clinics were created at a few medical schools, including the one at Yale University, where psychotherapist Dorothy Greenfeld was hired to monitor the mental health of patients and make sure they weren't driven crazy by treatment. The Jones clinic had a waiting list as soon as it opened, and Yale found itself inundated by patients who had to wait as long as eighteen months for an initial appointment.

This was despite the fact that in the 1980s IVF success rates were still excruciatingly low. As few as 5 or 6 percent of "cycles"—single rounds of treatment—resulted in a child. There were many reasons for this. One was that doctors were at the mercy of the female body rhythm. After their own failures using synthetic hormones, Steptoe and Edwards maintained—wrongly—that IVF would not work if the woman took drugs to boost her egg production. So doctors followed the natural ovulatory cycle of every patient: taking her temperature, doing blood tests, and otherwise endeavoring to detect that unseen moment when a single, mature egg slips out of the ovary. If they tried to retrieve the egg too soon, it would be immature and unusable. Too late, and it would have disappeared. This meant coming in the middle of the night, slogging in during snowstorms, banging on the hospital door to be admitted. Eventually, Georgeanna Jones decided to try using gonadotropins. She found that Steptoe and Edwards were wrong, and that drugs did work for IVF. This made it easier to control the release of the eggs, which meant that egg retrievals could be done during the day, when, as one doctor put it, "lab staff were fully awake."

Then there was the challenge of equipment: everything—everything—had to be invented or improvised. When Steptoe and Edwards were conducting their initial IVF experiments, they had to transport human eggs, human sperm, and human embryos back and forth between Oldham, where the hospital was, and Cambridge, where the lab was. They did this by putting a specimen in a sealed tube, and tucking the tube into a pouch that had been surgically created in the skin of a living rabbit. "It was pretty crude by today's standards," says Arthur Leader, who knew Steptoe and Edwards and went on to help develop the field in Canada. "It's a testimony to the hardiness of the human eggs and embryo that they survived." Even when incubators became available, scientists had to teach themselves how much carbon dioxide, oxygen, and nitrogen were needed to approximate the makeup of the female body. The same was true of culture media, the viscous mix in which specimens are contained and grown: it was trial and error, figuring out how to duplicate the rich, ever-changing environment an embryo experiences as it travels from the Fallopian tubes into the uterus.

In the beginning, nothing was known for certain. What size needle best retrieves eggs? How thin is too thin? Doctors in some cases made their own tools, while embryologists worried about every substance every instrument was made of. They kept the lab dark, says Lucinda Veeck Gosden, because they knew that light can affect the DNA in many cells. "We were always worried about toxic substances, anything you could smell, perfumes, construction fumes, somebody opens the door, you smell isopropyl alcohol. They can be extraordinarily toxic."

One of the most profound retardants was the fact that in the United States, there was no funding for human embryo research from the federal government, the world's most powerful engine of scientific and medical advancement. There was no way to test the effect of equipment or medications or light on the embryos that science was creating. Instead, doctors had to put the embryos into the uterus, and wait until a child was or was not born. IVF itself *was* the experiment. In Britain, the situation was more coherent: there pro-life groups also protested IVF, but government and scientific leaders made a serious effort to resolve the conflicting viewpoints, and found a way to regulate the field while letting it move forward. Members of Parliament were invited to peer through microscopes at eight-cell embryos. One member, Baroness Mary Warnock, was commissioned to draw up regulations. In Britain, it was decided that research could be done on human embryos for fourteen days after the embryo was created and that every IVF lab must be licensed, every procedure approved. A list of clinics and the procedures they perform is published annually. "That did an enormous amount, I think, to satisfy both MPs and the public that there weren't horrific Frankenstein scientists hiding in dark corners, making human-ape hybrids and I don't know what else," recalls Dame Anne McLaren, an eminent biologist whom I interviewed in a medical building at Cambridge University while she, tiny and amiable and birdlike, sipped water from a laboratory beaker.

In the United States, IVF got its start during the presidency of Ronald Reagan, a conservative politician presiding over an administration in which little money was available for any reproductive research. As for *human embryos*: well, bioethicists met; commissions agonized. Prolife organizations quietly lobbied. Women's groups were otherwise occupied. In a weird stalemate, it was decided that embryo research might be federally funded, if approved by a certain advisory board. The board was formed, briefly, but soon was deliberately disbanded. You could apply, but there was nobody to apply to. As early as the mid-1970s, a scientist named Pierre Soupart submitted a request to the National Institutes of Health (NIH), asking for federal funding to research IVF embryos in his lab at Vanderbilt University. He died without receiving a reply. It was truly a Kafkaesque situation, until in 1995 two Republican members of Congress, Roger Wicker of Mississippi and Jay Dickey of Arkansas, proposed a formal law that no government funding could be used for research that involved the destruction or endangerment of human embryos. Jay Dickey is no longer in office, but the so-called Dickey-Wicker amendment is passed every year as part of the Department of Health and Human Services appropriations bill; it has become a conservative touchstone.

In late 2005, Phyllis Leppert, a veteran NIH staffer, stood before a group of doctors and tried to find a diplomatic way to describe the effect of that ban. This group of doctors was interested in the still open question of the health of IVF children; under discussion was the effect of culture media on embryos that are living and excreting and growing and developing in it. Nobody—nobody—knows what the effect of the medium might be on the children conceived therein. "Anything funded by the federal government—that includes CDC, NIH, FDA—we have to follow the Dickey-Wicker amendment, which says that we cannot fund any research on human embryos, and that has been an incredible— what shall I say—*constraint*," said Leppert, a sensible-looking woman who rose to speak with real passion, the frustration of decades fueling her outburst. "And make no mistake about it, Congress watches, every day."

This ban on embryo research meant that no federally funded experiments could be conducted on the safety or efficacy of IVF, even as the field itself was surging ahead, unfunded and unregulated. Or, one should say, self-regulated: many among the still small circle of IVF doctors were members of the American Fertility Society, which would later become the American Society for Reproductive Medicine. Using money from the drug company Serono, they started an unofficial, unpublicized, ad hoc registry of the children conceived by IVF. The purpose was simple: to stave off malpractice suits. "We wanted a registry and Serono wanted a registry in case there was a bad outcome; in case your kid had a cleft lip, and you would sue the clinic, this way we would know that the number of cleft lips were three percent of the country," says Alan DeCherney, a cheerful, forthright man and former president of the society. According to DeCherney, at one point the National Institutes of Health approached the doctors about maintaining a government registry tracking child health, but "the only way they could do it was if the same pediatrician examined every kid, which was impractical and too expensive."

Getting at the Germ Line

But in the early days of IVF, what was really holding the field back was a technical problem having to do with women's tendency to protect their genetic offspring. Men, the thinking goes, like to spread their genes around indiscriminately, while women like to make sure their genes, and their genetic offspring, are well protected and carefully nurtured. At least, this is the really pretty reductive view of human behavior set forth by evolutionary psychology, a school of thought that sees many human actions as explainable by the drive to ensure the survival of the individual gene line.

And it's true: in women, that trove of genetic material known as the human germ line is protected to the point of being excruciatingly hard to get at. Human beings have two kinds of body cells. One kind are somatic cells: the ordinary body cells that make up the skin, the heart, the eyeballs, the blood, the hair. The other kind are germ cells. Germ cells are the unique, specialized cells devoted to human reproduction; they are the only cells capable of dividing their woven strand of forty-six paired chromosomes into twenty-three single ones, then fusing with another germ cell from another human body to develop a new human being.

Germ cells are, in short, the sperm and the egg.

For the purposes of IVF—for any purpose, really—the male germ line is absurdly easy to get at. In fertility clinics, men are given a pornographic magazine and sent to a small room to produce a sperm sample, which is processed in the lab. A small amount is then dropped into a culture dish in the vicinity of a waiting egg, which the sperm set upon and push around for some time, harrying the egg to and fro in their frantic effort to penetrate it, continuing the contest long after a single sperm has won. Meanwhile, getting at the female germ line is—or was—hell. For most of a woman's life, eggs, known as oocytes, are immature cells that lie resting in the ovary, a small organ, well protected, nestled behind other abdominal organs. One and sometimes two eggs come to maturity once a month, at which point the follicle, or pocket, in which the egg is growing swells with fluid. Although eggs are the largest cell in the body, they are, nevertheless, cells, and invisible to the naked eye. A doctor cannot simply open the belly, spot an egg, and tweeze it out.

What doctors had to do in the late 1970s for women with pelvic scarring was to inflate the woman's abdomen with carbon dioxide gas to perform a surgery—laparoscopy—under general anesthesia. Passing a telescope into the abdomen, the doctor would free the ovary from its scar tissue, insert a retrieval needle through a cut near the belly button, and, looking directly at the ovary and the follicle, puncture the follicle and remove its fluid. In the early 1980s, pelvic ultrasound allowed the ovary and follicle to be seen from outside. Once the results were brought to the lab, the staff would hope to find a pearly sphere—the oocyte—standing out against the bloody mess that is follicular fluid. Because of all the scar tissue from previous infections, half the time there would be no egg; "a woman went through a surgical procedure and anesthetic and got nothing to fertilize," recalls Arthur Leader.

To advance the field, what was needed was a better way of getting at that stubbornly inaccessible female germ line. It was a little like coal mining or oil exploration. Different extraction methods were tried. At one point a small, spring-loaded gun was in vogue. "You would measure the distance from the skin to the ovary," Leader remembers, "and you would fire a needle, and hopefully hit only the ovary. It didn't last very long. Nobody got hurt, thank God, but people didn't feel comfortable firing this needle into anybody."

Because these procedures required that the patient be put under anesthetic, they had to be done in an operating room. In hospitals, IVF labs had to fight for space with specialties that were better established.

Lab staff ended up rushing around with dishes of culture media, from the operating room to wherever the lab happened to be grudgingly located by the hospital administration, frantically pushing the buttons on elevator panels, terrified lest the egg be lost or dropped.

In the mid-1980s, doctors began to suspect it might be possible to get at the germ line from within—to take a fifth-column approach to egg retrieval. They wondered if it might be possible to insert a needle up through the wall of the vagina, rather than down through the abdomen, while using an ultrasound wand inserted into the vagina (vaginal ultrasound was developed in Europe in 1985) to get a more complete view of the ovaries. It made so much sense. When the ovary is stimulated with fertility drugs, it's larger and heavier than normal, and lies almost on top of the vagina, meaning that if you could travel through the back of the vagina with your needle, you had only about a half inch to go.

The prospect of putting a needle through the vaginal wall made some doctors nervous, because of the potential for infection, but nevertheless it was tried. The technique worked—so well that by 1987, it had taken off. In many ways this medical advance was the tipping point for fertility medicine. It transformed the field, gave it momentum, made so much else possible. It enabled not only more reliable retrieval of eggs, but also the donation of eggs from one woman to another. And the donation of eggs to science: vaginal retrieval would help advance therapeutic cloning and with it stem-cell research. Before too very long, the female germ line would become an engine driving not just fertility medicine but medicine, period. The female germ line would become one of the world's most sought-after natural reserves, essential to stem-cell research, one of the most hyped and controversial fields of medical and scientific endeavor. One day, it is hoped, the human egg, which exists to create a child, may become the vehicle for curing the diseases that accompany old age. It would therefore become a potential and potentially quite profitable fountain of youth: coveted, contested, bought, sold, shipped, argued over, freely sacrificed, stolen, traded, and sometimes obtained by ruse.

In the short term, what vaginal egg retrieval did was enable the explosive growth of IVF. Suddenly, human eggs were so much easier to get at, and as fertility drugs became more powerful and refined, there were more and more eggs to get at simultaneously. The beauty of it was that with vaginal retrieval you didn't have to put a woman under heavy anesthetic. You weren't really doing surgery. A woman could be given local anesthetic, and the egg retrieval could be carried out in ten minutes. IVF could be done—well, IVF could be done anywhere. It could become an outpatient procedure. Fertility clinics no longer had to be located in hospitals. All you needed was a couple of offices, some equipment, and a small lab, which could be attached to the retrieval room, with a window for passing specimens back and forth. "What it did was, it took IVF out of the operating room," says Leader. "It could be done now in any facility. It could be in an office building. A mall."

A mall! Why not? Fertility clinics were indeed set up in malls, discreetly, so as not to attract pro-life protests. There they remain. Human life can be created in labs in the most nondescript locations. There is one clinic near my house in Arlington, Virginia, located on the upper floor of an office building in a small strip mall, not far from a cluster of ethnic restaurants, a dollar store, and a Goodwill charity outlet. Clinics could now be run as private, profit-making enterprises. They could be set up as chains. They could advertise their services directly to patients, a development many academic, hospital-based doctors deplored, and still do, as diminishing the dignity of the profession. "The money ruined a lot of things," says Robert Nachtigall, lamenting colleagues who have left hospital practices affiliated with medical schools, and the research they promote and sponsor, in favor of the lucrative private sector. Somewhat ironically, the very profitability of the field served to impede its development: What research is done by private practices often is not shared, because of the competitive nature of the marketplace. Private practices that hit upon successful protocols guard them, "much like the forceps that [ob-gyns] guarded back in the eighteenth century to keep their competitive advantage," as Arthur Leader observes. And without the U.S. government funding and directing research, there was less of an effort—compared to fields such as cancer research—to develop standards that could improve success rates for all. Patients were paying for procedures whose quality varied markedly from provider to provider.

And they were paying a lot. Just over $12,000, on average, for a single procedure, though at some sought-after clinics, the bill could be closer to $50,000. Because infertile women—and men—were willing to pay anything. They were willing to go through IVF over and over again, submit to an unfolding series of procedures that doctors began to experiment with in the next several years. In 1986, there were 41 IVF clinics in the United States, according to an early industry registry. By 1996, that number would rise to more than 300, an increase of more than 700 percent. Ten years later there would be more than 400 clinics. Between 1968 and 1990, the number of annual office visits for infertility would increase from 600,000 a year to 2 million. Reported cases of impaired fertility would rise, from 4.9 million in 1988 to 7.3 million in 2002.

"Age of the Female Partner, Age of the Female Partner, Age of the Female Partner"

Part of the reason for the explosion was the steady improvement in treatment; part was pent-up demand from patients with infertility of a traditional sort. But part was the fact in the late 1980s and the '90s, a new kind of patient began to make her presence known. By the 1980s, feminism was well established, if eternally under assault. Professional aspirations for women were normalized; women graduating from college often were expected to work, attend graduate school, or both. Employers depended on these women, and they expected women to stick around once they were hired. Maternity leave was by no means a given. Neither were flexible workdays. Women responded by not taking maternity leave and by not asking for flexible workdays. They responded—to a complex set of incentives, including the fact that their partners and husbands were really liking this extra paycheck thing, this high-earning spouse, this chance to endlessly dine out and travel—by not having children, yet.

The new arsenal of reproductive technologies—the ones invented to help women *not* have children—would in some ways create a natural demand for the next reproductive technology, the one invented to help them have children once they were ready. "Controlling your reproduction" would become, for some, a three-decade continuum. Between 1973 and 1980, according to the Alan Guttmacher Institute, the rate of abortion in America almost doubled, rising from 16.3 to 29.3 abortions per 1,000 women. It would remain there for all of the 1980s. As for the Pill, as early as 1967, more than 6 million women were using it. In 1960, the average American woman had 3.6 children. The fertility rate thereafter began to fall steadily, until in 1970 the average women had 2.5 children. In 1975, the average American woman had just 1.77 children. Not until 1989 would the fertility rate edge back up to 2.0, where it has hovered ever since.

It's not that women weren't having babies. Well, some weren't; rates of childlessness have also increased markedly. But the ones having children were having fewer, and—crucially—they were starting their families later. At this time the age of first childbirth also began an inexorable climb. Since the mid-1990s, each year the average age of the first-time American mother has been higher than it was the year before, a pattern that has continued to this day and that has been replicated in other first-world countries. For some women the new paradigm worked; when desired, the children obediently materialized. For others it did not. And so doctors began seeing women like Melanie: bright, hardworking women who had wanted children all their lives; who had struggled to achieve in a workplace that was not geared to working mothers; and whose infertility was due to the fact that when the time finally seemed right, the timing itself was wrong. Now there were women sitting in fertility clinics who were fit; who ran marathons; who did yoga; women astonished to find out that getting pregnant was going to be even harder than making partner in their law firm had been. Now, thirty-eight is the average age at the Jones Institute, where the cutting-off point once was thirty-five. There is no cutting-off point anymore.

And the thing was, you couldn't tell who was going to end up in a clinic and who wasn't. Doctors couldn't tell—still can't tell—which women are going to become infertile and at what age. For sure, scientists began trying to get some handle on it: They studied the Hutterites, a religious sect in which contraception was not practiced and women had children until they couldn't have children any longer. What they found is this: In her late twenties, a woman's fertility undergoes an initial, modest decline. There is a steeper decline at about thirty-seven. Or rather, there *can* be a steep decline at thirty-seven. The decline is far more severe in some than in others, and it starts in some earlier than in others. Before thirty-five, most women are fertile. After thirty-five, women enter a period of extreme variability. A woman may remain fertile for ten years, or she may undergo a precipitous drop in her ability to conceive; her childbearing days may be over. As a rough gauge, doctors assume that infertility usually sets in ten years before menopause, which begins, on average, at age fifty-one.

According to one study published in *Human Reproduction,* 75 percent of women who begin trying to conceive naturally at age thirty will succeed within a year. At age thirty-five, about 66 percent will conceive within a year, and 44 percent at age forty. Failure is by no means the norm, even after forty. It just becomes much more likely. After forty, women undergo a terrible division into two camps: those who can, and those who can't. You can see these two camps eyeing each other warily in offices and coffee shops and street corners everywhere: the happily pregnant, and the miserably still-trying. At forty-five, 87 percent of women are infertile.

Why does female fertility drop, and why is it so variable? Scientists everywhere are trying to come up with the cause of reproductive aging, as well as a reliable test to predict when it will happen and even, some-day, to reverse the process. All signs point to the egg. According to the most widely accepted scientific theory, a woman is born with all the oocytes she will ever have. She has the most eggs while still in utero: at twenty weeks, a female fetus has about six to seven million oocytes in her tiny, newly formed ovaries. Massive numbers are quickly lost, however, so that at birth, about one to two million oocytes remain. The process continues; eggs die off, so that at thirty-seven a woman has something like 25,000 oocytes resting in her ovaries, and by the time she is fifty-one, she has only about 1,000 left.

A thousand may seem like a lot. But the older the egg, the less likely it is to work. Patricia Hunt, a reproductive biologist who studies aging, suspects the problem has to do with the way an egg undergoes meiosis, which is the process of dividing its forty-six chromosomes—the stumpy, loglike structures on which the genes are located—in half. Meiosis is something both sperm and egg must do before they come together. For reasons no one knows, meiosis in eggs begins almost immediately, when a girl-child is still in utero and her oocytes have only recently come into existence. The process then stops, and doesn't resume until the girl-child is mature and the egg is fertilized, years and years later. So eggs sit in suspended animation, their chromosomes awaiting the chance to sort themselves, for many years, even many decades. The older the egg, the less able it seems to be to finish dividing properly. And if an egg doesn't divide properly, the number of chromo-somes in the resulting embryo may be unequal: the embryo may have forty-five or forty-seven chromosomes, instead of the normal human complement of forty-six. Where there should be a pair of chromosomes there may be one or three, a condition known as aneuploidy that is either bad or fatal. This is why the chances of having a child with Down syn-drome—an extra chromosome 21—increase with maternal age. It's also why the chances of miscarriage rise as a woman gets older.

And more and more women are suffering miscarriage: according to the National Center for Health Statistics, the United States has seen a rise in miscarriage rates. Up to now, scientists have assumed that miscarriage rates are pretty much constant worldwide; while early pregnancy loss is notoriously hard to detect, rightly or wrongly it has always been estimated that roughly the same number of pregnancies miscarry in different countries and cultures. Some miscarriages are caused by infection, smoking, and other external influences, but the majority, scientists believe, are due to chromosomal abnormalities that doom the pregnancy from the start. Miscarriages are "nature's way" of ending a pregnancy that was never meant to be, as doctors invariably tell the devastated women undergoing them. For the first time, the miscarriage rate among American women has begun to rise, as more pregnancies are begun at later ages and more embryos are chromosomally compromised. More and more women, like Melanie, are experiencing the disappointment of getting pregnant only to lose the pregnancy, over and over.

And so it is that reproductive endocrinologists find themselves sitting across from patients, drawing the same dreary chart every day: the curve of female fertility. Little drop at twenty-nine, big drop at thirty-five or thirty-seven. They find themselves sitting across from women who, while smart and well educated, are surprised to learn that just because you are still menstruating doesn't mean you are still ovulating, and just because you are still ovulating—that is, producing a mature egg every month—doesn't mean the egg will work. They are seeing more and more women whose diagnosis is "diminished ovarian reserves," or—what often amounts to the same thing—"unexplained."

Keep in mind that older women are *not* the women most likely to be infertile. That distinction still belongs to poorer women and less well-educated ones, and to minorities. "Non-Hispanic blacks and other race women, high school dropouts, and high school graduates are significantly more likely to be infertile, and significantly less likely to have ever sought medical treatment to get pregnant or prevent miscarriage," notes one report. In fact, according to the National Survey of Family Growth, while infertility among women over thirty-five increased by 6 percent between 1996 and 2002, women under twenty-five reported a striking 42 percent rise in impaired fertility, for reasons nobody understands, from 4.3 percent of women to 6.1 percent.

But older women tend to be the ones able to afford IVF treatment, or to have insurance coverage, or both, so they are overrepresented among patients. They also attract more than their share of controversy. It is inevitably assumed that women over forty find themselves infertile because of a series of deliberate choices. If you do an online search for "women" and "delay childbirth," you will find any number of news articles and op-eds in which it is assumed that pregnancy delay is something women have single-handedly brought about—something women have *chosen*—and that it has nothing to do with, for example, reproductive dawdling on the part of indecisive male partners or active discouragement from bosses. The age of first childbearing among males has risen at the same rate as among women, yet this trend of men choosing to delay child-fathering has attracted little public hand-wringing.

The great problem is that older women are the least likely to be helped by treatment. They are the ones most likely to be spending $12,000 for an IVF cycle and getting nothing for their money. "You know the old real estate adage: location, location, location. Well, in IVF, it's age of the female partner, age of the female partner, age of the female partner," says Marcelle Cedars, a reproductive endocrinologist who has been working in the field since before IVF began. Doctors can almost gauge how treatable a situation is by looking at the woman's birth date. If a couple comes in and the woman is thirty-two and suffering from endometriosis, that's better than if the woman is forty-one and has no ailment anybody can identify. Studies show that among ART patients who are forty years old and using their own eggs, there is a 25 percent chance of pregnancy over the course of three IVF cycles. The chances diminish to around 18 percent at forty-one and forty-two, 10 percent at forty-three, and zero at forty-six.

In 2005, a group of doctors at Cornell surveyed IVF patients over forty-five who had attempted to conceive using their own eggs. Among women between forty-six and forty-nine, not one got pregnant using her own eggs.

Not one.

"Fat, Slutty, Old, Smokers"

And yet aging has remained a weirdly taboo topic among women's reproductive rights groups, who have not served their constituency well in this area. There has been an energetic muddying of the waters by some old-line women's groups, which have preferred either to ignore the facts or, in some cases, to deny them. When it comes to the consequences of aging, the same groups who encouraged women to use mirrors to collectively inspect their cervixes, and even taught them how to cobble together jars and aquarium tubing to pump out their menses, seem to regard rather differently any effort to acquaint themselves with the details of other reproductive organs.

The most notorious clash came in 2001, when the American Society for Reproductive Medicine (ASRM) decided to devote $50,000 to an infertility awareness campaign. The goal was to point out to men and women alike the effects of any number of lifestyle issues: smoking, obesity, sexually transmitted disease, and age (though, in the eternally unfair way of the world, age does not affect male fertility nearly as much as it affects female fertility). It was a pretty delicate message to sell. "Infertility patients thought we were saying they were fat, slutty, old, smokers," says ASRM spokesman Sean Tipton. But by far, the most controversial message was that of age.

This was surprising to the architects of the campaign. You wouldn't think sharing a few facts would be that inflammatory, says Marcelle Cedars, who practices at the world-class infertility clinic at the University of California, San Francisco, and who was in charge of the panel on reproduction and aging. "We knew it would be a hot potato," says Cedars, a working mother with a deep commitment to reproductive health. "We could tell that it was going to be so controversial, there was a push from some people on the committee to drop it. I'm like, 'You can't drop it. It's the most important factor in infertility for people of this generation. We can't *not* talk about it.'" So the committee approved an advertisement that showed a baby bottle in the shape of an hourglass, to communicate the idea of time running out. It wasn't a big campaign. The ads ran on metro buses in four cities. But *Newsweek* picked up on the topic with a cover story on reproductive aging; and all hell broke loose.

The opposition came from the National Organization for Women (NOW), co-founded by the late Betty Friedan. It should be pointed out that Friedan was no enemy of family life. In *The Feminine Mystique*, she posits the reasonable suggestion that women deserve a place in public life, and that they also deserve the intimacy and love afforded by children. But the current NOW president, Kim Gandy, saw the ASRM campaign as an attempt to bully women into having babies early. "I don't think we need to be putting that pressure on younger women," Gandy told CNBC. To Katie Couric, then host of the *Today* show, Gandy said that "what is essentially a scare campaign is extraordinarily ill-advised."

Fertility doctors were nonplussed.

"We were shocked. We thought [NOW] would be allies, you know, because we're empowering women," says Robert Stillman, a doctor at Shady Grove Fertility who was also on the committee, and who quickly realized that all in all the outrage was a good thing: it leveraged the tiny publicity campaign into something much bigger. Ultimately, ASRM was so energized that it decided to run the ads again, on a larger scale. "Women are, once again, made to feel anxious about their bodies and guilty about their choices," Gandy wrote in an editorial. But she eventually stopped commenting, out of a reluctance, a NOW spokesperson said, to provide the fertility awareness campaign any more free press.

Still, the conflict underlines the fact that while doctors who advocated for better contraception and abortion rights are seen by many feminist groups as natural allies, fertility doctors have been marginalized. Any effort to educate women—and men—about consequences of delayed childbearing is seen by some feminist leaders as tyrannical, oppressive, retrograde. Just as the hullaboo over the ASRM campaign was dying down, *Creating a Life: Professional Women and the Quest for Children,* Sylvia Ann Hewlett's 2002 book about the reproductive consequences of aging, was similarly attacked, from roughly the same quarters, as antifeminist. Increasingly, this is a challenge for the reproductive rights movement: there is a formidable swathe of women, now, who are lying on gynecological examining tables with their feet in stirrups undergoing the hysterosalpingogram—a terrifically painful procedure in which dye is forcibly injected into the Fallopian tubes to see if they are open—and rethinking everything they were led to believe about the ability to choose when to conceive.

"Boy, are they pissed," says Robert Nachtigall succinctly, after thirty years of working with fertility patients in San Francisco. Or as one fortysomething patient put it, speaking from behind the closed door of her San Francisco law firm, contemplating the birth of her own IVF-conceived, egg-donor child: "I didn't grow up thinking that I'll have a baby this way."

"All we were trying to do was give women information," says Marcelle Cedars, still stung by the way a major women's group turned on her. "We're not *telling* them to have children. We're not telling you should have children before having a career. We're telling them that if having children is a priority for you, it's something you should factor in. What I hate is women who show up in my office at forty, and say, I never knew this."

Why Don't You Just Adopt?

During the same period, a traditional solution for infertility, domestic adoption, was undergoing changes of its own. Adoption is a complicated topic, as fraught and controversial as anything else involving family making, and a full exploration is beyond the scope of this book. But it seems fair to say that during the second half of the twentieth century, developments were taking place that would propel more would-be parents away from adoption agencies and into medical clinics. The pool of U.S. children available for adoption was shrinking and the process for adopting them becoming more expensive, laborious, and daunting, making fertility treatment that much more appealing, at least as a solution of first resort. More bluntly: there were fewer white babies, and those who were available became harder to get.

According to Marsh and Ronner, there have never been enough adoptable children in the United States to meet the demand of infertile couples. Even so, the supply used to be larger than it is now, and adopting them used to be, for better or worse, more informal and streamlined. In generations past, parentless children were sometimes seen as miniature servants; for infertile couples in colonial times, the void was filled, in part, by taking in children as apprentices and household helpers. In the late nineteenth century there were actual "orphan trains" that carried children around the country, stopping at cities where orphans would be "put up" for adoption, displayed on platforms for viewing by members of the community, who often regarded them as sources of farm and household help, preferred boys to girls, and didn't have to jump through many hoops to take them home. In the early twentieth century the process became more formal, with the creation of maternity homes where unmarried women could bear their children, receive medical care, and surrender the newborns after birth. Couples who wanted to adopt a child had to apply, but "the application was literally one page long," recalls Paige McCoy Smith, a spokeswoman for the Gladney Center for Adoption, one of the oldest U.S. adoption agencies.

The 1970s brought a major change in adoption procedures, and changes, too, in the number of babies available. In prior decades, young unmarried mothers were often pressured by their families, and by social services, to relinquish their babies even when they wanted to keep them. That painful phenomenon is movingly explored in Ann Fessler's 2006 book, *The Girls Who Went Away: The Hidden History of Women Who Surrendered Children for Adoption in the Decades Before Roe v. Wade.* After *Roe,* women could more easily end problematic pregnancies; moreover, the pressure to relinquish lessened as single motherhood became more accepted. With abortion legal and birth control safer, more reliable, and readily available, fewer unplanned pregnancies came to term, and the babies born were more likely to be kept by the women who bore them. "What we started seeing in the eighties and nineties is a decline in the number of women making the decision to place a child for adoption," McCoy Smith says. "A dramatic decline."

A government study confirms this. In the early 1970s, 9 percent of unmarried mothers relinquished their children for adoption. By 1988, the rate was just 2 percent. The decline was most marked among white women, who—contrary to popular belief—have always provided the majority of adoptable babies.

According to the study, before 1973, 19 percent of children born to single white women were placed for adoption. That percentage fell to 8 percent in the period between 1973 and 1981, and it had dropped to just 3 percent by 1988. The relinquishment rate among black women has never been high; it has hovered around 2 percent. Hispanic women have never been very willing to relinquish babies. According to government researchers, women are more likely to relinquish children if there is a high "opportunity cost" to raising that child: if the child endangers some other part of her life—such as a career, education, marriage, or social status—the woman is more likely to give the child up for adoption. At least she used to be. According to the Evan B. Donaldson Adoption Institute, just 13,000–14,000 U.S. women now voluntarily relinquish infants for adoption each year.

At the same time, the number of parents seeking adoptive children was beginning to rise. Demand was increasing even as supply was on the downswing: "Now, for every one call from a birth mother, I receive six to ten phone calls from prospective adoptive parents," says McCoy Smith, who stresses that her agency is always—eventually—able to find a child for qualified parents. And the ethics were evolving. Before the 1950s, adoption was carried out secretly, and adoptive children were rarely told the truth of their origins. After midcentury a generation of adoptees began to protest. Adults who had been adopted as children began to argue that secrecy was traumatic and corrosive; that there was, or should be, no stigma surrounding adoption; that children deserved to know the truth of their origins and if possible the identity of their birth parents.

That movement had an enormous impact on the way domestic adoption happens. Now openness is the norm. Now the birth mother has much more control over where, and with whom, her child is placed. "In the past, when the birth mother would deliver the child, the adoptive parents were already selected by the agency," says McCoy Smith. "Now the birth mother has the option of selecting her parents based on criteria she provides." It's a better system, most believe; a birth mother who feels good about the adoption is less likely to try and reclaim the child afterward. But, McCoy Smith allows, it makes some parents nervous. It also makes the process more expensive. The average Gladney adoption takes a year, and costs at least $20,000. Even with agencies that can carry out adoption more cheaply, competition cannot be avoided. Would-be parents compete with other would-be parents; they assemble scrapbooks, describe their marriages and homes, pay for the birth mother's health care and, in some states, expenses. Some eschew agencies altogether, placing ads in newspapers and striking up independent agreements with a birth mother. All of these arrangements are closely monitored by the courts. For a judge to release a child for adoption, most states require that intended parents undergo a criminal background check, a home study by a licensed clinical social worker, and other formal evaluations. Part of the purpose is to find out how parents plan to deal with the fact of the adoption when talking to the child.

If a parent said in the home study that they did not intend to tell the child of his or her adoption, that, according to McCoy Smith, "would be a real red flag. That tells us: what's your issue with adoption?"

So Why Don't You Just Half Adopt? The Dawn of Egg Donation

Meanwhile, surprisingly early on fertility medicine proved able to deliver a compromise solution: sort of like adoption, and sort of like having your own biological offspring. Assisted reproduction could offer, to this group of older women, a way to salvage something from an unforeseen situation. The chance to—well, to adopt half a baby, in a sense, and gestate that baby and experience childbirth and breast-feeding, and control one's diet and alcohol intake and cigarette smoking and all the other things that cannot be controlled, or not so easily, if the child is being gestated by another. Fertility medicine offered the woman who cannot conceive using her own eggs the opportunity to conceive using the eggs of another.

Egg donation is one of the fertility procedures that turned out to be more doable than anyone had anticipated. As early as 1984—just three years after the birth of the first U.S. IVF baby, Elizabeth Carr—Sanford Rosenberg, a doctor practicing in Richmond, Virginia, was confronted by a patient suffering from premature ovarian failure, meaning that her eggs had stopped being viable well before the normal age. The patient asked whether her sister's eggs could be used in the IVF procedure instead of hers.

Rosenberg had no idea whether they could or they couldn't. Nobody knew what was possible in terms of trading eggs between women. Rosenberg called Zev Rosenwaks, a leader in the field; according to Rosenberg, Rosenwaks replied that there seemed to be no reason not to try. A baby was born, genetically the child of one sister and gestated by the other. Elsewhere, doctors performed scattered procedures using eggs scrounged from patients who were undergoing their own infertility treatment and had excess eggs, or women who had undergone hysterectomies and happened to have a mature egg in their ovaries when they were removed. Somewhat to everyone's surprise, many of these procedures worked. An infertile woman could rather easily be given drugs to prepare her uterine lining to receive an embryo conceived from the eggs of a different woman.

"Who knew that all you would need would be estrogen and progesterone to get the uterus to work?" says Richard Paulson, chief of the fertility clinic at the University of Southern California medical school and a pioneer of what has become a fast-growing area of fertility treatment: egg donation to older women. Who knew that a woman's body would willingly take in a little genetic alien? That the womb itself was willing to adopt? Could such a thing be done on a wider scale? Not right away. When eggs were being retrieved under surgical conditions, it was not practical, or ethical, to have a stranger undergo surgery for another stranger. At the outset, you couldn't do egg donation en masse.

And then, all at once, you could. Once again, it was vaginal egg retrieval—the new ease of getting at the female germ line—that allowed egg donation to become routine treatment. In 1987, Paulson and several other doctors in Southern California recruited a group of women to be formal egg donors. These were early adopters in the high-tech birth arena, open-minded women willing to allow their eggs to be removed and fused with sperm from a stranger, to produce children who, while related to them and their own children, would be raised in households other than their own. These women weren't birth mothers. They did not fit the social profile of someone relinquishing a living child. They were married women with children—women considerably older than most egg donors now—who knew how important their own families were to them and were glad to help someone else attain the same happy state. They were, as the pleasant saying went, "women helping women."

Doctors were making up the rules as they went along. There was no medical playbook here. There were no government guidelines. What was normal? What was wrong? Who knew? Paulson's group decided that egg donors, to be accepted into their program, had to be finished with their own childbearing. At the time, it was unknown whether egg donation might affect a donor's own fertility, or interfere somehow with a later pregnancy, so they admitted only donors who did not want more children, in case there were adverse effects. Another rule was that egg donors were not paid, or not much, because to pay them seemed coercive, and felt too much like buying a human being. These early donors "were altruistic, motivated," says Paulson, who sometimes shares a PowerPoint presentation on the history of egg donation with medical students for whom the early days must seem as remote, and as faintly absurd, as the U.S. invasion of Grenada.

Another rule, at first, was that Paulson's group would not accept infertile patients older than forty. Nobody knew whether the body of a fortysomething infertile woman could properly gestate a fetus. If a woman's own eggs had deteriorated, what about the rest of her reproductive apparatus? Instead, egg donation was performed on younger women who had suffered premature ovarian failure. Pretty quickly, though, the rule was relaxed. When some women waiting for egg donors turned forty-one, Paulson and his group decided to see what would happen if these women were allowed to stay in the program. They found that in this case age hardly mattered. Infertile women over forty could get pregnant; thanks to egg donation, they could gestate a baby as successfully as anybody. Women who were forty-two, forty-three, forty-four, forty-five. "Lo and behold, they got pregnant at the same rate as women under forty," says Paulson. "This was the light bulb. Oh, I get it: The problem must not be in the uterus. The problem must be in the egg." Egg donation underscored what doctors were already learning about female fertility: what mattered was not the age of the patient, but the age of the egg.

In 1990, Paulson and colleagues published a paper in the *New En-*

gland Journal of Medicine, "A Preliminary Report on Oocyte Donation Extending Reproductive Potential to Women over 40." This paper prompted a snowstorm of controversy in the medical academy and the popular press. Concerns centered not on egg donation per se, but on the ethics of older motherhood. Regular old-fashioned mainstream ob-gyns were concerned, pragmatically, that older mothers might suffer more complications in childbirth, which in fact they do. In papers with titles such as "Increased Maternal Age and the Risk of Fetal Death" and "Advanced Maternal Age—How Old Is Too Old?" doctors noted that the older the mother, the more likely she is to experience preeclampsia, gestational diabetes, hypertension. The older the mother, the more likely her child is to be premature, small, or—this does happen—dead.

And then there was the reaction of reproductive rights advocates. The feminist camp was split. There has always been a school of feminism that tends to view fertility medicine as one more way in which an oppressive patriarchy forces women to bear children against their will. These theorists, whom Richard Paulson refers to as "lunatic ultra feminists," had an easy time of it. Like many conservative skeptics, they could and did reject egg donation to older mothers as unnatural: "violations of nature" are often a deal breaker for Far Right and Far Left alike. But there is another line of feminist reasoning that says that egg donation to older women makes women more like men and therefore is a good and liberating breakthrough. Enabling women to become parents in their late forties is a kind of Title IX for the body, a gender equity measure that permits women to become parents at the age when many men do. After all, actuarial tables show that women live longer than men do, so isn't it more ethical for a woman to become a parent at forty-eight than for a man to do so? This is Paulson's view: "I am a feminist, and as a feminist I take great issue with the fact that [the lunatic ultra feminists] call themselves feminists," he said during an interview in his academic office.

Controversy or no controversy, Paulson and his group kept going. It was kind of the opposite of the limbo: how high can you go? In 1993 they published a paper reporting egg-donation pregnancies in women over fifty. In these cases, it was often necessary to give women drugs to reverse menopause and restore the hormones for supporting a pregnancy. At first, Paulson says, their fiftysomething patients tended to be women who had been infertile their whole adult lives, women who were always one step behind—or rather ahead of—the technology. He offers the hypothetical example of a woman born in 1943 whose tubes were blocked at age ten by a ruptured appendix and who for most of her life was regarded as hopelessly infertile. "When she was thirty-five years old in 1978, Louise Brown was born, and she thought perhaps there would be hope for me yet," Paulson explains, describing the sort of woman who did in fact make up about 50 percent of his older patients. "Then in 1983 when she was forty, IVF technology was starting to percolate into clinics in the States, but they were only taking women under forty. In 1988, she was 45 and her eggs were not going to work. Then she turns fifty and the technology of egg donation was now available. She finds that technology has finally caught up with her, and at the age of fifty, she can go in and have a baby."

This, however, is one of the many areas in reproductive medicine where the borders of "normal" are very hard to locate. Because the other 50 percent of his patients had not been infertile all their lives. They had just gotten started late. As more older women presented themselves for treatment, not just in Southern California but everywhere, clinics had to develop mathematical formulas and ad hoc age-related guidelines, which they are constantly revising and making exceptions to. Some practices have decided that they will not accept any couple whose combined age is more than, oh, say, 100. Others are willing to go further: "The combined age should not exceed 110, and I wouldn't do a single woman over 50, though God knows I might change that in a few years," one Los Angeles doctor, Vicken Sahakian, told me in 2004, snatching a conversation between treatments in his Wilshire Boulevard practice, where there is a separate waiting room for celebrities, a kind of reproductive green room. Sahakian acknowledged that he once treated a couple where the woman was sixty-four and the man was sixty-eight (they were using a surrogate). His staff were mad at him, he says, and when that pregnancy miscarried, "I refused to do it again with the frozen embryos. I said, 'I really can't do this anymore.' They took the embryos and went somewhere else."

Paulson and his group decided that fifty-five was a quasi-physiological limit for a female patient carrying her own pregnancy, the age at which the risks outweigh the benefits. Even within those parameters, the experiments produced extraordinary results. As early as 1993, Paulson and colleagues published an article in *Human Reproduction,* "Quadruplet Pregnancy in a 51-Year-Old Menopausal Woman Following Oocyte Donation." Just fifteen years after the birth of Louise Brown, assisted reproduction was able to make a menopausal woman pregnant with four babies simultaneously. According to the paper, the woman underwent "selective reduction" to reduce her pregnancy to twins.

The new normal: a fifty-one-year-old having twins.

And it was. It was the new normal. Around the world, a certain species of doctor/scientist/self-promoter was learning that a sure way to garner headlines—aside from fake claims of having cloned a baby—was to impregnate the "world's oldest mother." In 1994, Severino Antinori, an Italian embryologist practicing at a time when Italian fertility medicine was notoriously permissive, reported getting sixty-two-year-old Rosanna Della Corte pregnant. Taking pregnancy into the seventh decade! Clearly, that was going too far. "Antinori is a crazy guy," says Paulson, who himself was working well within the confines of what seemed acceptable: just sitting there in Long Beach, adhering to his first-half-of-the-sixth-decade limit. Then, one day, a woman named Arceli Keh walked into Paulson's office. Keh said she was fifty. Paulson thought she looked older but figured that every person ages differently. So he treated her, successfully, using donated eggs. He released her to a gynecologist, who presently called to let him know that Keh was in fact over sixty. "I said, watch her like a hawk," says Paulson, who was relieved when Keh, who developed gestational diabetes, delivered a healthy baby by C-section.

Paulson felt that ethically he had to write the case up for publication in a medical journal. But he needed to know the scope of his achievement. So he contacted Antinori to find out exactly how old Della Corte had been. Turned out that Della Corte was sixty-two years, six months, and two weeks on the day of delivery, while Arceli Keh was sixty-three years old and nine months. Turns out that he, Richard Paulson, feminist, normal guy, bona fide academic, had created the world's oldest mother! By mistake! His old mother was older than Antinori's old mother! So he wrote for the journal *Fertility and Sterility* an article titled "Successful Pregnancy in a 63-Year-Old Woman," which described the case and followed with a discussion of the difficulties of setting reasonable age limits on motherhood. After raising the question of whether there should be a law declaring how old is too old for a woman to give birth, Paulson reviewed the reasons why there should not. If there were a law, he pointed out, doctors would have to card their patients, and you know how easy fake IDs are to make. Plus, what would that do to doctor-patient trust? Plus, what if a fraud were committed: would the doctor be sent to jail? How unimaginable is that? He concluded, inevitably, that "the decision as to when and how to procreate is best left to the patients and their physicians."

In short, like so many others in every arena of reproductive technology, he fell back on choice.

The New Old Mother

Before long, "world's oldest mother" would become a staple of news bites, a reliably interesting and borderline sick category, like "worst-dressed actress at the Oscars" or "most-divorced country singer." "Retired lecturer, 67, set to be oldest mother, and it's twins!" said a 2004 article in *The Times* of London, describing the Romanian woman who may or may not be the actual "world's oldest mother" (though it turned out to be a singleton, not twins). This would be Adriana Iliescu, who in her Web photos really does look startlingly old, frail and wrinkled and bleak-looking, possibly because Botox and hair highlights are harder to come by in Eastern Europe than they are in Los Angeles. Before long, to make headlines a woman who was moderately old had to possess some other distinction, as in "California woman in her 50s gives birth to second child," a headline that appeared in 2005 in the *Orange County Register,* which, like many newspapers, adopted a jolly, who'd-a-thunk-it tone in describing Carolyn Pelcak, who, having given birth via IVF at fifty-two, decided to do the same at fifty-five, after which, the article noted, "the single mom is loving every minute of it."

Then there was "Woman, 55, Gives Birth to Grandchildren," the headline in the *Washington Post*, profiling a woman who delivered triplets in Richmond, Virginia, while acting as a surrogate mother for her own daughter. Or "Great-Grandmother Becomes One of the Oldest Women in the World to Successfully Give Birth," the careful headline on the 2006 press release in which an online casino called Nine.com awarded its "Gambler of the Week" award to Janise Wulf, of Redding, California, who at sixty-two—diabetic, blind from birth, the mother of eleven children, grandmother of twenty, and great-grandmother of three—used IVF to conceive her twelfth child. Her husband was forty-eight. Nine.com treated the feat as funny, noting that "Gambler of the Week" awards are reserved for "those who take risks by doing something unusual, risky, heroic or foolish."

Small wonder that by now the categories are finely parsed, the superlatives elaborately qualified. In 2004, when Aleta St. James, the "energy healer" whose newborn twins are, of course, featured on the website that offers her up for speaking engagements, gave birth at fifty-seven, she had to settle for being "the second oldest woman in the United States to give birth to twins."

As for over-forty moms: who cares anymore? A little more than ten years after the publication of Paulson's "preliminary report" on fortysomething mothers, a parade of high-profile women, many of them actresses facing the professional dry spell that Hollywood imposes on the talented but no longer nubile, have made fortysomething mother-hood seem almost—natural. Jane Seymour, forty-four. Susan Sarandon, forty-six. Geena Davis, twins at forty-seven. Holly Hunter, twins at forty-seven. The late Wendy Wasserstein, a premature daughter at forty-eight. The woman in the cubicle next to you—well, giving birth at forty-five just feels normal now, doesn't it?

You can conceive at forty-eight, can't you? Using your own eggs?

This is what many people now assume. Ironically, the parade of ART-assisted older mothers has served more than anything to cloud the waters of reproductive accuracy. Infertility awareness campaigns not-withstanding, the medical profession is hard at work blurring its own message. Even now that "biological clock" has become a grim cliché, women can be forgiven for not knowing exactly when the alarm bell goes off.

The real ethical issue isn't older mothers. It's the fact that many celebrity older mothers, who inevitably benefit from the publicity surrounding their darling newborn, or newborns, tend to leave out the pesky detail of having had help. Aleta St. James, bless her self-promotional heart, thanks her egg donor in the course of tirelessly telling her story of New Age determination and late-in-life childbearing. But she's the exception. How should one think about the don't-tellers? Should closeted egg-donor moms be outed? For the sake of collective education and reproductive truth telling? What would be the thinking feminist position here? Example A in this bioethical essay question might be Joan Lunden, the trim former anchor of *Good Morning America*, who, after raising one family in the usual window of time, remarried a younger man and decided to do it all again. (This is a common pattern: many fiftysomething mothers are women who have married a younger man and want to "give him a child," a situation that in and of itself is a little hard to know what to think about.) In so doing, Lunden secured her place on magazine covers not one but two years in a row. Lunden "had" twins in 2003 at fifty-two; two years later she had "twins again!" as *Good Housekeeping* trumpeted on its cover. The double whammy of coverage must have been a welcome boost to Lunden's career, which includes a book on childhood nutrition, and, inevitably, a stint hosting a reality show.

In a series of interviews on her "miracle babies," Lunden acknowledged using a surrogate to carry the babies, a woman she warmly thanked. She would not, however, comment on whether she also used an egg donor, despite the fact that she clearly—unless these truly were miracle children—must have done so. Her reticence was for the benefit of "all the other people who are calling and writing me now, wanting to do this," she said piously when *Ladies' Home Journal* had the temerity to ask her whether she used an egg donor. "I don't want them to feel that they can't achieve what we have if they can't produce their own eggs. I want everybody to understand that however they make their families doesn't make any difference."

Huh? If doesn't make a difference how you make your family, why not go ahead and say how you made your family? Could it be that not acknowledging egg donation is something a public woman does because she wants the public to think of her as young?

And how to feel about that warm and likable late-in-life mom Elizabeth Edwards, wife of former North Carolina senator and Democratic presidential candidate John Edwards? During the 2004 campaign, when he was running for vice-president, the Edwardses' young children, Jack and Emma Claire, were one of the few reliably cheerful sights on the campaign trail. Charmingly reminiscent of John-John and Caroline, they enhanced John Edwards's Kennedyesque air of youth and vigor. And yet their means of conception was nothing if not edgily modern: After the Edwardses' teenage son, Wade, was killed in a car accident in 1996, the grieving Edwardses, who also had an adult daughter, decided to conceive again. When she bore Emma Claire and Jack, Elizabeth Edwards was forty-eight and fifty, respectively. In 2004, an article in *Slate* speculated on what doctors had been discussing as a certainty: the younger Edwards children were most likely conceived using egg donation. Elizabeth Edwards has declined to address this or to acknowledge using IVF, saying only that she had some help from "shots," and that to say more would not be "ladylike." John Edwards, in keeping with the way of the world, does not appear to have been asked.

The *Slate* article provoked a barrage of e-mails to the Fray, the magazine's tetchy repository for reader response. Many thought it was wrong to invade the Edwards family's privacy. But was it? And if so, why? It's true, families who use fertility medicine are not obliged to reveal every detail of treatment, and if public people choose not to comment on family matters, maybe they have good reason. Maybe they don't want their children to suffer any stigma. Then again, wasn't it precisely "stigma" that provided the rationale for secrecy back in the days of hiding the truth about adoption? Wasn't it just this idea of stigma that adoptive children were battling? While hardly rising to the level of a campaign issue, the question did offer Edwards the opportunity to be of service to other women, and to their children, and dodging it may have misled some into thinking a fiftysomething pregnancy can be accomplished easily.

"You need a huge section [in the book] talking about the biology of reproduction," I was earnestly told by Fady Sharara, a Virginia reproductive endocrinologist who gets very exercised on the subject of well-known women who are in the closet about having used an egg donor. Sharara was one of the doctors who went on record as saying Elizabeth Edwards likely used egg donation to have those lovely, much-loved children. To Sharara, honesty in these matters *matters*, because he is the one left cleaning up the mess created by minor little half-truths and ladylike omissions. "Women are clueless, *clueless*," he said. "They come in at forty-four, forty-seven, and don't know why they aren't getting pregnant." The day I spoke with him he said that a fiftysomething woman had recently called him from rural southwestern Virginia, expecting that fertility medicine could produce for her a biological child. She was incensed to learn that it can't. She felt he had wronged her. In just thirty years, ART itself has become part of the problem: in persuading women that they can safely wait, assisted reproduction is creating for itself a new patient base.

"If It's Not Going to Be Mine, Couldn't It Not Be Both of Ours?": The Marital Politics of Egg Donation

"I wish I had known some of this ahead of time," says Melanie over lunch, still trying to make sense of her own uncertain journey into the hardest parts of treatment. Having once sworn that she would never even try IVF, Melanie is considering using a donor egg. This is the kind of "choice" now facing women: When should I abandon all hope of a bio-logical connection to my baby? When should I give up on ever seeing my own features reflected in the face of my child? At forty-two? Forty-three? Forty-four? Melanie is looking for any small hint, any piece of data, any test result that might offer an answer. Meanwhile, more and more fertility practices are offering egg donation as the natural next step of treatment, an automatic transition in an ongoing medical process. "We've seen over the past couple of years, moving couples quickly into donor egg, almost as a standard of care," worries one therapist. "Donor egg for these couples is a last-minute switcheroo, snatching victory from the jaws of defeat," agrees Robert Nachtigall.

And patients are left wondering: What will it feel like, giving birth to a child who is genetically related to my husband, but not genetically related to me?

"I think about this all the time," says Melanie, who, after she had endured the second blighted ovum—or was it the third? There have been so many—decided that old-fashioned adoption would suit her fine. No more injections. No more bruises. No more blood tests. No more miscarriages. Tired of the physical toll treatment was taking, she decided that she would be glad to adopt a child domestically. She would be glad to adopt a child internationally. She would willingly adopt a special-needs child. She and her husband had the means; they could afford to give a child, any child, plenty of resources and love. At least, this was the way Melanie saw it, and so she came home one day and said to her hus-band: "Let's adopt."

"We haven't exhausted our options," her husband replied.

Who knew that Melanie's husband would be the one propelling them toward egg donation? "If it's not going to be mine, couldn't it not be both of ours?" she asked when they were discussing what to do next. Her hus-band replied, "I can't believe you wouldn't want to at least try to have a child that's ours."

Ours?

"That," she says now, "shocked me more than anything."

And so now it's Melanie's husband who is urging her to continue with egg donation. While genetic connection matters to him, he cannot see, she says, that it should matter so much to her. After all, she'll be carrying the child. She'll bear the child, deliver the child, breast-feed the child, bond to the child. Melanie is not so sure it will be that easy. She doesn't think she would have the same ambivalence about adoption. Adoption, she thinks, would be clear-cut. Adoption would be straightforward. With adoption, she and her husband would be equally unrelated to the child. Moreover, adoption seems more honest. She worries that egg donation would feel—to her—like a public lie.

"I would feel like a fraud, like I'm going through this whole process with something that's not really mine," she speculates. "This wouldn't be my biological child, so why am I going through this whole charade?" What, she worries, if she doesn't love the child? What if she does love the child, but in a different way than she might have loved a child related to her? Would such a child feel like her child? Would it feel like more of her husband's? Would he have more authority? Would he have more say?

What about their marriage? Melanie and her husband have had to work through a disagreement that many couples never imagine confronting. "I don't think he's ever going to be comfortable with adopting, and I have to respect that. That's where his boundary is," she says now. Their marriage was strained, but emerged, she thinks, stronger. Adoption: out, for now. Egg donation: possible. So they talked to the fertility practice about their available egg donors.

What they found was that the logistics and ethics of egg donation have changed enormously since Richard Paulson and his colleagues were making up the rules in Long Beach, California. "Women helping women" is no longer an ethical guideline; it's an advertising hook in every college newspaper and neighborhood pennysaver. The initial guidelines have been jettisoned. Now egg donors are young; the majority have not begun their own families. Now egg donors are paid. They're paid a lot. And because they are paid a lot—because there's good money to be made—private commercial agencies have gotten involved in procuring donors. Unlike adoption agencies, egg-donation agencies do not have to be licensed. They are not charitable. They are not run by churches or nonprofits. They are businesses. Yet unlike real estate agents, egg brokers do not have to take written exams or obtain certification. Real estate agents are dealing with houses; egg brokers are dealing with the essence of life, the most sought-after human cell in the world. And they are eagerly recruiting donors, with campaigns so slick that many medical practices can no longer hold their own against the Tiny Treasures and Loving Donations and Creating New Generations of the world. "We're out-advertised, out-hustled," says Richard Paulson, acknowledging that even USC now often must get its egg donors through profit-making agencies. Some of these donors are young enough to give Paulson pause. How young is too young to surrender your eggs to another? Paulson personally draws the line, or tries to, at twenty-one.

"I struggle with these things," says Paulson. "I would so like this to be women helping women."

Melanie struggled too. At first, she and her husband looked through the thin notebook provided by their hospital practice, which recruits many of its Caucasian donors from rural Pennsylvania, there being a shortage of white donors in Washington, D.C., a majority African-American city with a high demand for donors of all races. Problem was, there weren't many donors in the book, and they weren't very well presented. Whereas every Sunday newspaper carries advertisements from private clinics and for-profit agencies that maintain fancy databases of donors. So Melanie and her husband checked out one of these private clinics with an enormous egg-donor program. "Each one was numbered, and had a little bio," Melanie remembers, and there was a wealth of information about each donor's achievements and aspirations.

Melanie had more relationship surprises in store. She and her husband had different donor-selection priorities. While Melanie was looking at donors' weight and hair color, hoping for a child who would resemble her and her husband, her husband was drawn to one thing: height. "The donor he was completely obsessed with was about six feet tall," Melanie says now, laughing. "I'm like 'Look at me! I'm not six feet tall!' He's like 'What if we have a boy?' I'm like 'What if we have a girl?' We start calling her the Green Giant. We'd look at other profiles, and then I'd be like 'Oh, we're back with the Green Giant.'"

So who gets to choose?

Both. Neither. Melanie talked her husband out of the Green Giant and they picked a donor who satisfied both of them. They paid about $20,000 to get treatment started. Only after they had paid did the fancy private clinic initiate psychological and physical testing of the donor. This was Melanie's introduction to the fact that many for-profit egg brokerages don't do the major testing of donors—testing for disease, family medical history, and fertility potential—until that donor has been selected. It isn't cost-effective to test a donor who won't ever get chosen. In truth, most huge databases of donors aren't as huge as they seem. Often, as in Melanie's case, the first donor doesn't pass some aspect of screening. The donor they had selected had alcoholism in her family, parents who were serious, active drinkers, and now that donor was no longer a candidate. Would they like to pick another?

Melanie and her husband now find themselves back at square one, trying to figure out whether to go with a different donor, or try to get their money back and seek out a different agency, or what. Ironically, Melanie is taking the birth control pill. One thing she could not live with is another natural pregnancy followed by another blighted ovum, another miscarriage. And that, she feels, would be the inevitable result of a natural pregnancy at forty-two. "I do feel like I was betrayed. Maybe you can have it all, but you need to be thinking long and hard from the beginning," says Melanie, picking at her pasta.

"I could have pressed hard earlier. I feel so stupid. I just didn't think it was going to be an issue."

CHAPTER THREE

EVERY MAN A FATHER, EVERY MAN INFERTILE

Invisible Man

Paul Turek is mad at me. Well, not mad at me personally, but mad at the state of the world. And not mad, really, as much as chronically irritated. Just now, Turek is irritated because I have shown up in his office to talk about men only after doing months of interviews, with women's doctors, about women. Women! To Turek, a urologist and a passionate if somewhat embittered advocate of men's health, this goes to show what men are up against. "I'm the guy who's been here ten years, and you've just discovered me," says Turek glumly. He is a dark-haired, intense male infertility specialist at the University of California, San Francisco. "That's exactly what's going on. The men's health movement is way behind the women's health movement."

For proof, Turek describes a situation he encountered not long ago, a case that did not have to do with infertility, exactly, but seems to him powerfully illustrative of how neglected men's reproductive needs are. The example involved a patient who had lost a testicle to testicular cancer. Now, if you are a *woman* and you have lost a breast to breast cancer, after treatment is finished you will have no problem finding a breast implant, if you would like your body restored to what it looked like before. Turek wanted to do for this man what is routinely done for women: he wanted to put an implant where the excised body part used to be.

The problem was that testicular implants for humans didn't exist. The only thing Turek had found was a testicular implant for a *dog*. The

FDA wouldn't approve its use in a human, so Turek had had to literally make an implant and get it approved, and once implanted it made a huge difference to this man's self-esteem, and why is it that nobody ever thought of that before, when women are offered many cosmetic solutions to cancer-related disfigurement?

He asks accusingly, looking at me.

While Turek was talking I became momentarily distracted by the question of why testicular implants for dogs exist. I can't ask Turek because he is still venting, still for-exampling, still talking about the unsung men who end up in his office, including, first and foremost, men suffering—really suffering, as much as women do—from infertility. Rich men. Poor men. Men with misshaped sperm. Men with physical blockages that prevent sperm from emerging. Refugees from African civil wars who have had their testicles shot to pieces, their gonads shredded, no joke, by gunfire. Men who want a child, just as women do, and who will, like many women, go to extremes to try to have one. Nobody pays much attention to these men because everybody is obsessively thinking about women.

"Men are encouraged to work hard, shut up, be culturally invisible, and get the work done, and not complain, and deal with it," Turek says. "And so in a man's world, you see several reactions to infertility. You see deep depression, self-blame, and guilt. Lots of guilt. Or you see men brush it under the rug and have no idea of how to deal with it, and it comes back in a destructive way such as a divorce or something bad. When I give lectures about male infertility, I always say: 'Remember the underdog.'"

So okay!

Men and infertility: A terrible thing.

Men and infertility: An affliction that is possibly getting more common.

Men and infertility: Medical progress so complete and so successful that it's deeply alarming—especially, as it happens, to Paul Turek.

"People Don't Like to Ship Bitches"

"Come in," says Rob Ginis, a pleasant investment banker in his early forties, opening the door and turning to maneuver his wheelchair back toward the dining room of the Marin County home where he and his wife, Hillary Fox, and his older daughter, Marnie, are lingering over a Friday night Shabbat dinner. I follow him down the hallway, trying to be quiet, knowing that somewhere in the house a baby is sleeping. But the parents are not worried about noise. This is a great baby. This is a good sleeper. This, for any family, is a priceless moment of domestic contentment. The challah is half eaten, sitting on a breadboard on the table. There is wine, still, in the wineglasses. Ten-year-old Marnie is at the table fully engaged in conversation with the adults. The golden retriever is on the floor playing with a tennis ball. The household has that palpable air of happiness enjoyed by a family coming together after the week's work and schooling are through.

For Rob Ginis, it is a particularly hard-won home life. In 1985 Ginis was a senior at the University of Michigan, planning a career in business and blissfully unaware of advances in reproductive science. For spring break that year, Ginis and a friend took a trip to Southern California. There was a car accident. Ginis woke to find he was a paraplegic, paralyzed from the waist down. Several vertebrae in his neck were fractured, and his spinal cord was completely severed at the T12 vertebra, farther down his spine. The first thing doctors told him was that he would never walk again. The second thing they told him was that he would never father children. The second piece of news was worse by far than the first. He wasn't ready for children yet, Ginis says, but "I knew I wanted to be able to" have them.

In one way, his accident happened at a good time. When Ginis was flown back to Michigan for rehabilitation, he encountered a doctor who told him about an experimental technique to help men with spinal-cord injuries have biological children, a technique that sounds unbearable but isn't, at least not compared to what a paralyzed man has already gone through. The technique is called electroejaculation. Simply put, electroejaculation involves inserting a probe into the patient's rectum, and turning the power on. The device emits an electrical current, which stimulates nerves and causes the paraplegic body to do one thing a paraplegic body usually cannot do: ejaculate semen. In most paraplegics, the procedure doesn't hurt. In Rob Ginis—"because of a quirk of anatomy"—it did.

But he was glad to have it, deeply glad. The device was invented by an Irish veterinarian named Stephen Seager, whose achievement is an example of the unlikely sources of advances in treatment for male infertility, and the way in which these advances, without coordination or planning, have resulted in near total victory in the battle against male reproductive failure. In the 1960s, Seager was a veterinarian working at the University of Oregon medical school, where he produced the world's first litter of puppies conceived using frozen dog semen. To animal breeders this was a welcome breakthrough: up to then dog owners who wanted to breed a female had to subject her to a car trip or a plane flight to visit a stud, stressful experiences during which valuable animals can be lost. "People don't like to ship bitches," explains Seager, and thanks to him, people no longer had to: frozen semen could be shipped to the bitch's owner.

Seager began to get involved in wildlife breeding. A self-described "zoo nut" with an affinity for gorillas, big cats, and other charismatic megafauna, he joined the international effort to preserve endangered species by breeding them. Traditionally, an impediment to wildlife breeding had been the fact that wildlife are wild. Before a female can be artificially inseminated, semen must be obtained from a male, which is difficult, for different but obvious reasons, whether the animal is awake or asleep. Seager began tinkering with an electroejaculation device, which he first used, successfully, on a sedated Canadian timber wolf. He began traveling the world, electroejaculating the snow leopard, the cheetah, the Sumatran tiger. He also used his expertise on zoo animals, which can be famously reluctant to mate. "You would have a pair of leopards," says Seager, "and they would just sit there and look at each other."

During this work Seager joined the Center for Comparative Medicine, an effort at Baylor College of Medicine to find ways to use animal models to cure human disease. There Seager began to wonder whether electroejaculation might help another captive population: men with spinal-cord injuries.

These are men whose infertility, like that of older women, in a sense could be described as self-inflicted, as well as particularly unfortunate for the species, because it tends to happen to those members who are most ambitious, most high-spirited, most risk-taking. Many spinal-cord injuries occur in younger men, who are prone to drive fast, rock climb, parachute, scuba dive, bodysurf, fight, and stand on their heads while drinking from the tap of a beer keg at a fraternity party. When any of these endeavors results in a paralyzing spinal-cord injury, the patient has to confront a life deprived of normal movement—a life deprived, often, of children. It is a terrible punishment for exuberance. When Seager called to offer his services to a Michigan doctor who was doing research on spinal injuries, the doctor said, "I've been waiting for sixteen years for someone like you to call."

In 1986, Seager used electroejaculation to achieve the first pregnancy using sperm from a spinal-cord-injured patient. "Since then my life hasn't been the same," he told me. "It totally changed my life." He refined the device, for which there are now promotional brochures; got approval from the U.S. Food and Drug Administration to market it for use in humans; went around the world, teaching urologists how to perform the procedure. There was a deeply grateful market. "Most of my guys are so happy to see their sperm. Some haven't seen it for twenty to thirty years, and both they and their wife break down and cry. You can't realize the problems other people have, until you stand in their shoes," said Seager, an avid polo player who understands the appeal, and the possible consequences, of extreme physical risk.

Seager also saw that his invention could have a broader use: the same device might be used to get sperm from men who were impotent owing to diabetes, spina bifida, multiple sclerosis; men who had undergone pelvic surgery and suffered damage to fragile, important nerves. In all of these situations, the male germ line can be unusually hard to get at. "Ejaculation and urination are two of the most complex things in the body," explained Seager, who is justly proud of the fact that thousands of babies have been born thanks to his invention, and hundreds of doctors have been personally trained by him worldwide. In Washington, D.C., where Seager ended up, the National Rehabilitation Hospital has a formal fertility program for men with spinal-cord injuries, a population that is always finding ways to renew itself. "No sooner does mankind invent air bags and safety belts than there are other ways to become paralyzed," Seager reflected. "In Washington, you get a lot of gunshot wounds. In Basel, Switzerland, they don't have gunshot wounds, but the day I came in, there were three guys with hang-gliding injuries. There are between ten and fifteen thousand spinal-cord injuries [in the U.S.] a year, and that doesn't count the casualties back from Iraq and Afghanistan."

Seager was working at the National Rehabilitation Hospital in 1994 when Rob Ginis—now married, working as a banker in Washington, and living in Falls Church, Virginia—came in for treatment. Seager performed the procedure, and the happy result was Marnie, the hundredth child born as a result of his device. Ginis and his wife later divorced; Ginis moved to the Bay Area and remarried. Once he and Hillary Fox were ready to have a child, Seager flew out and trained Paul Turek. Hence the blessed wonder of baby Lucy, sleeping in a bedroom off the dining room as we sit and talk about her conception. In Ginis's case, it was not even necessary to use IVF. The retrieved sperm could be directly injected into the cervix. It was so easy to get both of the girls. "Marnie is truly a miracle baby," says Ginis, using the term parents inevitably adopt to describe a child who is in truth the result of profound advances in medicine and science, plus plain old mechanical tinkering.

"Maybe I am," says Marnie, who sits at the table listening intently to a story riveting and familiar: how her parents had to carry the sperm sample from appointment to appointment, how it took seven times till it worked, all the clinical details that are just—well, they're just part of her story, the story of how Marnie came into being. "Maybe I'm not."

"After I've Been Talking for an Hour You'll Have Several Billion New Sperm"

Spinal-cord injuries are a small subset of infertile men, but it's striking how many doctors take real satisfaction—and rightly so—in treating the men who have suffered them. The success of this treatment also illustrates how completely, and how quickly, men have benefited from a branch of medicine intended to help women. Thanks to an assortment of sometimes unlikely experts, tinkering with contraptions and refining procedures in various places for various reasons, medical science has all but solved the problem of infertility in men. At many clinics a quiet sea-change has taken place. Male patients are now as numerous as women, and in some cases—as at the Cornell clinic in Manhattan—men out-number women. In Western Europe, procedures to overcome male infertility are more frequently used than those that help women. When this was announced at an international conference in the summer of 2005, there was widespread discussion of the possibility that male infertility may be on the rise.

To understand the significance, and consequences, of these developments, it helps to understand sperm production, which unlike egg production continues for most of a man's adult life. Men, like women, have a germ line, a reserve of special stem cells that arise early in fetal development and travel to the gonads in an instinctual migration (one of the many wonderful things about cells is that most of them can move) and whose express purpose is the perpetuation of the species. When a boy reaches puberty, his testicles turn into ceaseless, tireless sperm factories; sperm stem cells begin to proliferate through a process called mitosis, doubling and bubbling like soap spilling out of a dishwasher. The process of spermatogenesis, or sperm manufacture, takes sixty-four days as the sperm moves through a series of passageways, undergoing discrete stages of development on the way to maturity. A new round of spermatogenesis is initiated every sixteen days, so that there are always sperm in different stages in different parts of the testis. "Every time your heart beats, you make a thousand sperm," a veterinarian named Ina Dobrinski told male scientists sitting in a classroom at the Marine Biological Laboratory in Woods Hole, Massachusetts, in the summer of 2005. "So that if nothing else, after I've been talking for an hour you'll have several billion new sperm."

Sperm manufacture begins in tiny compartments of the testes called the seminiferous tubules; once made, they flow into the also tiny epididymis, and from there into the vas deferens, an 8-inch tube, from which they pass into the urethra and out of the body. As is also the case with oocytes, sperm cells are shadowed by somatic cells: regular body cells whose job it is to nourish the human germ line, talk to it, nurture it, raise it—like parents, or teachers, or coaches, or mentors—and guide it into maturity. Somatic cells load sperm cells with information, passing messages along at the right time, helping them develop properly. During maturation the sperm grows a tail, which will propel it forward; it assumes a missile shape, and it acquires the acrosome, which produces enzymes that will enable it to penetrate an egg. The sperm, like the egg, must undergo meiosis: it must divide its forty-six chromosomes into twenty-three, something the sperm accomplishes by splitting. After the split, one sperm gets twenty-two regular chromosomes plus the X chromosome that will result in a female baby; the other gets twenty-two chromosomes, plus the Y that will result in a male. The chromosomes draw together and become compacted in the head.

After a sperm is ejaculated, it truly is like a swimmer, organized for maximum velocity: streamlined, shaved down, sleek. During normal intercourse, the sperm enters the vagina, where it must first make it past the viscous barrier of the cervical mucus and then through the uterus and into the Fallopian tubes, an arduous passage during which it continues to develop, undergoing capacitation, which means it acquires the ability to bore a hole into the egg's outer covering, or zona pellucida. The sperm knows where to go because it follows the chemical path of released follicular fluid that surrounded the egg before ovulation. The penetration of the winning sperm triggers, in the egg, the cortical reaction, a burst of calcium that releases enzymes causing the egg's outer membrane to instantly harden, to prevent more sperm from entering. At this point the winning sperm causes the egg to resume its own meiosis, or chromosomal splitting. (Half the egg's chromosomes are directed into the polar body, which drifts away outside the egg and disintegrates.) In turn, the egg opens up the sperm and encourages the sperm's compacted DNA to express itself. The sperm's energy being spent, and its set of chromosomes appropriately delivered, the egg takes over the development of the embryo.

Anybody who studies this process finds it amazing that conception ever goes right. And when it goes wrong, there can be any number of causes. In some men, infertility is caused by a varicocele, a varicose vein that presses against the scrotum, causing sperm to overheat and die. Sometimes, the problem is a missing passageway, the vas deferens, which means the sperm cannot exit the body. In some cases, sperm die en masse, or are not produced to begin with, resulting in a low or a zero sperm count. In some cases the sperm suffer from low motility: they lack the engine to make the passage and penetrate the egg. In some cases they are misshapen, with two heads, or two tails, or worse: UCSF reproductive endocrinologist Linda Giudice has a vivid chart showing the grotesque ways sperm can be malformed, which she likes to show at conferences.

In men as in women, disease and infection once played a large part in causing infertility; in third-world countries they still do. Among the culprits are smallpox, mumps, and gonorrhea, all of which can cause scarring and blockage of male tubules. Most of these have been eradicated in first-world countries, where they have been replaced by "lifestyle" causes: smoking, for example, has a terrible impact on chromosomal organization. Stress is bad for sperm. So are sitting and heat. So is sitting while exposed to heat: studies have confirmed that sperm counts are diminished by hot tubs, by long-distance driving, by laptops. Happily, for men, some problems can be reversed by putting the laptop on the desk, or quitting smoking, or taking a long vacation, and waiting sixty-four days for a new batch of sperm to be brewed. The ability of sperm to rebound is shown by a study that found men who work outdoors are less fertile in the summer and more fertile in cooler seasons, a phenomenon that may partly account for the fact that in hot climates, fewer babies are born in spring.

Then there is the inescapable problem of being human. Compared to other animals, even fertile men have a mortifyingly low sperm count, existing, as humans apparently do, on the windy edge of extinction. "Human males are just infertile, period, compared to any other animal," cheerfully notes Sherman Silber, a urologist who heads the infertility clinic at St. Luke's Hospital in St. Louis. "Humans have the worst sperm except for gorillas and ganders of any animal on the planet. If you look at almost any animal, cockroach or elephant, the average sperm production is about 25 million sperm per gram of testicular tissue. That's pretty constant. In humans it's about 4 million. We have, in general, about one-sixth the sperm production per testicular weight."

The reason for this appears to be the false security induced by female monogamy. The tendency of women to be faithful, at least in the short term, has enabled the man to rest on his reproductive laurels. In many animal species, a female may be mounted by a number of males, whose sperm then enter a race to the finish, a genetic swim meet in which the fastest and/or most numerous are most likely to procreate and pass their genes along. "The moose with the biggest horns and the biggest muscles is going to win," Silber explains, because that moose will also have the biggest and most productive testicles. Male humans can get by with fewer sperm, because it is rare that one man's sperm must race any other man's to reach the egg.

So things are pretty bad in humans, sperm-countwise, in the best of circumstances. A real question is whether they are getting worse. In 1992, a Danish scientist published a study suggesting that between 1938 and 1990 sperm counts declined by about 1 percent a year—a striking fall—from about 113 million sperm per milliliter to 66 million. (A sperm count of less than 20 million per milliliter is associated with substandard fertility.) This set off a great deal of controversy; as inevitably happens in science, other studies challenged the methodology of that first study. Then in 2000, an American scientist, Shanna Swan, duplicated the Danish findings, showing that sperm counts are dropping by about 1.5 percent a year in the United States, and 3 percent in Europe and Australia, though they do not appear to be falling in the less developed world. This may not sound like a lot, but cumulatively—like compound interest—a drop of 1 percent has a big effect. Swan further showed that there is a regional variation in sperm counts: they tend to be low in rural areas and higher in cities, suggesting the possible impact of agricultural chemicals.

This finding coincides with what may be a subtle feminization of male babies, a disturbing series of alterations in the male reproductive organs. In the United States, scientists have reported an increase in male babies of hypospadias, a birth defect of the penis, in which the urethral opening is located on the shaft rather than at the tip. A rise has also been noted in cryptorchidism, or undescended testicles, which also can cause fertility problems. Some scientists suspect that many chemicals, including those found in pesticides and plastics, are acting like estrogen, a major female hormone, disrupting the development of the male reproductive organs. Shanna Swan has done experiments showing that in male babies, the ano-genital distance is decreasing; if you measure the distance from a baby's anus to the genitals, the distance in males seems to be getting shorter, closer to that of females. These findings suggest a possible new interpretation of statistics showing that young women are having more difficulty conceiving. Maybe it's not the young women who are infertile. Maybe it's their partners.

There is even talk that a celebrated development like the decline in teenage pregnancy may not be altogether a good thing: rather than a sign of more responsible sexual behavior, the decrease may signal that young men are less able to get their partners pregnant. In the United States, there is no government-sponsored effort to track male fertility; researchers interrogate women about their efforts to conceive, but rarely approach men with clipboard in hand to gather similar information. Trends in male-factor infertility can only be inferred from a catchall category called "impaired fecundity," a relatively new statistical category that includes any woman, of any age, who is trying to get pregnant and not having any luck. Since 1992, the category of "impaired fecundity" has seen, in the United States, a steady rise. That the rise has been most striking in younger women is something researchers find significant when trying to ascertain trends among men.

In a letter to the journal *Family Planning Perspectives,* Shanna Swan reasonably proposed "that the role of the male be considered in this equation." Maybe the rise in impaired fecundity is taking place not in younger women, but in younger men. Maybe it isn't just abstinence education and better birth control that are leading to a decline in teenage childbearing. Maybe it's males who are less fertile than their fathers. If sperm counts drop each year, then younger men will have lower sperm counts than older ones. This last point was suggested, strongly, by a group of Danish researchers. In a study published in *Human Reproduction* in 2002, they noted that teenage birth rates have also dropped in Denmark. Unlike in the United States, in Denmark there have been no changes in outreach efforts to encourage responsible behavior in teens. There has not been a rise in abortion rates either. The researchers propose that the real reason could be falling sperm counts in younger men.

"The use of contraception and education levels among Danish teenagers during the past 10 years has remained largely unchanged," the study noted. "Neither is it likely that diminished sexual activity is the cause. It therefore seems particularly important to younger cohorts to see whether their future fertility rates are affected." And the Danish government has done so, launching a program in which the sperm counts of six hundred young Danes will be measured annually. Scientists are even asking whether the worldwide drop in birth rates might be due, in part, to rising overall infertility. "The decreasing trends in fertility rates in many industrialized countries are now so dramatic that they deserve much more scientific attention," noted one 2006 study in the *International Journal of Andrology.*

In other words, having fewer children may not be something that we first-worlders are choosing as a lifestyle decision. It may be something that is happening to us as a species, unbidden.

"Are You Feeling Any Pain from Any of This?"

For men, as for women, infertility is harrowing; studies have shown that men feel the stigma much more when they know that they, not their partner, are the cause. Infertile men feel "profound loss," one study notes, adding, "not only is fertility taken for granted by men; it is central to men's gender identity."

Infertility is also not something men like to talk about. Infertility-related websites are full of chat rooms and message boards, which are populated, without exception, by women sharing pain and pointers. When I was beginning research on this book, I looked at Yahoo chat groups to see what kind of talk was going on. There were hundreds of groups in which women discuss infertility. There was one—count it, one—for men, specifically Christian men who wanted to share their distress at being unable to carry out the biblical directive to procreate. Men willing to air their feelings publicly are few, though they do exist. In a Washington, D.C., conference room in early 2006, a young man named Jeffrey Virostko talked to public health advocates about his sperm motility problems, which have caused him and his wife much anguish, and which he suspects may have something to with the mercury levels in well water on their Indiana farm. "We are the kind of folk who understand that life is not fair," said Virostko, pleading for more intensive research. "But we are also the types not to stand by if there's something we can do about it."

Another man, James Simpson, spoke in an interview about what it feels like to be unable to conceive the child you long for; his story also showed how motivated men can be in pursuing treatment. An air-conditioning installer, Simpson lives in southern Maryland, a rural area that is increasingly suburbanized. When James married his wife, Teri, she had a daughter from her first marriage. Simpson loved having a child in his life, and he and Teri wanted to have a child together. When no baby was forthcoming, they sought the services of a Virginia doctor who found that Teri suffered from tubal problems, while James had low sperm count and motility. To afford IVF, James sold a new truck on which he was making monthly payments. IVF resulted in a positive pregnancy test, but when they joyously went for the sonogram, they were greeted by a blighted ovum. They sat in the examining room, stunned and devastated. Now the couple had no truck, no baby, and payments to make on—nothing.

"We don't have another vehicle to sell," James Simpson said. Their doctor gave them a quiet discount on a second procedure. They put the remainder on their credit card. The result was a cherished daughter. During this period, Simpson was honest with co-workers about what he was going through. Some of them offered sympathy along the lines of "Can't knock up your old lady?" and "Want me to do it?"

Given comments like these, it's small wonder that, now and historically, many men, when possible, have endeavored to deny their infertility and/or pin the blame on women. George Washington is said to have held Martha responsible for the fact that the United States was the only thing he was ever father of, despite the rather glaring evidence of her fertility, provided by four children from a recent prior marriage. (It has been argued that George Washington was infertile owing to a childhood bout of tuberculosis.) And the wives of Henry VIII knew all too well how women tend to be blamed for male-factor issues.

For most of history, really nothing could be done to alleviate or treat male infertility. Then in 1884 a Philadelphia doctor, William Pancoast, decided to experiment on a female patient whose husband was infertile. While the woman was under anesthesia for what she thought was a routine procedure, the doctor took sperm contributed by a medical student he considered an apt specimen of manhood, and—curious to see whether it would work—inseminated the woman without asking her permission, or her husband's. This is the first known case of artificial insemination using donor sperm. The doctor eventually told the husbands what he had done, but the woman went to her deathbed thinking her husband was their child's biological father.

Thus was born a procedure that for decades was the only reliable "treatment" for male infertility: artificial insemination using sperm donated by another man. By the early 1900s, sperm donation was practiced by doctors in New York and in London. After World War II, the baby boom was putting pressure on couples to reproduce even as some men were returning from the war physically and psychologically damaged by combat, not to mention infected by STDs contracted in Europe. Donor sperm became, at this point, widely if surreptitiously used: obgyns would perform the simple procedure using fresh sperm obtained from anonymous sources. Sometimes the donor was the doctor himself. Often it was a medical student at a nearby teaching hospital. "Let's say if you were a tall medical student or if you were a tall ob-gyn student, you were going to be a sperm donor. That's just how it was," recalls one veteran doctor. It was a loose, unregulated system. Little attempt was made to record the identity of sperm donors. In some cases the blurring of truth went further: sperm from the intended father would be mixed with sperm from the donor, a ritual whose purpose was to preserve the fiction that the intended father might be the biological father. The result was an estimated 30,000 children a year born as the result of a procedure whose existence was hardly acknowledged.

Then everything changed. In the past twenty years, extraordinary advances have taken place in two areas: finding sperm, and using the sperm that's found. Even as Stephen Seager was refining his device for paralyzed men, urologists were perfecting surgical methods to free up the germ line in men infertile for other reasons. They were learning to plumb the reserves of the sperm-making apparatus to locate the tiniest reserves of usable sperm. In men with low or zero sperm counts or physical obstructions, there are often pockets of sperm somewhere; they can be found by moving backward in the sperm-manufacture assembly line. You can watch how readily sperm can be obtained by searching the Web for a video in which the urologist Sherman Silber retrieves sperm from a patient who lacks a vas deferens, which prevents his sperm from emerging naturally. Silber, a blunt, humorous man wearing surgical scrubs, performs "sperm aspiration" on the patient, who has sportingly permitted the operation to be webcast. First Silber injects a needle into the man's scrotum, giving him anesthetic to numb the area. Presently he plunges another needle in, and aspirates sperm into a catheter.

"Are you feeling any pain from any of this?" Silber asks the patient.

"No, I'm not," says the man cheerfully.

"The patient is able to see exactly what he has within a few minutes of the procedure starting," Silber tells the camera. To prove it, Silber puts the sperm in a dish under a microscope, which is then projected on a screen. For the first time in his life, this fully awake patient can see his own sperm, wiggling and swimming. "You never thought you'd have sperm!" Silber congratulates him. "After all these years of no sperm!"

"Nope!" says the patient, who is wearing glasses and a scrub cap, speaking in a flat midwestern accent, lying on his back looking up at the screen. "That's really fantastic."

It is a routine procedure, yet revolutionary. Urologists like Silber and Paul Turek have refined microsurgery to the point where if a man has a pocket of motile sperm anywhere—if, for example, the majority of his sperm are dead but there is live sperm in one tubule—they can retrieve it and use it. They're like the SWAT team of reproductive surgeons, trained to get the hostage out safely. (In military hospitals, these are actually called "commando extractions.") They can suck the sperm up with a needle—or if there are not enough sperm to do that, they can carve out a "divot" of testis tissue, digest the skin down with enzymes, and probe dish after dish of tissue to find a few sperm lurking. Turek has even developed a technique of "mapping" the testicles to find fragmentary reserves that might otherwise be missed, a technique that he uses on cancer patients as well as on his African civil war victims.

But much of this retrieved sperm is problematic in some way. It may be deformed, or immature. It may be energyless. It may be scarce. It may be confused. In these cases, IVF alone is not enough to achieve fertilization. "It was a tremendous problem to deal with the male factor in the early days," says the embryologist Lucinda Veeck Gosden. During regular IVF, the sperm sample is plopped, from a pipette, around a waiting egg, but even this isn't enough for some sperm. "We did everything we could to concentrate sperm right around the egg," she says, but the sperm would sit there, helpless, hapless, fertilizationally challenged.

You've Got to Break Eggs to Make a Baby

So, even as surgeons were becoming adept at getting at the germ line, scientists had to figure out how to help the sperm get where it needed to go. At first, embryologists had no idea how much manhandling an oocyte could take, so they attempted to work the sperm into the outer membrane of the egg without breaking through. "We were so gentle," says Lucinda Veeck Gosden. "We were afraid to damage the egg." The person who achieved a literal breakthrough in this area was Gianpiero Palermo, an Italian doctor who in the early 1990s went to complete his training at the Free University in Brussels, a noted European IVF center. Palermo's mission was to refine his oocyte retrieval skills. But once he arrived in Brussels, Palermo was told that he also would need to find a laboratory research project. Palermo, who had trained as a clinician, says, "I had never seen a lab in my life."

He was fascinated by what he saw. One of the interesting things about reproductive science is that it requires a variety of skills, some high-tech and some quite humble. In addition to scientific and medical erudition, reproductive lab work requires practical skills, not unlike cooking, or woodworking, or knitting. Hand-eye coordination is important; you can be the world's expert on calcium oscillations in the oocyte, but if you flunked "cutting with scissors" in kindergarten, you're not going to shine in the lab. Handyman skills are helpful. Many scientists use the gas flames on lab burners to "pull" their own pipettes, for example—to alter the diameter and/or fine-tune the tip of the hollow glass needles that are used for all sorts of procedures, including aspirating, or sucking up, egg and sperm. Moving eggs from dish to dish also takes a peculiar skill: scientists often move eggs by sucking the egg into a pipette using their mouths, then spitting the egg into the next dish. There is a filter attached to the pipette, to keep them from swallowing the eggs, and to prevent any virus or bacteria from getting into or out of the dish.

Palermo discovered he had the kind of pragmatic sense that's perfect for a laboratory. "I've always been fascinated by screws and machines," he said. "I've been described by my parents as a missed mechanic."

In particular, Palermo was attracted to the micromanipulators: a pair of mechanical hands, sitting on a microscope, which are used to manipulate eggs and sperm. "I thought, what a ridiculous machine; very rough; very dangerous," Palermo recalled in an interview. He decided to tinker with it, and with other equipment. He practiced pulling his own pipettes; found a new way to grind the tip so that it had, at the end, a tiny spike; learned to whet the microgrinder so that there wouldn't be debris left on the tip after grinding. He tried remounting the micromanipulation arms on the microscope. When he got the equipment the way he wanted it, he set out to slip a sperm inside an egg better than anybody had ever slipped a sperm inside an egg before.

Under a microscope, the egg is a lovely thing. It is big, and round, and of course living; inside is a grayish substance, the cytoplasm, which looks undifferentiated but is in fact teeming with tiny objects called organelles; and with mitochondria, the energy-generating pack that drives the egg and will eventually power the embryo. Also inside are the chromosomes, the nuclear DNA that, upon merging with DNA from the sperm, will provide a unique genetic blueprint to build a unique baby. Enclosing all of this is the zona pellucida, the rubbery membrane that a normal sperm penetrates by dissolving it with digestive enzymes. Palermo's goal was to help a not-so-normal sperm accomplish the same feat. He experimented with every part of the process. Eggs are three-dimensional, and, like globes, can be turned this way and that. Palermo decided a good injection spot was at around three o'clock. At first Palermo tried to inject the sperm "subzonally": like others who had experimented with sperm injection, he wanted to create a tiny pouch within the zona, a depression, really, to enable the sperm to have closer contact. But during one procedure he pushed too far, and pop, the zona broke. The sperm was inside the egg, not just nestled near it.

Palermo put a "?" on that culture dish, expecting fertilization to fail in the egg he had mistakenly punctured. Instead, the opposite happened: the egg whose membrane had been breached produced the only embryo to develop into a viable pregnancy. For the first time ever, perhaps, a human being was born as the result of a lab mistake.

When he realized that the dish with the question mark on it was the one that had produced the pregnancy, Palermo started sticking eggs on purpose. Eggs, it turned out, were tough. They were durable. He achieved a second pregnancy, and a third. He went to his lab director, who, he said, accused him of being "excitable" and exaggerating, or mistaking, what had happened. But in 1992, after four sperm-injection pregnancies, Palermo and colleagues published in the British journal *The Lancet* an article called "Pregnancies After Intracytoplasmic Injection of Single Spermatozoon into an Oocyte."

Palermo called the new technique ICSI, or intracytoplasmic sperm injection.

"I thought intracytoplasmic was two separate words," he says, laughing.

Looking back, he thinks the most important skill he brought to lab science was industrial, almost old world. "The success of ICSI, if you want to reduce to one word, I would call it the pipette," says Palermo, who is now at Cornell's Weill Medical College clinic. "That was the reason ICSI worked. The pulling is one thing; then you need to grind, then you have to make the spike at the tip. The major problem with the previous pipettes was the large shank at the tip. It would move a lot of fluid quickly and damage the egg. You have to think all that by yourself. It was about a nine months project to put everything together."

The result of a better pipette was hundreds of thousands of children born to amazed and gratified fathers. Thanks to the glass pulling skills of Gianpiero Palermo, almost any man can become a biological father. If a man has a few sperm in his body—no matter how remote, how malformed, how weak, how misshapen—urologists like Paul Turek can often retrieve them. Embryologists can take one, and inject it into the egg. Presto: a pregnancy. Every doctor agrees that next to IVF, ICSI has been the greatest, if much less heralded, advance in fertility medicine.

"When I started doing IVF, we used to say, 'I'm sorry, your sperm is so bad, IVF won't be able to help,'" recalls Peter Brinsden, director of the Bourn Hall clinic in England, the first IVF clinic in the world, founded by Steptoe and Edwards. "Now we get one sperm and say, 'I think I have something that can help you.'"

The problem is, since ICSI was an accident, it had not been tested before it was performed in humans. Nobody knew what kind of babies would result from a substandard sperm being injected into a broken egg. In the United States, the Food and Drug Administration regulates devices—like testicular implants—that are put in the human body, and it also, of course, regulates drugs. But the FDA does not regulate scientific techniques, even if the technique is being used to create a human being. There is nothing to stop anybody from doing ICSI. I could do it in my basement. You could do it in your garage. And now everybody does it: some IVF labs perform ICSI for all patients, regardless of the cause of infertility. Yet scientists immediately suspected that genetic problems in a father—problems that might be responsible for the infertility—might now start showing up in the sons. "The benefit of this technique was shown almost immediately," says Paul Turek. "However, the demonstration of side effects and complications takes much longer, and they are likely to be very interesting ones."

"Now we can take a man with zero sperm—zero sperm—in his ejaculate, not even one," marvels the embryologist Veeck Gosden. "We can go into his testes with a fine needle, or we just cut out a little biopsy, take it to the lab, digest all that tissue down, search literally for hours, through dish after dish after dish, and come up with one or two or three sperm, and inject these sperm with ICSI, one at a time, and come out with wonderful babies. Now, the downside to that of course is that some of these men have very good reasons why they are infertile. We are possibly setting up the next generation" of infertility patients.

Every Man Infertile

Not possibly. Certainly. Because not all male infertility is created by infection, or laptops, or hot tubs. A substantial portion is genetic, caused by a problem in the chromosomes of the sperm. As Sherman Silber explains it, genetic infertility is something that arises spontaneously in every generation. What happens is that a normal fertile man manufactures mostly normal sperm. But about 1 percent of the sperm a man makes are going to carry some little mistake, some chromosomal glitch. If the problematic sperm is the one that fertilizes the egg, a boy born as a result will have a chromosomal problem with all his sperm, and will thus be infertile. Nature, it should be remembered, is not perfect. Mistakes are made. Genetic infertility is nature's way of making sure the same mistake does not happen twice. Genetic infertility is nature's levee, if you will, holding back a flood of chromosomal mishaps.

And that levee has now been breached.

"All of a sudden, every reproductive endocrinologist who sees a guy with a low sperm count is sweating," says Turek. "If he knows what he's doing, he's sweating."

For sure, ICSI is permitting infertility to be transmitted. One of Turek's colleagues at UCSF, Renee Reijo Pera, has pioneered the study of the Y chromosome, the sex chromosome that is the defining quality of men, who inherit an X chromosome from their mother, and a Y from their father. She found that in 15 percent of percent of men with azoospermia (almost no sperm) and 6 to 9 percent of men with low sperm counts, infertility is caused by a deletion on the Y chromosome—a skip, like the kind that happens to CDs and record albums. Up to now, a father who had a Y deletion could not pass the deletion to offspring, because he could not have offspring. Now he can, and does. In 1999 scientists tested four ICSI boys. Without exception, those boys exhibited their father's Y deletion. This was proof that ICSI has enabled infertile men to father infertile sons.

So what else is passing into the gene line? Who knows? Chromosomal deletions are just one cause of genetic infertility. Another is cystic fibrosis, the single most common genetic disease. CF is a recessive disease, meaning that if an unaffected man who is a carrier conceives a child with an unaffected woman who is a carrier, their offspring has a 25 percent chance of being afflicted. Cystic fibrosis is also a complex disease; there are hundreds of mutations, not all of which can be detected through testing. One physical problem associated with some CF is congenital bilateral absence of the vas deferens: in men who carry a certain mutation, one or both passageways through which sperm is ejaculated are missing. In these men, the absence of the vas deferens acts as a natural vasectomy, a kink in the hose of reproduction, a way of preventing CF from being transmitted. Now that sperm can be obtained from the areas where it is manufactured, the hose has been opened, the kink uncoiled. Without genetic testing, some male carriers could now pass along CF to a child, thanks to ICSI.

As well as cancer, maybe. Turek and colleagues have also found that some infertile men are incapable of repairing the DNA in their testicles. Most of us, when our DNA is damaged by, say, sun rays, or ordinary wear and tear, possess an enzyme that will zip along the chromosomes and restore them to stability and order. An inability to make this repair is associated with some forms of cancer. "This was an accidental and quite profound finding," Turek says. "It made us worry: if their testicles can't repair their DNA, maybe the individuals conceived from that have a higher cancer risk."

In addition, men who carry a genetic defect called a Robertsonian translocation, a rearrangement of chromosomes that also causes infertility, are at risk for causing pregnancies that miscarry, and for fathering children with mental retardation and birth defects.

The ultimate impact remains unknown. They could be minor; they could be dramatic. Sherman Silber and colleagues published a paper showing that, thanks to ICSI, in 10,000 years every human male could be infertile. If 1 percent of men are genetically infertile in every generation, and every one of those men were to reproduce through ICSI, then 1.1 percent of the next generation would be infertile: the ICSI babies, plus the new 1 percent of infertile males that arise spontaneously. And so on and so on: infertility would be magnified; like compound interest, it would grow, and grow, and grow, until all men were sterile. In 2004 Silber presented these whimsical findings at an ASRM presentation, to knowing looks and scattered laughter. Every man infertile! How preposterous! Yet every doctor accepts that they are literally making future infertility patients.

And not everyone is laughing. Paul Turek certainly isn't.

"I have enormous respect for genetic infertility," says Turek. "When a man is infertile for genetic reasons, maybe that's a large statement by someone who knows more than we do that it's not good for the species for that individual to reproduce. The reasoning for that is, you are put on this earth to eat and reproduce, and if you can't reproduce, that's kind of important for the species. Is male infertility the ultimate medical problem? Is it a species problem? It's funny: on a day-to-day basis, infertility does not affect the way you drive, or your job. Yet I'm telling you that it might be the most important thing that ever happened to the species."

At the University of California, San Francisco, Turek and others have created the Program in the Genetics of Infertility, or Progeni: an attempt to monitor what is happening to the human gene line as a result of male infertility treatment. As part of that program, patients are asked for a full family history and offered extensive genetic testing; they are counseled about risks associated with their cause of infertility. "It makes me feel much better, when the big bad thing happens—if the big bad thing happens, genetically—that there is something awful being transmitted that we are not clear about, at least we have systems in place to deal with it," Turek says. But industry-wide, there is no rule compelling clinics to conduct genetic testing of ICSI patients, or even to counsel them about the consequences of treatment.

Up to now, much of the ethical discussion surrounding reproductive technology has focused on the possibility that genetic testing may someday enable the creation of enhanced babies. Little attention is being paid to the fact that the same technology is clearly, here and now, enabling genetic glitches to be perpetuated and possibly, in some cases, magnified in the children of ART. Assisted reproduction may someday make the human race better. Then again, it may make the human race worse. Who knows? And why aren't more people asking this question? Could it be because men are the ones benefiting?

Okay, Okay, Okay

There is another problem associated with ICSI, one that has nothing to do with the fate of the human gene line and everything to do with the fact that when ICSI is performed for male-factor infertility, the person who bears the brunt of treatment is the woman. In 2005, I spent a week with David Keefe, a reproductive endocrinologist who at the time was director of the fertility clinics at Women and Infants Hospital in Providence, Rhode Island, and Tufts–New England Medical Center in Boston. (Keefe now heads the clinic at the University of South Florida.) Both Massachusetts and Rhode Island require that health insurers cover IVF. As a result, many patients in these states are working-class. In Rhode Island, a state that was a birthplace of the U.S. industrial revolution and continues to host a lot of industry, Keefe suspects chemicals may be the source of much of the infertility doctors see.

One morning Keefe, a former Harvard rower with an air of athletic vigor despite an arthritic limp he has developed in midlife, walked into a strip-mall satellite clinic that Women and Infants Hospital maintains in Woonsocket, about an hour from the main facility in Providence. In an examining room, Keefe found himself facing a cheerful, strawberry blond patient who was accompanied, for that morning's introductory visit, by her cheerful, strawberry blond sister. The woman's husband had not accompanied her. Testing had showed that he had a very low sperm count, far below the normal threshold.

"Was that suspected?" Keefe asked the woman, sitting down to review her chart.

"No," she said.

"How did he handle the news?" he asked, delicately.

"It wasn't good for a while," she said.

"It's important to separate masculinity from sperm count," advised Keefe; looking at the patient's chart, he saw that her husband's sperm had been evaluated in the lab, and did not display any chromosomal problems. "Probably the best way to handle this is to treat you," Keefe said, warming up to a discussion of ICSI.

"Okay," the woman said.

"Guys are the fragile sex," Keefe added, explaining that the treatment for male-factor infertility was to take a sperm from her husband and to inject it directly into her egg.

"Okay," the woman said.

He followed with a detailed explanation of what ICSI would involve for her: taking hormone shots; having her eggs retrieved surgically; then, after fertilization, transferring embryos into her uterus. He explained that at her age—she was in her mid-thirties—there could be issues with her eggs, but that it looked like the problem was male factor. He also explained that they would probably get more eggs from her than they could use, and possibly end up with excess embryos. He explained that if they transferred more than one embryo, she could end up with more than one child—with, say, twins.

"Okay," the woman said.

Taking a moment to talk more at length about embryo transfer, Keefe reviewed the consequences of multiples. In addition to enumerating risks and statistics, he shared a vivid anecdote about a man he treated who appeared to have Mafia connections, and whose second wife wanted to have as many children as his first wife had, which meant that she wanted triplets. Meanwhile the guy was hoping that the treatment would fail, so he surreptitiously asked Keefe if he could just shoot, you know, blanks into the eggs. When they looked at the ultrasound screen and saw three babies growing, Keefe was a little nervous about the father's reaction, and though the father did not have Keefe whacked, the many varied problems of a triplet pregnancy are why he recommends, in most cases, putting back no more than two embryos.

"Okay," the woman said.

He explained that the drugs she would take are very natural, so natural that if she took them by mouth they would be digested by the body, which is why they must be administered by a shot.

"Okay," the woman said.

He explained that if the shots hurt, they could give her a local anesthetic cream.

"Okay," the woman said.

He explained how egg retrieval works: the growth of her follicles would be monitored using an ultrasound wand inserted into the vagina—"it's a little embarrassing at first; the nurses do it"—and then once the eggs mature, they would give her general sedation and use ultrasound to guide the needle through the vagina, which doesn't hurt, he explained, because there are few nerves there. They would then take the eggs and fertilize them, and transfer the embryos, and freeze the rest if there were any, "and if it doesn't take the first time, you can thaw the frozen ones and put them in."

"Okay," the woman said.

He explained how hard it can be, waiting to find out whether a pregnancy has resulted. There followed a conversation about infection from the egg retrieval, which happens about 1 percent of the time; hyperstimulation of the ovaries from the drugs; whether she could work out while pregnant; whether she was taking prenatal vitamins. He asked the woman, who was French Canadian and at risk of being a carrier for Tay-Sachs, whether she would like to be genetically tested.

"Yep," the woman said, willing to undergo everything Keefe had described, in order to have a baby with a husband who didn't show up for the appointment with her.

"It's better this way," she said of her husband's absence.

"I Just Like to Have the Bits and Pieces Passed Back to Me."

Issues like these have not stopped ICSI from being widely adopted. According to the U.S. Centers for Disease Control and Prevention, more than 50,000 ICSI procedures were carried out in 2003—testament to the fact that men, like women, will do what it takes to have biological children. ("Desperate" is rarely an epithet used to describe infertile men, while it is routinely applied to women.) Testament, too, to the lasting impact of Gianpiero Palermo's accidental puncture. "Male factor is hugely more of an issue than it was in the past," the Richmond doctor, Sanford Rosenberg, told me. "It's not so much that more men are infertile—although that may be the case—but it used to be that about 20 to 30 percent of patients had male factor, and now it's easily half. I would say that the ability to deal with abnormal sperm has brought many abnormal sperm patients out of the woodwork, where before, the only option was donor sperm. I used to do a ton of donor sperm in the early days, and now the only time we do it is when there is virtually no sperm, zero. We had a pregnancy where we only had five sperm. It took the embryologist several hours to go through the ejaculate. ICSI has virtually eliminated male factor as a significant infertility issue. Ninety-nine percent of my donor sperm now goes to lesbian couples."

It is a cliché, of course, but truly it is wondrous what medical science can achieve. I watched ICSI being performed one day in the lab of a doctor named Geoffrey Sher, and have found it impossible to forget. Sher is one of fertility medicine's more flamboyant practitioners. A member of the early group of doctors who began offering IVF in this country, Sher, a charismatic and somewhat larger-than-life South African, separated early on, and with great fanfare, from the academic establishment, breaking with the American Fertility Society and striking out on his own to found a private IVF clinic. Sher helped pioneer a "consumer" approach to fertility medicine, advertising to and actively recruiting patients, rather than relying on such niceties as professional referrals. One doctor I interviewed referred to Sher's sessions with patients as "dog and pony shows." Sher now runs a chain of clinics, the Sher Institutes for Reproductive Medicine, with offices in Los Angeles, Las Vegas, Manhattan, and several other cities. He travels among them, performing retrievals and transfers. His patients fly in from around the world, sometimes actually staying in the clinic: a kind of fertility camp. While some doctors scoff at his tactics, there are patient groups who admire his vigor, accessibility, willingness to try new techniques, and, they say, commitment to his patients.

One thing that can certainly be said of him is that he does not mind visitors.

The day I was in his Las Vegas clinic, Sher let me put on scrubs and spend time in his lab, which was located not far from a wall showing photographs of his patients' many pets. ("People who are trying to have babies usually have pets," one nurse told me.) In fertility clinics, labs can be very small; embryos don't take up much space, and neither do the incubators and microscopes. Labs are usually supervised by one bona fide PhD; other staffers may be degreed scientists, or they may be trained employees who manipulate gametes for a living. The day I was there, an embryologist was sitting at a micromanipulator. In front of him was a microscope that had on its stage a dish of sperm. The microscope image was projected onto a wall screen. Under a microscope, sperm look exactly as you think they would: little tadpole things darting in a herky-jerky motion. Using the tip of the right-hand pipette, a thin hollow needle controlled by a computerized joystick, the embryologist pinned a sperm to the bottom of the dish. Then, using the tip (ICSI pipettes are now commercially produced), he sliced off the sperm's tail, to immobilize it. Tail-slicing is another thing scientists were at first loath to do. Palermo was afraid of damaging something vital, but after watching a younger embryologist slice tails like a sushi chef, Palermo says he realized that no harm seemed to be done, and it seemed to improve success rates.

Every now and then in the Sher clinic, the pinning and tail-slicing would result in a sperm's getting stuck to the bottom of the culture dish, making it impossible for the embryologist to aspirate the sperm into the needle. Rather than pry the stuck sperm off the bottom, he moved on to another one that looked good. There were thousands to choose from. He would slice the new one's tail and then aspirate it. The stuck sperm would stay immobilized at the bottom, its chance at a future life lost.

I don't know why, but to me this seemed moving. Sperm may be little more than DNA delivery systems, but it was strange to think one might-have-been baby would now not be born, and another baby would be, all because one sperm got mashed to the bottom of a dish, and another sperm managed not to be. I watched the poor stuck ones, mesmerized by the randomness of what was happening.

The embryologist didn't sit around philosophizing. He brought to his task all the emotion of a car mechanic changing a spark plug. Once he had the sperm in his needle, he moved to the egg dish. Holding an egg, by suction, to a wider pipette tip, the embryologist pushed the needle with the sperm against the surface of the egg, which had been prepared by removing the "cumulus" of somatic cells that surround it. He pushed, and pushed, and pushed. For a while the zona pellucida stretched inward, like a balloon. And then pop, it punctured, and the pipette was inside. The embryologist injected the sperm slowly into the core of the egg, where it could now be seen. What existed was not, strictly speaking, an embryo. It would be a day or so before the two nuclei would draw together and fuse, forming what's known as a pre-embryo.

I could have sat there watching ICSI over and over and over. Not just the creation, but the somewhat ad hoc *selection* of human life, taking place about a mile from the Bellagio and the MGM and the teeming, sordid pedestrian activity of the Las Vegas Strip. It all seemed like gambling. This was what, for most of human existence, God has done, or natural selection, or whoever chooses which gametes become humans. There is so much about natural conception that involves selection: the growth of sperm and oocyte (millions of gametes are lost growing to maturity; only the select survive); the passage of sperm to egg, which like all epic journeys requires energy and resourcefulness, as well as luck; the way a single sperm is drawn into the orbit of the egg, which appears in some profound chemical way to exercise choice over the sperm that enters. All this was lost—all of it. All of it had been taken over by us. And yet, I reminded myself, it's important not to fetishize nature. Natural conception itself also involves randomness and chance.

As I was watching, the lab technicians chatted about this and that, as co-workers working in close quarters will do. One staffer was still annoyed with a patient who earlier that day had been pushy and demanding. "Do you ever go up to a heart surgeon and tell them how you want it done?" the embryologist asked, incensed.

"I don't like to talk to the patients at all," said another staffer. "I just like to have the bits and pieces passed back to me."

CHAPTER FOUR

IT TAKES A VILLAGE
TO MAKE A CHILD:
ART AND THE EVOLVING HUMAN FAMILY

Standing on Ceremony

At 10 a.m. on May 22, 2005, Kendra Vanderipe was sitting on the outdoor patio of the Bay Landing Hotel in Burlingame, California, lingering over breakfast and watching the planes descend into San Francisco International. Her luxuriant hair was freshly styled and very blond, her pedicure red and unchipped in high-heeled, open-toed sandals. She was wearing a crisp beige-and-white-print cotton skirt, with a beige knit top that hugged her figure without appearing tight or unseemly. Stretching in front of her were the choppy waters of San Francisco Bay, flanked by a promenade on which people were calling to one another in words that were not quite audible from where Kendra was sitting.

"If only they could be shouting in a foreign language," said Elizabeth, Kendra's best friend, sitting next to her, wearing a strapless dress in a green-and-white tropical floral print. "*Any* foreign language."

Kendra laughed. It was a cloudless morning. Spring in northern California was warmer than spring in Aurora, Colorado, where Kendra and Elizabeth, both flight attendants, lived in a rented townhouse. Sitting here you almost could believe you were in the Caribbean, or the south of France. But the Bay Area was itself a fine weekend destination; yesterday after arriving the two women had driven to Fisherman's Wharf and spent the afternoon exploring. In the evening they'd gone to visit Laura and Hector Ramirez, hospitable and consumed by child care in their tiny Bay Area home.

It was the first time Kendra Vanderipe had seen the Ramirez triplets, conceived from her egg donation more than a year earlier. She'd had no idea what to expect from the meeting. What would it be like, encountering three children who were genetically hers, and yet practically and legally not hers but someone else's? "None of us has been here before; you don't really know until you walk through it," she had said sunnily in a telephone interview, not long before setting out. Once there, she saw that at six months the boys were already their own persons. Edward had the medium-brown skin of his father, while Preston and Hunter were fairer, with complexions that came from what might be called Kendra's side of the family. The smallest, Preston, moved around a lot, trying to keep up with his brothers. Preston had a lot of energy, a trait that Kendra, who tends to be restless and peripatetic, thought might also come from her. Other than that—well, other than that, they looked like babies. "They look nothing like me," Kendra said, musing on the comfortable sense of disconnect she felt, contemplating her genetic offspring. "It's like: did I have anything to do with this? I don't think I did. At this point, they're just babies." She said that she felt more of a connection with Laura than with the babies.

More than a year earlier Kendra had signed with a California egg-donation agency, thinking it would be "cool" to help another woman have children. It was her first egg donation, though not her first experience with third-party family-building. In her early twenties she had become pregnant at a time when she felt unready to raise a child, and rather than aborting the fetus, she had given birth and offered the baby, a boy, for adoption. Far from a searing experience, Kendra found being a birth mother uniquely gratifying. There was real pleasure, she discovered, in helping a person who couldn't have a family have one. "Gosh, to be able to do something for somebody that they can't do for themselves—I'm wired a certain way and I'm wired to be able to do that. I understand that a lot of women aren't; they feel like, 'Those are my babies.' I look at it as 'We're all brothers and sisters on the planet.'" Kendra has an open relationship with the adoptive family, and sees her birth son several times a year. Finding that experience "super-rewarding," Kendra became interested in "helping people have babies" more deliberately. "I'm super-passionate about it." She looked into surrogacy—her pregnancy had been easy, and she would have been glad to carry a child for a woman who couldn't—but was told by agencies that to qualify she needed to have a child living at home, as proof that she would be emotionally capable of relinquishing any other baby she carried. It seemed ironic, given that relinquishing was what had gotten her into this in the first place, but there it was. So if she couldn't contribute the womb, she'd contribute the oocyte.

Going online, Kendra filled out egg-donor applications, answering questions about her height and weight and education and health background. She was most interested in the Egg Donor Program, a California brokerage run by Shelley Smith. What appealed to Kendra was that Smith was willing to permit her to be identified to a recipient family. Most egg donation in the United States, like most sperm donation, is done in permanent anonymity, which is reassuring for some parents and convenient for the agencies; setting up donor-recipient meetings is time-consuming and emotionally fraught, requiring a level of record-keeping and professional counseling that egg agencies aren't in the business of providing. But Smith didn't mind Kendra being open if Kendra wanted to be open. Kendra wrote into her profile that she would be willing to be known to any offspring, and within a few months she was selected by a family. After taking medications to stimulate her ovaries, she flew to California for the retrieval. Which is where she was, recovering on a gurney, when a dark-haired woman materialized, bearing a bouquet of flowers.

The woman was Laura Ramirez, who was known, in the parlance of the trade, as the "intended mother." Laura had tried to send flowers via the doctor, who, unwilling to get in the middle of such a charged and unpredictable encounter, had waved her into the room. At thirty-seven, Laura was nine years older than Kendra, who was twenty-eight. Good-natured and down-to-earth, Laura had gone to college, gotten a master's in psychology, done an internship with the police, and found her calling. While walking her beat Laura had run into Hector Ramirez, an attractive Guatemalan immigrant who worked as a security guard. Hector, whom she had met before, at a San Francisco dance club, was on crutches. The two stood chatting about what had happened to him—he'd been grazed by a car after stepping off a curb—and one thing led to another. Within a year, they were married. Their efforts to become pregnant resulted in a series of miscarriages, and Laura was diagnosed with premature ovarian failure. "We just wanted children," says Hector, who like Laura was willing to consider any solution, pretty much, including an egg donor. Laura started trolling the Web for a donor willing to be known to the intended family. While Hector worried that an open arrangement might make the children "confused about who was the mommy," Laura, who had seen too many dysfunctional families with too many uncommunicated secrets—"as a police officer, you see so much stuff"—didn't want any mysteries lurking in her own. After a lot of clicking and scrolling she had located Kendra's profile. Now, at the hospital bed, the two women greeted each other warmly. Kendra had also brought along a gift: a necklace. Into the box she had tucked her name and address. "Call me anytime," she said.

And Laura did. Once Kendra's eggs were fertilized with Hector's sperm and the embryos were transferred into Laura's womb, Laura called Kendra to let her know the pregnancy had taken. After the birth of the triplets, which was horrific and nearly fatal—Laura hemorrhaged badly after the triple C-section, losing an enormous amount of blood—Laura e-mailed Kendra photos of the newborns. Kendra put the pictures up in her townhouse. She e-mailed them to friends. Kendra's own mom put photos of the babies up in her cubicle at work. Operation Triplet, they started calling it. Along the way, it became apparent to Kendra that Laura Ramirez was not only no longer a stranger; she had become a friend, a mentor even, somebody warm and grounded to whom Kendra could turn for advice on her own life as it unfolded. When Kendra was thinking about changing jobs, Laura wrote letters of recommendation. When Laura invited her for the baptism, Kendra accepted.

And so here Kendra was, with Elizabeth for companionship and road-map navigation. They weren't planning to stay long: just go to the ceremony, watch the boys be baptized, and fly back to Denver. Kendra wasn't sure what her role would be. She thought it might involve pouring water. As flight attendants, she and Elizabeth were professionally punctual; promptly at 10:30 they rose from the patio table, got in the car they had rented at the airport, and headed for San Mateo. On the other side of the freeway they stopped for a camera battery; went the wrong way on El Camino Real; realized their mistake; turned around; found St. Matthew's, the church where the baptism was taking place; parked at a curbside meter; picked their way on party shoes toward the entrance; and made it thirty minutes early. They were the first guests to arrive.

Almost immediately they ran into the Reverend Beth Parab, who had finished the regular morning service and was crossing the courtyard in her ministerial robes. "I'm Kendra," said Kendra, a little tentative.

"Oh, you're *Kendra*," said Reverend Beth, smiling warmly. "I'm so glad to meet you!"

Writing the prayer for a baptism involving three infants, six godparents, two parents, and an egg donor had turned out to be unexpectedly easy. Reverend Beth had sat awhile, contemplating her assignment. She had put herself in Laura Ramirez's place, trying to imagine how she would feel if she had conceived children with vital help from another woman. How would she want her children to regard the woman who had given them the gift of their existence? Inspiration struck, and fifteen minutes later, the prayer was written. Rereading it, Reverend Beth figured—as she always did, when writing flowed like that—that "it must have been a God thing."

Now Reverend Beth chatted with Kendra, leading the women toward the church office to see if the Ramirez family had arrived and was waiting there. As they walked, Beth asked Kendra if she had known the Ramirezes for long. "No," said Kendra, without going into the whole story. A boy glided by wearing Heelys, the shoes that convert into roller skates by means of wheels tucked into the heels. "Did he have *wheels* on his shoes?" asked Kendra, who had never seen these before, turning to watch, delighted, as he skated past. "I am *so* going to get a pair of those!"

As they skirted the parking lot, they could see the Ramirez family emerging from their car, a cumbersome process. After the boys' birth, Laura had been incapacitated and Hector had adroitly stepped into the vacuum, changing the boys and feeding them and doing their laundry and bathing them and getting them on a schedule. "I was more intimidated by triplet care than he was," says Laura. Eventually it would be Hector who stayed at home, and Laura—the one with the pension plan and the health insurance—who had to return to work. It was hard to say who had the more arduous duty: Hector, at home with three newborns, or Laura, who had to walk out the door and leave the family she had worked so hard to achieve. Laura, emerging from the car, wearing brown pants, a turquoise sweater, and brown and turquoise shoes, had the happy, dazed look common to new mothers. Hector Ramirez, fit and smiling, was wearing a suit and his dark hair was gelled; he was carrying a backpack into which he had loaded the milk and diapers and the video-camera. Friends and relatives were now milling around as the Ramirez boys, dressed in white cotton outfits complete with little white socks, little white shoes, and little white hats, were removed and loaded into an enormous triple decker stroller. Laura, embracing Kendra, introduced her to the guests, saying, "This is Kendra."

Everybody seemed to know who Kendra was. They looked at her with discreet curiosity, surveying the woman who had supplied half the DNA that had resulted in these three little beings, who just now were looking wide-eyed at the toys dangling from the stroller bars. "This is Kendra," Laura said to her own father, who approached them. "Nice to meet you," Laura's father said to Kendra, and then—as everyone was trying to think of what to say next—it was time to make their way into the church.

Inside the coolness and tranquility of St. Matthew's, Kendra and Elizabeth stood to one side, a little apart from the others. Kendra was unsuccessfully hunting for a Kleenex, sensing already, from the dim, sanctified atmosphere of the church, that the ceremony was going to be more intense than she had anticipated. All at once a small brown-haired woman was ecstatically hugging her. "I would like to thank you eternally for the beautiful children!" said the woman, who introduced herself as Laura's mother. Looking at Kendra with affection and an unmistakable air of interest, she continued, "I don't know if you know how happy you have made my daughter!" Kendra murmured a polite acknowledgment and then the ceremony started, godparents and parents gathering around the baptismal font. Reverend Beth began talking, pointing out that Trinity Sunday seemed a delightfully appropriate day to baptize triplets. She segued into the announcement that "we get to celebrate Kendra, who helped make this possible."

Kendra, who was preparing to take a photograph, was surprised to hear herself called forward. "Put that camera down," said Reverend Beth affably, so Kendra did. As the group's attention shifted to her, she approached the font and stood immobile as Reverend Beth put her hand on Kendra's head.

"Dear Lord," said Beth, beginning the prayer she had written, "we thank you for Kendra, and her selfless and generous gift. From her loving act, the miracle of new life has become a reality and the dream of a family has been realized." Kendra continued standing as Beth continued praying: "Now we have come to the day of baptism" of Edward, Preston, and Hunter, "who couldn't have been born without the compassionate help of Kendra. Bless her and keep her, Lord, and make your face to shine upon her always, and let her sharing in the life-giving story of these baby boys always be honored and remembered, that others may also be inspired to live with hearts that overflow with grace and generosity toward their fellow brothers and sisters."

Kendra, listening to this tribute, started to weep. Laura, standing on the other side of the font, also was weeping. The two women stood facing each other while the babies waited in the arms of their godparents, cheerful and quiet. "Lord Jesus," Beth prayed, asking that Edward, Preston, and Hunter "always keep a special place in their hearts for their angel, Kendra, who by your abiding grace helped to place them in the arms of their parents. Amen."

Kendra wobbled back to her place. Reverend Beth had given her the piece of paper on which the prayer had been written, and she was clutching it. Surreptitiously wiping her face with the back of her hand, she tried to recover her composure while the godparents presented the children for baptism. But hardly any time had gone by before Reverend Beth was saying, "Kendra, can you come and pour the water?" And so Kendra did, taking a silver pitcher and pouring water into the font as Reverend Beth said, "We thank you, God, for the gift of water." The children were baptized one by one, blinking with mild surprise as Beth dipped a silver seashell into the font, and trickled water onto their now bare, fuzz-covered heads. "I baptize you in the name of the Father, the Son and the Holy Spirit," said Reverend Beth three times, and suddenly Kendra wasn't comfortably disconnected anymore from what was taking place.

Standing while the three Ramirez children were blessed and celebrated and welcomed, she felt a sense of relatedness she had never expected. She was, she realized, fully part of these proceedings. She wasn't an observer. She wasn't on the periphery. She wasn't a minor player. She was part of this family. She had something to do with these babies, something real and crucial. She was here, at the center of what was happening.

"I'm not on the outside looking in. I'm on *the inside,*" she marveled afterward, talking rapidly to Elizabeth, overcome and more than a little rattled. "I didn't know I was going to feel this way. Like, I felt this *bond* with them." Disconcerted by so much prolonged attention, she added, "When I was up there, I worried that people were looking at my nose. Is there something wrong with my nose?"

Transforming the Family

As I watched the baptism, which the family was kind enough to let me attend, it seemed to me that Kendra Vanderipe's emotional passage—detachment and uncertainty giving way to an overwhelming sense of connectedness to a family of which she is, undeniably, a part—was a good example of the transformation that assisted reproduction technology has wrought on the human family itself, which is being reshaped and re-formed, literally reconceived in ways that are still unpredictable. It seemed a good example, too, of the fortitude required of anybody forging new ties using new technologies—fortitude required not just of Kendra Vanderipe, standing at the baptismal font for all to see and scrutinize, but of parents like Laura and Hector Ramirez, who are building a family that honors both the power of genetic connection and the central importance of love and parental commitment. For Laura Ramirez, building a relationship with her boys' genetic parent is part of being a mother: it's that simple. Even as assisted reproduction technology has enabled many men and women to have children, period, it has enabled them to have children in novel combinations, with the help of a third, a fourth, sometimes a fifth party. It is now possible, and no longer uncommon, for a child to have two or even three people involved in what should perhaps now be called the mothering process: the egg donor, possibly a surrogate, and the mother herself. As of now, two is the maximum number of types of fathers—sperm donor and rearing father— science can attain. In the industry, procreating with help from a friend or stranger is called "collaborative reproduction." It is creating unprecedented kinship patterns, altering our sense of who parents are and what parents do.

In the United States alone, at least 15,000 IVF cycles are performed every year using a donated egg, a number that grows by more than 20 percent annually, as women seeking fertility treatment are increasingly encouraged to consider borrowing the reproductive capacity of another woman. According to the most commonly cited industry figure, 30,000 children are born in the United States annually from sperm donation, but the number could be higher; the FDA estimates that between 80,000 and 100,000 donor sperm inseminations are carried out each year. At least a thousand children are born each year from surrogacy, in which a woman carries a child for another woman, or—in more and more cases—for two gay men. Most of these are situations in which people who want children cannot have them unless they enlist the help of someone who will have a direct genetic connection to the child, now and forever. As one group of psychologists put it, making these families work for everybody—parents, children, shadowy other players—takes "courage, conviction, and strategy."

"The whole concept of family is so wide open now," observes Lori Maze, director of a pro-life organization, Snowflakes, that helps parents sign over surplus frozen IVF embryos to other infertile couples. People who just a few years ago were trying to have a family are now poring over biographical profiles to select the families who will receive their own excess potential offspring. Full siblings are being raised in separate households, sometimes getting together for barbecues and holidays, sometimes unaware of one another's existence. In the alternative universe that assisted reproduction can be, conservative Christians can be every bit as willing as the most progressive, most left-wing couples to consider new forms of family-making. "Family is not just that little nuclear genetic family that it was in the 1950s."

Of course, family has never been exclusively a genetic concept. The nuclear family has always been the theme on which variations have been improvised as the human family has evolved over time, surviving migration, war, adultery, epidemics, childbed fever, remarriage, slavery, harems, stepparents, increasing human longevity, social upheaval, and dramatic shifts in gender roles. We know already that family can take different forms, and that different kinds of people can serve as parents. We know, thanks to the greater openness of conventional adoption, that motherhood is not something automatically bequeathed by blood or genetic connection; that "birth mother" does not necessarily mean "real mother," yet birth mothers do matter and adopted children, if possible, should have access to them or know who they are. We know, in the wake of the rise in divorce in the 1970s and '80s, that "family" can mean something split and then reunited and cobbled together, comprising ex-spouses and stepparents as well as children hammering out relationships with stepsiblings and often half siblings who may or may not be close to the same age.

But science has given us something new: families that are designed, from the start, to have only a single parent; to have quite a few parents; to have two parents, only one of whom is biologically related to the child, the other of whom is biologically unrelated, with a third party out there who is biologically related but, often, unknown. Families with these qualities have spontaneously arisen in the past, and still do, of course, but now they are being consciously formed. They are being planned in advance and meticulously constructed, sometimes in exhaustive detail. Children are being conceived with the parents knowing that the child will be genetically related to one parent but not to the other. Parental roles are being divided up and divvied out, outsourced and reshuffled and even deleted. In addition to enabling the creation of relatively "traditional" families headed by a heterosexual couple like Laura and Hector, reproductive technology has fueled the creation of families headed by same-sex couples, challenging our understanding of what a mother is and does; what a father brings to the table; and precisely what significance these old-fashioned terms "mother" and "father" still have. Dads are the new mothers. Single moms are the new combo breadwinner/nurturers. In the United States, there are an estimated 1 to 9 million children living with a gay or lesbian parent. Often these are lesbian partnerships in which one or both women conceive using sperm donation, or gay male couples who enlist an egg donor and/or surrogate. "The average surrogate who carries for men prefers it," says Gail Taylor, a lesbian mother who founded a surrogacy agency, Growing Generations, which serves gay men using ART to have families. Gay men tend to be generous, attentive, unthreatened, and really happy about this whole family-building thing. "We don't resent [surrogates]; it's our first choice, and we treat them really well," says Doug Okun, gay father of surrogacy-enabled twins.

The possibilities are endless. One woman I interviewed, Kristine Cicak, was able to enter into a co-parenting arrangement with her best friend, Lisa Thornberry. The two women had been co-valedictorians of their high school and were lifelong friends. Both were single, both were straight, but they lived in Florida around Cape Canaveral, where all the men seemed to be terminally uninterested, or constantly being transferred. Lisa had always wanted a child, so, as she approached her fortieth birthday, she decided to select a sperm donor, have a baby through IVF, and enlist Kristine as co-mother. They would live in the same house and raise the girls together. Lisa bore twins, both girls. Then Lisa was killed along with one of the twins, broadsided by a car as they were turning into a fast-food restaurant parking lot to get a drink. Now this collaboratively reproduced family consists of Kristine and the surviving twin, to whom she has no biological connection but of whom she considers herself—she had to work to convince a Florida judge of this—the mother. There are online co-parenting bulletin boards, and some infertility websites include an "arranged parenting" classified section whereby a man or woman can advertise for a partner to have children with, using IVF. It's online-assisted shared custody, without the dating, the marriage, the sex, or the divorce.

Then there are co-parenting situations where a lesbian couple raises children together with a gay couple, one or both of whom donates sperm to one or both of the women, resulting in children with two biological and two "social" parents, a term that has evolved to mean a non-biological, but caretaking, parent. At a 2005 session convened by a gay and lesbian advocacy group called Family Pride, one teenager described being conceived by a lesbian couple as well as two gay men who later split up and found new partners. The upshot was that he ended up with two mothers and four fathers, something he enjoys revealing to classmates at the cafeteria lunch table, because the six-parent thing always trumps whatever another kid is bragging about. "It became, like being double-jointed, the thing I save for the last moment, when I needed extra attention," said the young man. "Once you realize it's something to be proud of, you're set to go," he added; the qualifier, "once you realize," perhaps implied that coming to this realization did involve some struggle. Along the way, ART has expanded the definition of not only "mother" and "father" but also "uncle," which is often the preferred designation for sperm donors, straight or gay, who want to be weekend dads to their offspring, enjoying many of the perks of biological relatedness, you could say, without having to clean up the puke. "It has really changed the way we view the family," says Dorothy Greenfeld, a Yale psychotherapist who has been studying IVF families since they began. "The idea that you can have so many parents."

The More People Who Love Them, the Better

After the Ramirez baptism concluded, everybody settled down and became more relaxed; there was the kind of friendly mingling that occurs after a formal ceremony is over and everybody has performed well, not just the parents and the egg donor but also—especially—the babies. Laura's father called the participants up to the font, and took photographs with St. Matthew's stained glass Nativity windows for background, the California sunshine throwing a soft colored glow over the proceedings. He photographed the godparents; he photographed the parents; then, with Laura and Hector standing there, he summoned Kendra. The three stood at the font: mom, dad, donor, each adult holding a baby.

"Are you coming to the lunch?" Laura asked Kendra. A celebratory meal was being held after the service. Kendra hesitated. She had been planning to just slip into the baptism, slip out, and call it a day. She had to return the rental car within twenty-four hours or pay extra; she and Elizabeth had to make an afternoon flight to Denver so Elizabeth could fly out for work the next morning. Not to mention that so much attention from strangers was getting stressful. It's hard to be a public egg donor for a prolonged period of time. At a certain point you want to get back into a place where people aren't assessing the slant of your eyebrows and the curve of your chin.

Then again, there was Laura, who had worked so hard to make these relationships work on every level. Laura, who had drawn Kendra into the family fold and kept her there; shown her that she mattered to the boys and their parents, both as a progenitor and as a person. Laura, who had created this family from, it sometimes seemed, pure force of will: enduring pregnancy loss after pregnancy loss, being comforted by her grief-stricken husband even as he tried to manage his own disappointment. Laura, the architect and general contractor for a family built out of disparate but surprisingly sturdy pieces. Laura, determined to locate a donor her boys could grow up to know. Laura, enduring the pregnancy itself; during the delivery she had hemorrhaged so severely that she needed blood transfusions. "She's an incredibly strong person," Kendra said in wonder.

So: yes. Yes, she'd go to the lunch, she told Laura.

Laura explained what she hoped her boys would realize someday. "I hope they will understand that they were very wanted," said Laura. She was still moving a little stiffly. Six months is not a long time to recover from a triple C-section and the loss of half the blood in your body. "A lot of work went into having them, and the more people that love them, the better."

Stronger Families

This is an important point: how deeply loved all of these children are. Back when IVF was getting started, it was feared that any family created through assisted reproduction—even plain IVF—would be stressed, maybe dysfunctional, because the parents had gone through so much to achieve them. It was thought IVF parents might be scarred; that fathers might feel alienated from the child, and mothers pathologically overprotective. Instead, psychologists found that families achieved with assisted reproduction tend to be stronger and more highly functioning than naturally conceived ones, because the parents are so motivated to have children, and so gratified once they arrive. "Children conceived via assisted reproduction are not disadvantaged," found Susan Golombok, a psychologist who has been tracking the welfare of IVF families, first at City University in London, and now at Cambridge University, where she directs the Centre for Family Research. Golombok has found that assisted reproduction parents are "more involved with their children and less stressed by parenthood," at least when parenting only singletons. This was true of the first families she looked at, who had conceived using IVF. Then, as she began to study families who had used egg and sperm donation, she found that these families consisted of people like Laura and Hector Ramirez, who were driven to make them work.

"These are the most motivated people I've ever met in my life," says Robert Nachtigall, the San Francisco reproductive endocrinologist who has directed studies of families formed by egg and sperm donation. "And what they want is a good relationship with their children." These are parents who so value their children that they tell them constantly how deeply they are loved. "When you've had to work so hard at having children, you talk to them about how much you value them," says Fay Johnson, a staff member of the Center for Surrogate Parenting and herself a mother by surrogacy. "You've learned that lesson that you need to tell the people you love that you love them."

Still, there is no question that many of these parents feel anxiety over their child's origins. Assisted reproduction is shaping, and complicating, the way we think about genetic relationship and its impact on family ties. In a major and largely unnoticed irony, in vitro fertilization—developed explicitly as a way to help infertile couples have their own biological children—now makes possible families in which the power of biological relationship is often both affirmed and denied, the importance of genetic inheritance simultaneously embraced and rejected.

In almost every way, the Kendra-Laura union is unusual. Many families using collaborative reproduction do not tell their children the truth of their origins, and most do not know the identity of the egg or sperm donor. Many don't want to. "If I never meet her, it won't feel like my husband is having a child by her," I was told by one mother conceiving with the help of egg donation, who did not plan to tell her child the truth. Parents sometimes fear the power of genetic connection so acutely that, if they can get away with denying it altogether—or rather, asserting their own fictional status as genetic parent—they often will try to do so.

The industry itself is shot through with inconsistency and self-contradiction. Parents facing the task of selecting an egg or sperm donor are bombarded with the most minute pieces of information. On websites for sperm banks with high-tech names like Xytex and Cryos and egg brokerages with sentimental monikers like Loving Donations and Tiny Treasures, prospective parents can sort donors based on ethnicity, personality, College Board scores, the shade of their skin and the kink of their hair; in fantasy they can construct a child as they would build a customized car. In the age of the genome, the message is that genetics are paramount in the *formation* of your child—and yet at the same time genetics are nothing in the formation of your *relationship* with your child. Your child won't be related to you, but will still love you! No problem! Every day, families are formed by parents trying to hold in their heads the competing notions that genes, while important, aren't. "We still have a very deep-seated cultural ambivalence about families not based on genetic ties," points out Jean Benward, a psychotherapist who specializes in adoption and has also done work with families formed by sperm donation. "Many years ago I heard a speech from Margaret Mead saying fundamentally, culturally, we did not believe it would work."

Prior to ART, there were two ways, in modern society, for two people to build a family from scratch. They could do this naturally, through sex, or legally, through adoption. As Benward points out, the two, in a way, are mirror images. In the first, both parents are genetically related to the child, while in the second, both parents are genetically unrelated to the child. Either way, both parents have equal standing. With collaborative reproduction, what you get is a sex-adoption hybrid, and this, for some families, is a source of real, unresolved anxiety. One reason parents don't tell children about egg and sperm donation is that they fear genetics will trump conscientious parenting. They fear that genetics will trump love. They fear genetic bonds are so fundamental and overpowering that the child, if told the truth, might reject the unrelated parent. "Parents are afraid that if they tell the child, the child will love them less," says Fay Johnson. "Period. It is a fear of being loved less than the other parent."

Moreover, adoption, which is socially accepted, facilitated by non-profit agencies and administered by the courts, creates what family law attorneys call a "bright line" of parenthood. Once a child is relinquished by a birth parent, the adopting parents are the legal parents, not the birth parent. With collaborative reproduction through ART there is no "bright line" between parenthood and nonparenthood. There is no legal relinquishment of an actual, living baby. Instead, egg and sperm are sold, the rights to their genetic contents and reproductive energy legally transferred. Yet many parents continue to worry about the not-well-articulated claim of this third party. Every parent conceiving collaboratively has to decide whether genetic relationship matters, or doesn't matter, or matters only somewhat; whether it matters just to the parents selecting the characteristics of a donor, or to the child created as a result.

Nor does it get easier. As is the case with adoption, parents who do tell their children about their origins usually create a happy narrative that minimizes the importance of genetic connection ("Every family can be made differently!" "You were a much-wanted baby, and that's what matters!"). But there is no guarantee offspring will agree. "We have a culture that talks about genes and genetics, but the offspring are not supposed to think that's important," Jean Benward says. For most of human history, as one adoption attorney has pointed out, "being a father was a matter of conjecture, and being a mother was a matter of fact." Now nothing can be known for sure.

For counselors trying to help these families, the unresolved question is: Do they resemble adoptive families, or do they resemble families created through natural procreation? Should there be screening for parental fitness and counseling afterward? There is furious disagreement. "I've always looked at this as adoption that is run by the medical profession," says Bill Cordray, an adult offspring of sperm donation who believes donor-conceived individuals have a moral right to know the truth of their parentage and that the industry needs to be reformed in the direction of encouraging disclosure. And yet: "I do see it as very, very different from adoption; it's a purchase of cells," I was told by a pregnant egg-donor mother, who did not intend to disclose to her child the truth of how the child was conceived. "It's like a blood donation," insists Gail Taylor, the surrogacy broker for gay men, who also runs an egg donation agency, and is embarking on a sperm bank. Taylor argues that sperm and egg donors are "genetically contributing to the conception process" and do not need to be identified to a child. Even those who support telling a child the truth do not necessarily see gamete donation as analogous to adoption. In adoption there is loss. For the birth mother there is surrendering; for the children there is the possibility of feeling given away or abandoned. In collaborative reproduction, it's argued, there is none of this. "There is no loss; nobody is giving up anything," argues Kim Bergman, a psychologist who works with Growing Generations, portray-. ing collaborative reproduction as a purely win-win-win situation.

"Adoption starts in a moment of incredible agony: some woman is going to give up a child, and that child is going to live the rest of his life thinking why did she give me up, and that's a wound that's hard to heal— for some people, that's a wound," agrees Robert Nachtigall. "Donor gametes isn't the same thing by any means. Nobody's unhappy. Every-body's happy to be doing it. Nobody's being abandoned. The donors— none of the donors are looking at this like these are offspring."

Collaborative reproduction also exposes the contradictions between several popular theories of child development. As Susan Golombok has pointed out, same-sex families created through assisted reproduction have "challenged deeply rooted beliefs in child psychology about the processes by which parents influence children." For more than a cen-tury, child-rearing experts have debated the ideal role of mother and father. Should mothers be strict? Firm? Warm? What about fathers? Authoritative? Loving? The assumption, implicit until recently, is that children would have—should have—one of each. Freudian theory, to take just one example, holds that a child must work through a series of conflicting feelings toward parents of both sexes—longing, resentment, sexual attraction, murderous urges—in order to emerge, sane and intact, on the other side of all that turmoil.

More recently, an "attachment" model of parenting has emerged: the idea that what a child needs most is a strong connection to a loving and attentive grown-up. A warm, fully bonded relationship with a loving adult—any loving adult, or six loving adults, or one, or two, male or female—is what enables a child to grow up well-adjusted and secure. "There is a growing body of empirical evidence to show that the course of a child's social and emotional development is related to the quality of a child's attachment to a parent, and that secure attachment relationships are fostered through parents' sensitive responding to the child," Golom-bok has written. You could call this the Harry Potter theory of child development: the idea that a parent's warm, unconditional love, dis-pensed early and often, provides a magical protection against many and varied trials of later life. This persuasive theory—it is intuitive, of course, that a well-loved child is more likely to be a healthy child— minimizes the importance of having parents of both genders, and it also minimizes the power of the genetic bond to trump all others.

But what, then, about genetic bewilderment? Genetic bewilderment is the term used to describe the confusion of a child who does not know the true identity of his genetic parent, and as a result cannot fashion a satisfactory identity of his or her own. The central importance of knowing one's biological parent—and being known by one's biological parent—is the narrative propelling the *Odyssey*, any number of Dickens novels, and—as is often pointed out, as if this somehow is a winning point—*Star Wars*. It is also a tenet of the way adoption is practiced now. In the 1970s, for an adopted child to search for his or her birth parents was considered pathological and maladjusted. Now it's considered normal, even necessary. Certainly, it is accepted that adopted children at the least should know they are adopted. Whether children of gamete donation have the same need is, however, both unresolved and furiously debated.

Regardless of their own cultural politics and beliefs, parents using collaborative reproduction often bounce back and forth unhappily between all of these theories. Parents using donor gametes in heterosexual relationships hope against hope that attachment will suffice; that children will realize that the mother who feeds and kisses and holds them is the real, the true, the important mother. At the same time, many fear that the genetic ties are the ones the child will secretly honor. The terror of many parents is that somewhere out there is a *replacement parent,* a *real parent,* and that the child may love this other parent more. "It's *not* all right. It's *not* all right with me," said one father, describing his reaction if his sons—who were created through both sperm and egg donation, and were playing just then in a very ordinary yard in a very ordinary suburban neighborhood—wanted to find their sperm donor. His wife, sitting with him on their deck, said she wouldn't mind if they wanted to find the egg donor; she no longer felt insecure and threatened, and thought the donor was someone they might be curious about. Her husband did not agree. "I'm the dad, damn it. To me, it's an indicator of whether the kids are happy or not." As for single mothers and gay and lesbian parents, it's a question many take seriously and want to know the answer to: whether and to what extent their children will be affected by the lack of a rearing parent of each gender. "All of us struggle with the concept of our kids not having mothers or fathers," acknowledged Gail Taylor, founder of Growing Generations, at a meeting of gay and lesbian parents.

"I Don't Know Any of These People"

At Celia's, a Mexican restaurant not far from St. Matthew's, a banquet room was overflowing with baptism celebrants, many of whom were already, by the time Kendra and Elizabeth got there, sitting at a long table, eating chips and ordering from the menu. "I don't know any of these people," Kendra said as they stood poised on the threshold, working up the nerve to enter the room. They took a place toward the center of the table, where there were still some empty seats. The two women were seated across from one of the godmothers, a friendly woman with long brown hair and magnificent gold highlights, who was sitting beside her husband, a police officer who was agreeable and built like a fire hydrant. Laura's mother materialized once again beside Kendra. "I will sit here, because I don't get to see Kendra very often," she said cozily, taking a seat nearby. Then she introduced a group of people sitting across the table and to her right, saying: "Kendra, this is my son, and this is his daughter, four-year-old Quinn." Kendra greeted them, and they greeted her, and I sat there trying to figure out if Kendra was related to the people she was meeting. It seemed, almost, possible. Four-year-old Quinn was the cousin of the Ramirez triplets, and Kendra was the biological mother of the Ramirez triplets, I kept thinking, so what relationship would Kendra have, if any, with, say, Laura's brother, down there nodding and saying hello? Aside from being the biological forebear of his child's three cousins?

Thinking about it made my head hurt—it was like one of those math word problems where a train is moving at 60 miles per hour and another train is approaching on a parallel track at 45 miles per hour, and if they are 3 miles away to begin with, how soon will they pass one another—so I studied the menu and ordered a burrito.

Meanwhile, one of Laura's law enforcement colleagues walked over and asked Kendra if she would like a margarita. He was holding a full pitcher, its contents slushy and cold and inviting. "Yeah!" she said, gratefully. It seemed precisely the sort of social situation for which alcohol was invented: You are an egg donor, seated in the midst of a large group of strangers looking at you with goodwill and curiosity, and yes indeed you would like a margarita, please very much and thank you. I understood now why Kendra had brought Elizabeth. It seemed a sensible and farsighted thing to do, having built-in moral support as well as, eventually, an excuse to leave. I also understood what a brave thing it had been for Kendra to come at all. "Is it snowing in Colorado?" the god-mother was saying to Elizabeth, and Elizabeth politely said no, it wasn't snowing, though it had snowed not long before, and they sat for a while, talking about the weather. At one end of the table was another group, chatting among themselves. Coming over to say hello, Laura introduced them to Kendra and Elizabeth as Hector's aunt, uncle, and cousin. "Some live in San Mateo, and some come from across the way, and some live in Napa."

As we talked, the triplets were being passed around from one willing lap to another. At some point someone handed Preston, the baby who Kendra thought most resembled her, over to Kendra. She bounced him on her knees. "Hey, babe, how ya doing?" she said cheerfully. She held him gladly but at a distance, bouncing him on her fresh-pressed full skirt. At the other end of the table, Hector was sitting with the backpack from which, Mary Poppins–like, he removed an endless series of child-care-related objects, including baby bottles covered in foil to keep them warm. He was trying to keep the boys on roughly the same feeding schedule, which was hard to do under the circumstances. Laura came back to sit with him, talking about people in the family and how they reacted to the egg-donation issue. Laura is kind and steady, exactly the police officer you would like to see approaching when you are broken down with a flat tire, or when you have heard a noise in your basement.

One of the issues with new reproductive technologies is that, even if you accept them, chances are somebody in your family will not. At first, Laura said, her mother didn't want to tell a certain uncle about the egg-donor aspect of the triplets' conception, fearing he would not embrace them into the family. Then word leaked out, and Laura frankly didn't care. Laura's own view is that if anyone is bothered by the lack of a genetic connection between her children and her, and is in any way unkind to her boys or treats them as lesser, she won't have anything to do with that person. As for being open, it seemed to her, based on instincts and training, by far the best way. "I didn't want any secrets. That's the worst thing that could happen: They get to be teenagers, and they already hate you, just because that's what teenagers do, and then they discover that you've been lying."

The margarita-serving colleague came round to chat with Laura. He was her partner, it turned out, and part of their dynamic seemed to be that he enjoyed giving Laura grief, needling her to try to puncture her air of affable poise. Speaking of Kendra, he said, enigmatically, "I didn't know she was a flight attendant." Talking about the fact that Kendra was suffering from a bad back that was keeping her off full flight attendant duty—she was working, for the time being, at the cosmetics counter in a department store—and that Laura just now had a bad knee keeping her off full police duty, he said: "That's even more that you two have in common! You're both on workmen's comp!"

"I've never been on workmen's comp in my life," said Laura evenly. Then he started teasing her about the baptism. "Something I've never seen before, [you] crying!"

Laura put up with the ribbing. It was clear that all of the police officers had a powerful bond. Laura explained that she had gone through police academy with the husband of the godmother sitting across from Kendra, and the godmother herself used to work for the police force. At one point, that godmother was talking about how they all—she, Laura, the others—used to go clubbing together; how now that they were married, with children, they never went out dancing anymore. It was too hard to find child care. They were old marrieds now. Kendra sat listening politely and checking her watch against the time she and Elizabeth had to leave and get back to their regularly scheduled lives, as single women living in a rented townhouse and flying, when they could, for weekend getaways. As they talked, it struck me that egg donors and women who use egg donors are actually the same people; they represent two stages of the way many women live now. There is the dancing/getaway weekend part of a woman's life, and there is the mothering part of a woman's life, and it's very hard to achieve both simultaneously. Now, "women helping women" can cut both ways: While reporting this book I heard of more than one egg donor who—having donated eggs as, say, a graduate student—found that once she got her degree, and spent years establishing herself in a profession, she, too, was obliged to seek the reproductive kindness of strangers. Or the kindness of her own former recipients: I interviewed one woman who was contemplating giving her frozen embryos to her egg donor. They had built a friendship over many years, and her egg donor, a lawyer, was married and now having fertility problems. When the woman tentatively suggested that these frozen embryos might be of use, her donor exclaimed, laughing, "That would be so *us.*"

The same thought had occurred to Kendra. The previous night, when she had met the triplets, she and Elizabeth had sat around the Ramirez house, chatting with Laura and Hector. At some point they came round to the topic of frozen embryos. Laura and Hector have seven embryos frozen, and at some point will have to decide what to do with them. The next morning, taking a shower in her room at the Bay Landing hotel, it occurred to Kendra Vanderipe that she herself wasn't getting any younger. She was almost thirty: old, by egg-donor standards. How long was her own reproductive window? Did she want children? Of her own? Ever? She goes back and forth. Right now, she is content with having offspring scattered around the country, children related to her whom she can visit. "I kind of feel like there are children out there who I'll kind of get to know and see them grow and be part of their life even in a small way," she says. Whether she had children of her own would depend on finding the right partner, and the right partner thus far had not materialized. Then it struck her. She could be her own egg donor! Those frozen embryos of Laura and Hector's! Before she left, she said, only half kidding: "They need to save some for me!"

CHAPTER FIVE

"SPERM BANK HELPS LESBIANS GET PREGNANT!":

HOW WOMEN CHANGED THE SPERM-BANKING INDUSTRY—AND THE MAKEUP OF THE FAMILY

"This Is a Sperm Bank?"

The Sperm Bank of California is an obscure, studiously nondescript place, like most sperm banks. It's located in well-worn upper-floor offices on Milvia Avenue in Berkeley, a block from the BART station and not far from the University of California. You push the button, and somebody buzzes you up. Tiny as it is, the Sperm Bank of California played an outsize role in transforming both the fertility industry and our collective thinking about who can and should be a parent. Its history and success show how ART has quietly but consistently offered a detour around social gatekeepers, a way for unconventional people to become parents and in the process to prove their parental fitness. Founded in the early 1980s, The Sperm Bank of California not only brought transparency to the sperm-banking business—a significant achievement in and of itself—but paved a wide and relatively smooth route to parenthood for a class of people who at the time were considered so aberrant, so profoundly unfit to be mothers, that more conventional means, such as adoption, often were barred to them. TSBC seized control of the very technology that had been forcibly inflicted upon an unconscious woman back in the nineteenth century, and enabled women to use that technology to their own personal and political advantage. It enabled lesbian women—lesbian women!—to become mothers. Together! To form lesbian-run households! Thanks in part to its lack of regulation and off-the-radar status, ART, then as now, has often enabled would-be parents to do an end run around family law judges, social workers, and other enforcers of normative thinking. It helped make the world safe for gay and lesbian parents, and thus it not only reflects social change but has actively and really quite powerfully driven that change. This is true of no place more than The Sperm Bank of California, founded more than twenty years ago by a women's health activist named Barbara Raboy. It was Raboy who helped inspire the kind of families being made today, though she didn't do it intentionally and had no idea when she was starting that she was setting in motion a revolution in how families are composed and our ideas about who can form them.

She just thought she was doing her job.

In 1981, Raboy was working as a health educator and counselor at the Oakland Feminist Women's Health Center. What the 1980s were about, for many women, was living out the freedom bestowed by *Roe* and other major legal and social victories. The eighties were about equal rights in the workplace; they were about women's health and female sexuality; they were about the Pill and the right to a legal, safe abortion. And that's what Barbara Raboy was about, too. As part of her job, Raboy gave classes on contraception. This being California, she also offered "fertility awareness classes" for women who wanted to contracept naturally, teaching them how to track their monthly cycle and gauge the window of time when they would be least likely to conceive. But then an unexpected thing happened: women started coming up privately after class, to ask Raboy how they could gauge the window of time when they would be *most* likely to conceive. Most of them were lesbians interested in using sperm donation, and wanting to maximize their chances of conception in an era before the availability of cheap at-home fertility kits. "They were saying that what they really wanted to do," Raboy remembered, "was *have* a child."

It was a radical concept in more ways than one. Social conservatives were of course against this new development—assuming they knew about it—but feminist leaders weren't exactly advancing the cause, either. Women's reproductive health in the 1980s wasn't primarily about helping women have children *now*. It was about helping women delay childbirth until they were ready. But they'd come to the right person: Raboy, who was feeling precisely the same urges, understood their instinct. If some of her clients wanted to have babies rather than avoid them, it was her job, as she saw it, to help them have babies.

Raboy flew to Los Angeles to visit a women's health center she had heard was quietly inseminating women. Raboy asked who their clients were, and was told that most were lesbian couples. She asked where they were getting their sperm. By now, a few sperm banks had opened, among them California Cryobank, as well as the infamous Repository for Germinal Choice—the so-called Nobel Prize sperm bank, a short-lived eugenics project to spread the seed of A-list donors, whose story is vividly told by David Plotz in his book *The Genius Factory: The Curious History of the Nobel Prize Sperm Bank*. The L.A. women's clinic used a small local bank, and somebody on staff drove Raboy over to meet the director. "I'm looking around thinking, 'This is a sperm bank?' Wow, this is pretty basic," says Raboy. "There was this little microscope, and a bathroom that had these pornographic photos," along with index cards on which sketchy information about donors was recorded. On the plane going back, Raboy wrote up her notes. She planned to tell her supervisor that to serve the needs of all their patients, the health center should start an insemination program, purchasing sperm from an existing bank. She didn't think they should get into sperm banking themselves: "We didn't have experience with the male population, and would have to create a scientific approach," says Raboy, whose science training consisted of using a microscope to look at yeast infections. She presented her recommendations to her supervisor, and "it took her one minute to say we should start a sperm bank."

Thus was born The Sperm Bank of Northern California (the name was shortened in 1988), created to provide sperm to women who wanted to conceive children and raise them in a father-free household. It was an extraordinary cultural moment: a radical change in how sperm banking was done, and, not coincidentally, the beginning of a new social movement. More concerned with the practical than the theoretical, Raboy obtained a list of sperm banks from the American Association of Tissue Banks and started calling banks to find out how they operated and whom they served. Invariably, she was told their clientele was limited to married couples.

And she was shocked—really shocked—at what married couples were willing to put up with. When she asked banks how they secured consent from donors, she found that there was no standard method. As for disease screening: "I was horrified," Raboy says. "There were a lot of banks unwilling to tell me what their criteria were for eligibility, and even those that shared it with me were doing almost nothing in terms of testing the donor. Some tested for gonorrhea; others didn't. Some tested for syphilis and Hepatitis B, and others didn't. I'm thinking to myself, 'Why am I even calling these banks?' I'm not a scientist, but I'm sort of stunned; there may be women out there who are acquiring gonorrhea." Raboy realized that here was an opportunity not only to start a sperm bank serving women, but to set a new industry standard. "And that," she says, "is when I started getting excited."

It was the summer of 1982: the tenth anniversary of the Supreme Court decision in *Roe v. Wade* was approaching. In the wake of *Roe,* the Oakland women's health center had developed a thorough system of counseling women seeking abortions. Now, she realized, The Sperm Bank of Northern California could do the same for men donating sperm. In a formal contract, they spelled out that these men would be surrendering all parental obligations and rights. They hired a sperm-bank director, and on the morning of October 5, 1982, called a press conference to salute the upcoming anniversary of *Roe* and announce the opening of The Sperm Bank of Northern California. The day before the event, the person they'd hired to direct the bank quit, saying he did not like working with an all-female staff. "No problem," Raboy's supervisor told her. "You can do this." Raboy was now the director, and the message she delivered was: "Any woman can come to the sperm bank for assistance getting pregnant."

"What do you mean?" Raboy remembers reporters asking. Any women? All women? What if a woman didn't have enough money to purchase a vial of sperm? "We said we'd work it out; we'd have a sliding scale," Raboy replied. "And they'd say: 'What about—they don't have to be married?' And we'd say, 'No, no, they can be single, they can be lesbians.'" To which, the reporter would say, "They can be lesbians—oh my God!"

"They asked, 'Is anybody else doing this?' and we said, 'We don't think so.'" Raboy filled several scrapbooks with articles. They had headlines like: "Life Without Father," "Farewell to the Turkey Baster," and—in the *National Enquirer*—"Sperm Bank Helps Lesbians Get Pregnant!"

Naturally, the controversy attracted more women: lesbians and some single women started pouring into the bank to be inseminated by strangers. Raboy started attending tissue-bank meetings to get a better feel for how the industry worked. She was appalled to learn that banks often tested only single donors for disease, and not married donors. She was shocked by how little information was available about the donors' family and health histories. It was Raboy's view that women had the right to know the health history of the donor whose sperm would go, and grow, inside their bodies. Up to then, sperm banks had been able to sell pretty much any sperm, trading on the fact that married couples were so ashamed of male infertility, they would keep things hush-hush and not ask too many pushy questions.

"What was starting in the 1980s was the idea that consumers have the right to know: about their donors, their donor's parents, their donor's grandparents, and what if anything does run in the family," Raboy says now. "They want to see the medical history. They want to know how does that sperm bank track the donor after the donor quits; what if their kid at ten years old is discovered to have some kind of ailment, and they want to retest the donor. What was going on in this period—and is still going on today—is that the paradigm was shifting: sperm banks were no longer allowed to be this cottage industry, behind closed doors. It was driven not by the FDA, but by people in the community saying: 'I'm not going to give you my business if you won't tell me this. You're asking me to put this tissue in my body and you won't tell me anything about it.' " In the mid-1980s the AIDS epidemic hit; soon the FDA would recommend that all sperm be tested for HIV and other communicable diseases, then frozen, and retested later. It was AIDS testing that fueled the creation of large-scale and eventually worldwide sperm banks; no longer could doctors use fresh samples that had been casually collected. Soon banks would begin providing all sorts of specialized information about donors. Nowadays, California Cryobank's website includes a monthly "featured donor"; one I looked at recently noted that in addition to experience umpiring baseball, and a taste for "seafood, chicken, fruit, pizza, and Italian food," this donor has "fair mechanical skills (enough to do things around the house)."

There is something else that sets The Sperm Bank of California apart from its peers: it was established as a nonprofit. Sperm donors were paid and women paid for sperm, but the middleman—or, in this case, the middlewoman—wasn't making millions of dollars. In this TSBC is set up more along the lines of Planned Parenthood, and other conventional nonprofit women's health groups. And TSBC has done well financially, earning its way without charitable contributions. They applied for grants from foundations, Raboy says, but "providers would say, you know, you're doing a great thing, providing access, but we have to give to homeless people before we can think about women getting pregnant."

Then lesbians transformed fertility medicine one more time. Once the bank opened, Raboy began giving classes on donor insemination. She alternated nights: one night would be for couples, another for unmarried women. It wasn't long before Raboy was approached with yet another question. "About two months into these classes—it was sometime around the holidays—a couple of women asked, in the women's orientation, if they could meet the donor."

Meet the donor? Meet the unknown, anonymous, carefully advised you-will-never-ever-have-access-to-this-child donor? "I sort of gasped," Raboy says. "Feminist that I was, I said, 'Wow, I don't think so, because that's the only protection we have.'" Donor anonymity was the only protection these mothers had against the ancient notion that genetic linkage carries with it the privilege, and responsibility, of being a father. Keep in mind, too, that genetic responsibility was, and remains, an important principle for women's-rights groups, in another context. Up to then, and now, a bedrock notion of women's rights has always been that any man who fathers a child, wittingly or unwittingly, has the responsibility of paying child support—precisely because he is genetically related to that child. Now women were saying: we want that notion of genetic responsibility to hold true, except when we don't want it to hold true. Would the distinction hold? What would protect women against a donor's coming back to seek custody or paternal privilege? It was hoped a watertight contract would suffice. But what if the woman wanted to know who the donor was? Was it possible for a woman to preserve custody of her donor-insemination child and still have access to the identity of the donor? Raboy didn't think so. "If there were to be some sort of paternity claim, the last thing we want to do is to put ourselves in the middle," she remembers thinking. "It was a huge liability for us. We didn't know what the future was going to look like as sperm banks evolved. I said, 'We feel really vulnerable.'"

Still, Raboy had the presence of mind to ask the women why they wanted to know the donor. They replied that they wanted to know more about him not for their own benefit, but for their children's. They wanted to be able to say to a child, "This is what he looked like, this is what he seemed like, his voice was like this and he carried himself this way, and his eyes were like that." Realizing this, Raboy says, "I just was beside myself: Oh my God, of course. You want to be open with your kid. You want to be honest."

So Raboy called the sperm bank's attorney, who by now must have been used to her novel queries, and "she said, 'Barb, you're right to faint when they want to meet donors. But the state defines an adult as eighteen years old: you could develop a policy so that the identity of the individual is revealed later on.'" If the offspring were to meet the donors as adults, there would be no custody contests. Like the doctors pioneering egg donation, the new female sperm brokers were making things up as they went along, figuring out both what seemed right and what seemed workable. "She said, 'I don't know of anyone who has ever done it, but let me draft up a new consent form. We already had the consent form for the donors that said you'll be anonymous forever. Now she drafted an identity-release consent form saying that when a child was eighteen, [the donor] could be contacted. And I said, 'Whoa, talk about a social experiment.'"

They called it "identity release." Rules were devised: Under no circumstances would a donor be allowed to change his mind. Nor would women be permitted to request a meeting before their child reached the age of eighteen. "It was scary to think that we were establishing an event that was going to happen, not only a ways down the road, but having nothing—nothing—to fall back on, nothing in the world to say, this is what it could be like. It was scary, and exciting."

At first, Raboy says, only a few women were interested in having an identity-release donor. Many women feared, still do fear, that a donor might try to seize custody of the child or interfere with the household. But by the late 1980s, "ID release just started becoming huge." Raboy realized that they needed access to a larger pool of open-minded donors. So The Sperm Bank of California moved to Berkeley, to be near the university, and that, Raboy says, "did exactly what I hoped it would do: It increased our pool" of men willing to say yes, when a child is eighteen, this child has the right to meet me. Other sperm banks would follow suit: there are always far more sperm donors who want to remain anonymous, but some of the large sperm banks have now created ID-release programs.

These donors became even more sought after in the 1990s, when ICSI abruptly wiped out a large part of sperm banks' traditional male clientele by offering infertile men a way to have their own biological offspring. In the wake of this technological tsunami, sperm banks might have been washed away had it not been for women. The percentage of sperm-bank users who were women soared to at least 60 percent. Ironically, the industry that initially shunned lesbian women soon found itself depending on them. Their patronage is considered routine and highly desirable. Lesbian patients are considered good clients and parents, not overinclined to engineer their children or become obsessively invested in minor attributes. "Lesbians are very accepting of whoever the child turns out to be," reflects Charles Sims, one of the founders of California Cryobank.

Ultimately The Sperm Bank of California achieved a number of significant changes. Practically speaking, it raised standards for testing and disclosure. It introduced the possibility that sperm (and, later, egg) donors could be known to children. And it enabled women to have children and raise them, deliberately, without men, inaugurating the age of the technology-assisted alternative family. It showed how a relatively simple fertility technology could catalyze massive social changed, creating, really, a whole new family unit. It showed how having children—maybe even more than not having children—can be a radical, cataclysmic act. It showed how ART offered some people who want to have families a way to make an end run around the people who don't want them to have one.

Father-Free Households

"We really wanted a gay donor. It would be neat if our kid is queer, too," said Gretchen Lee, sitting in the walk-up apartment in San Francisco's Castro district that she shares with her partner, Evie Leder, and their daughter, Rose Leder-Lee. Rose was ten months old at the time of our interview. Rose is a happy, round-faced girl with curly blond hair; her toes were bare and beautiful as she scooted around the sunny front room, chewing on her toys, playing with the household cat, a silky black beast who adores her, and clambering in and out of the arms of her mother. From time to time, Rose would demonstrate an impressive ability to stick her tongue out and curl it dramatically around, so that the bottom of her tongue faced the ceiling. Gretchen says the tongue-curling ability must have come from the donor; it's heritable, and she herself cannot do it.

Theirs is, in many respects, a typical two-parent household. Evie

Leder works full-time as a Web designer and filmmaker, and Gretchen Lee does part-time consulting and writing for a lesbian magazine, *Curve*, that lets her be a primary caregiver for Rose. Finding the right donor was a time-consuming process; finding the right man almost always is, even or maybe especially when you don't plan to meet him. Every lesbian couple, when they want to build a family through donor insemination, has to decide whether to select a "known donor" from among men they know, or go straight to a sperm bank. Leder and Lee had conversations with a gay friend, whose partner did not react well to the idea of his donating sperm. Two other gay men, a couple, were interested; they wanted a hands-off, "uncle" designation, which suited the women, but they also wanted an elaborate insemination process that would disguise who was the biological father. They suggested that their sperm be shaken together, in the old-fashioned surreptitious doctor's-office way, and fate could decide the rest. Lee and Leder were skeptical, and thought maybe they would put the sperm in two jars and pick one.

They ultimately opted for a sperm bank. They considered working with the Rainbow Flag Health Services, a small East Bay outfit that brokers arrangements between lesbians and gay men who want to co-parent together. They also wanted to use a midwife to perform a home insemination, thinking it would be "more organic" to conceive at home. But the midwife—like many things organic—turned out to be expensive, and their Kaiser health insurance offered office insemination services for just $38. To make a long story short, the Kaiser people did not work well with the Rainbow sperm bank people, so they found their way to The Sperm Bank of California.

There an interesting dynamic emerged: While Leder and Lee were inclined to select a "yes" donor—TSBC's term for a donor who agrees to be identified to an adult child—more important was a donor who physically resembled Evie Leder, the nonbiological mother of their child. In this, same-sex parents are strikingly like conventional ones. For years, heterosexual couples patronizing sperm banks have invariably selected donors that matched the appearance of the infertile father, the better to hide the truth of sperm donation and, with it, the fact of male infertility. But lesbian women often want to do the same thing: to match the donor to the unrelated parent. To a certain extent, same-sex couples do this because it's easier and circumvents awkward questions. But they also do it, I think, because of what having a child represents for any two people who love each other. During the week I spent with David Keefe, the reproductive endocrinologist, who also trained as a psychiatrist, Keefe talked of how couples use IVF out of a desire to "embody" their love for each other. Keefe was talking about heterosexual couples, who even if they have children from prior marriages often want a child to cement their union. Donor matching among gay and lesbian couples also seems to me a powerful example of a profound and often unacknowledged role of children: they are human love made concrete, human love given flesh, human love that takes form and is now moving about and growing, the living, breathing testament of our union with another person.

At TSBC, the women found an identity-release donor whose written description seemed to resemble Evie. But his sperm didn't deliver, so they went back to TSBC to pick a new one. They were in a hurry; Gretchen was approaching insemination time, and there was an anonymous donor whose description sounded close. "We were like, okay, he's a Jew, he'll do," Gretchen remembers. "Come to find out, after we'd picked him, we were back at the sperm bank and the counselor who was getting our order ready was real friendly, and said, 'Actually, this guy looks more like Evie than the first guy did.' When they purchased the "extended profile," it turned out that the donor had a gap between his two front teeth, another characteristic of Evie's. *Et voilà*: the result, just now trying to ascend her rocking horse, is lovely Rose, who at the time of the interview didn't have enough teeth to tell whether the gap would materialize. Gretchen speculated that perhaps Evie and the unknown donor share some genetic ancestry: "Maybe they're from the same village in the Ukraine."

In the wake of Rose's birth, Gretchen says, the two men they had considered as donors have stuck by them; they "have been really nice people to have around, in terms of taking an interest in Rose, and just like good friends." And that, she feels strongly, is enough male influence for her daughter. Of Rose's lack of a father, she says, "I really don't consider it a bad issue at all. She's got two mothers who are really devoted to her, two extended families who love her, and—I just think when she is grown up, it's going to be such a different world. She won't even believe us that it was a problem." And it's not as though Rose will be the only child she knows who has two mamas. "Anecdotally," Gretchen says, "almost every couple I know is trying to get pregnant."

As for Rose's eventual orientation: though at first they thought it would be good if their child turned out to be gay—"it's a nice life," Gretchen Lee says—they feel open to anything. "I would not have any problem if Rose turns out to be gay, nor would I have a problem if she turns out to be straight. I don't have any desire to prove anything by producing a straight child, which I think some parents do."

"Differences Are Not Deficits"

The ordinariness of the Leder-Lee household shows just how much has changed in three decades. Thirty years ago, it was a not only a fear but a widespread conviction that children of same-sex parents would be stigmatized and sexually maladjusted, and it was this fear that encouraged the idea that lesbians were unfit mothers. In the 1970s and '80s, the apprehension often played out in child custody cases; not only were lesbians barred from adopting children, but when they left heterosexual relationships, their biological children were sometimes taken away. Judges were willing to sever biological bonds altogether, for fear of the impact lesbians might have on their own offspring. In a 2006 lecture to gay and lesbian parents, psychologist Susan Golombok recalled the temper of that time. Holding up a 1976 copy of *Spare Rib* magazine, with a photo of three women on the cover, and the question, "Why could one of these women lose custody of her child?" she recalled the bias against lesbian mothers. In divorce cases, Golombok said, there would often be a Freudian psychoanalyst testifying on behalf of a father suing for custody, arguing that the combination of a lesbian mother and an absent father would create irrevocable problems for a child, who could not work through the necessary stages of attraction and separation. It was put forth that children of lesbian couples would be teased; that they would develop behavioral problems; that they would show atypical gender development. Golombok, who was involved with the women's movement in England, thought that it would be interesting—and that it was important—to try to devise a study to see whether or not this was true.

Lesbians using sperm donation to conceive would emerge as a popular study group. At the time, the same movement occurring in California was quietly making itself known in England. At the outset many doctors were unwilling to assist lesbians or single women seeking sperm donation, but in the late 1970s, a newspaper discovered a London doctor who was inseminating unmarried women with donor sperm. When the news broke, one member of Parliament thundered that "this evil must stop for the sake of the potential children and society, which both have enough

problems without the extension of this horrific practice." While apoca-
lyptic, this view was not uncommon: studies of "traditional" single
mother households had shown that a child without a father is more likely
to live in poverty, and to do less well in life.

To test whether this applied to children of lesbian women, a first
wave of research—including a longitudinal study conducted by Golombok and a colleague—looked at women who had had children in heterosexual marriages, had come out as lesbians, and been permitted to retain
custody. The number of households surveyed was small, something that
would remain a problem for later same-sex-parenting studies: because
so many gay men and lesbian women were not living openly as such, it
was (and to a certain extent, remains) hard for researchers to gather data
from a large, representative sample of randomly selected households.
Instead, researchers often use small "convenience" samples of households gleaned through social and community networks. The first wave of
data suggested that children of lesbian mothers, while sometimes stigmatized by their peers, were growing up well and healthy; that they were
impressively resilient; and that they fared much the same as children in
other households. This research was important; even today, the majority
of children living in same-sex households are living with gay and lesbian
parents who have divorced heterosexual partners, and come out.

A second wave of research, in the late 1990s and early twenty-first
century, began focusing on planned lesbian families created through
donor insemination, which are useful in that they enable researchers to
remove a number of variables—the negative impact of divorce, the
stress of having a parent come out, the reaction to re-partnering—
and focus more purely on child well-being under same-sex-parenting.
As sperm-bank policies became more relaxed and lesbian donor-
insemination (DI) households more widespread, they were studied by
psychologists in the United States, Britain, and Europe, who compared
them to heterosexual DI households, to single mothers using DI, and to
heterosexual couples with naturally conceived children. These studies
again suggested that the children were doing well. It was found that lesbian DI mothers, like many other assisted reproductive parents, are so
motivated that they in some ways are more committed to parenthood
than heterosexuals. The data, while still limited—most DI mothers, lesbian or straight, are white and middle-class, and tend to live in progressive regions—even suggested that the average lesbian mother wants
children *more* than the average heterosexual man, who is often led into

parenthood by a motivated woman. In lesbian households, one study showed, the co-mother who is biologically not related to the child does *more* hands-on parenting than the average heterosexual father. Studies also suggested that children of lesbian households grow up freer of conventional ideas about gender; in one, daughters of lesbians were more likely to envision themselves in professional careers than the daughters of heterosexual mothers. And, researchers repeatedly emphasized, children of lesbian parents showed no greater inclination to same-sex relationships.

This research did a great deal to legitimize same-sex families in the public and professional minds. As the sociological researcher Judith Stacey put it in a major 2001 review of the literature, research on gay and lesbian families is by no means an abstruse academic sideline. It is hard-fought, hotly contested, and of enormous social significance. Though still, as she puts it, a "youthful body of research," it is having a major impact on our views about how parents influence children, and about what is an acceptable environment for a child to grow up in. At the dawn of a new millennium, she points out, same-sex parenting and marriage are one of our chief areas of cultural controversy; struggles by gays and lesbians to "secure equal recognition and rights for the new family rela-tionships that they are now creating represent some of the most dra-matic and fiercely contested developments in Western family patterns."

Indeed. This research has been used in an ongoing and successful effort by gays and lesbians to gain the right to parent: by retaining cus-tody of their own biological children, by adopting, by serving as foster parents, by gaining greater access to reproductive technology. It's a radi-cal shift. Until 1973, homosexuality was included in the American Psychiatric Association's list of mental disorders. Now, the APA is one in a long list of organizations—the American Academy of Pediatrics, the Child Welfare League of America, and many others—that oppose discrimination against gay and lesbian parents.

Along the way, however, ideologues on both sides of America's cul-tural divide have debated the merits and ramifications of this research. As Stacey enumerates in her 2001 paper, a small but vocal body of con-servative researchers has stubbornly resisted the idea that children are unaffected by the sexual orientation of their parents. Their views are shared by many right-wing commentators: "When the State sanctions homosexual relationships and gives them its blessing, the younger gener-ation becomes confused about sexual identity and quickly loses its understanding of lifelong commitments," the conservative religious leader James Dobson asserts on the Focus on the Family website. In July 2006, the New York Court of Appeals ruled that gays and lesbians did not have a constitutional right to marry, saying that "Intuition and experience suggest that a child benefits from having before his or her eyes, every day, living models of what both a man and a woman are like." The irony of this position is it becomes counterproductive: gays and lesbians now are having families by the boatloads, through adoption and assisted reproduction both, and preventing them from marrying—denying them access to a legal recognition of precisely the "lifelong commitments" Dobson holds central to child well-being—mainly prevents their children from enjoying the considerable privileges of legitimacy.

Yet some cultural liberals, buoyed by the lesbians-parent-more-attentively-and-more-open-mindedly argument, have pushed those findings pretty far, going so far as to argue that not only are children *not* harmed by not having a father, but that a woman-headed household is in some ways superior to a household in which a father is present. The idea is that children raised by "maverick moms" are unencumbered by old gender stereotypes; that growing up without a father enables boys to adopt a new, less retrograde form of masculinity. This view can be—and has been—easily lampooned; while not having a father in the household may not be a bad thing, always, it's hard to accept the idea that it's actually an advantage. Stacey makes the point more moderately, noting that "lesbian parenting may free daughters and sons from a broad but uneven range of traditional gender prescriptions."

But Stacey and her research partner, Timothy Biblarz, also pointed out, eloquently, in their 2001 paper, that many researchers tend to sympathize with gay and lesbians, and that as a result, they have sometimes downplayed more controversial findings. Her 2001 review confirmed that the research suggests that lesbians are good parents, and that their children are well adjusted. But it has also shown—this is what has been downplayed—that being a child of gay and lesbian parents probably does result in a "broadening of children's gender and sexual repertoires." Golombok's longitudinal study, which tracked children of lesbian mothers into young adulthood, found that six of twenty-five children of lesbians had had a homoerotic relationship, compared to zero of twenty children raised by single mothers. With regard to their sexual partners in general, Golombok's study also showed that girls raised by lesbian mothers are more sexually active and "less chaste," as Stacey put it, whereas the sons of lesbians mothers evince the opposite pattern: more chaste, less sexually adventurous. Stacey concludes that children of lesbian parents do seem to "grow up more open to homoerotic relationships."

This report caused quite a stir. Stacey's arguments were seized upon by right-wing opponents of gay marriage, transformed into what she rue-fully calls "O'Reilly factoids." Persevering, Stacey has argued that aca-demics need to abandon their "defensive posture." She stresses that "differences are not deficits."

Like Golombok, Stacey spoke at a 2006 conference, "Real Families, Real Facts," convened by the gay and lesbian family advocacy orga-nization Family Pride. She looked a little tentative at first. But the audience—many of them same-sex parents—were certainly not hostile. They wanted to know what she was finding. Stacey reminded her audi-ence of the difficulties of doing this research: thanks to assisted repro-duction, methods of conception are changing so quickly that certain family-making arrangements are still unstudied. There are few studies of the increasingly popular arrangement known as co-mothering, in which one lesbian partner contributes the egg, which is fertilized with donor sperm through IVF, and transferred into the womb of the other woman. There are even fewer studies of gay men conceiving with the help of sur-rogates. And it's hard to measure the impact of stigma on children; atti-tudes toward gays and lesbians are changing so quickly that the fact that some children felt stigmatized, ten or even five years ago, might not hold true today. But despite these limitations, Stacey believes that research on different kinds of family arrangements, many of them technology-enabled, allows researchers to better understand parenting in general, by teasing out the impact that gender, biological relationship, and sheer desire to be a parent have on a person's ability to parent.

Looking at studies of parenting by heterosexual couples, single mothers, single fathers, and lesbian DI households, Stacey says the cumulative findings confirm that even today, women tend to do more childcare and household tasks; regardless of the amount of time women put into paid labor, "fathering involves more recreation and play, while mothering involves more caretaking." Stacey observed, however, that there does seem to be a "maverick dad" effect: when heterosexual men become single fathers, they parent more attentively, more like women. Future research on gay fathers will, she said, allow researchers to test this finding. Does being liberated from the presence of a mother in the house enable men to come into their own as warm, hands-on parents?

Stacey also emphasized that lesbian mothers—compared to heterosexual fathers—do spend more time caring for children; exhibit more warmth and physical affection; use less corporal punishment. They're very egalitarian. Overall, she said, being female seems to be associated with more attentive parenting. Gender still trumps biology, when it comes to warm and hands-on parenting. But biology also influences the parent-child relationship. In lesbian donor insemination families, she said, "the biological mother does tend on average to be the primary parent, to have a slightly closer relationship to her children, to feel more ownership and control." Biology cannot, in other words, be entirely discounted or wished away.

"Who Takes Out the Rat That the Cat Brought In?"

At a lunch after Judith Stacey's talk, Will Halm, a lawyer involved in gay family activism and the father of three children through surrogacy, was teasing Kim Bergman, a psychologist and lesbian mother. Both are on the staff of Growing Generations, the Los Angeles surrogacy agency founded to serve would-be gay fathers. Halm was joking about what he saw as another male-female pattern. Gay couples, he said, when they break up, tend to be easygoing and to remain on good terms, whereas lesbians break up in such a cataclysmic way that one woman "never wants the other woman to ever lay *eyes* on that child again." Which from his point of view is not a bad thing: "The worst lesbian breakups make for the best case law."

He was talking in part about a case that has dismayed the gay and lesbian family-building community even as it may have advanced their legal cause. In San Francisco, two women conceived by co-mothering. In this case, for reasons that are unclear—one suspects an overabundance of biological egalitarianism—the partner who contributed the egg signed a basic egg-donor agreement, giving away legal rights to her oocytes. But she intended to be a co-mother, and she participated fully in raising the child. After the couple broke up, the partner who carried the child tried to argue that the other partner had relinquished her right to custody. The Supreme Court of California rejected that argument.

Tellingly, the court did not focus on the fact that the woman suing for shared custody was in fact the child's genetic mother. Instead, the court ruled that the important thing was that this co-mother had *intended* to parent. Halm was pleased by the ruling; it fit his strategy of directing judges to look not at biology or sexuality but at *intent*. In deciding whether or not a person is a child's "real" parent, he argues, the only question that matters is whether that person meant to be the parent.

For Halm and other gay-family activists, wanting to be a parent—planning to be a parent—is the litmus test of whether somebody is the rightful parent. In California, this idea is now so widely accepted, thanks to precedent-making court cases, that both members of a gay or lesbian partnership may be listed as parents on a child's birth certificate, even though one parent has no biological relationship to the child. The idea of intent as a defining quality of parenthood is characteristic of assisted reproduction in general; in cases involving gamete-donation, judges have increasingly accepted the idea that the "real" parent is the person who intended to raise the child.

Perhaps needless to say, this development alarms those conservatives who are paying close attention to these potentially seismic changes in the legal definition of parenthood. In 2006, a group of conservative scholars, the Commission on Parenthood's Future, issued a report, "The Revolution in Parenthood: The Emerging Global Clash Between Adult Rights and Children's Needs." The report noted, with alarm, the varied ways in which, in states like California and a few foreign countries, assisted reproduction is encouraging judges to look at parenthood as "psychological" rather than biological. The principal author, the researcher and commentator Elizabeth Marquardt, sees this as an assault on the traditional idea of family as based on biological ties. She argues, with some justice, that in "law and culture, the two-natural-parent, mother-father model is falling away, replaced with the idea that children are fine with any one or more adults being called their parents, so long as the appointed parents are nice people." Marquardt argues, correctly, that technology is changing our very definition of a parent, and she is not wrong in arguing that gamete-donation scenarios sometimes privilege the desire of a parent—the desire to be considered the full, the rightful parent—over the child's own needs. On the other hand, thirty years ago, judges removing children from the custody of lesbian mothers weren't all that respectful of biology, either. And it seems to me that modern judges, in rather sensibly ruling in favor of a parent who intended to raise the child, who has raised the child, and to whom that child is attached, are acting primarily in the best interests of the child.

But it's certainly true that parenthood is being redefined, and how you feel about that probably does depend, in part, on your cultural politics. Sitting at lunch after Stacey's talk, Will Halm and Kim Bergman both urged me to accept the idea that it's meaningless—even politically incorrect—to even talk, anymore, about biological relationship. They want to get away entirely from the concept of "biological mother" or "biological father," because in gay and lesbian couples, such terms imply that the biologically related partner is the one with greater moral claim to the child. When I asked why, then, do so many lesbian couples choose a sperm donor who looks like the unrelated partner—why create the appearance of biological connection, if biological connection doesn't matter?—Bergman said that it's just something you do. If it were possible to have a child together, she allowed, of course a couple would want to do that.

Because in truth, biology cannot always be managed into irrelevance. "Genetics are a really strong thing, when you see them staring you back in the face," acknowledges Gail Taylor, founder of Growing Generations.

Gail Taylor, a lesbian and two-time mother by sperm donation, is honest about the difficulties she and her own partner encountered building their family, navigating what turned out to be unexpected biological landmines. The first time she and her partner, Trish, conceived together, Gail Taylor says that like many lesbian couples they wanted to come as close as possible to creating their own genetic child. Gail was planning to carry the baby, so they asked a close male relative of her partner to be the sperm donor, thinking that he was "this natural extension, genetically, of Trish . . . as close as we could come to doing it ourselves." A daughter, Chloe, was born, and what happened next, Gail says, is that the male relative "had more feelings about it than he thought he would." The male relative had a newborn of his own. The idea had been that he would be father to his own child, and "uncle" to Chloe. But the women felt he was beginning to regard himself as a father, claiming a say in how they raised their child based on the privilege of genetic connection. "That hit us just completely blindsided," Gail says. "We're like: 'I don't want to negate the gift you've given, but there's a big difference between what you do, and what Chloe knows fathers do.'"

When they decided to conceive again, they chose an anonymous donor.

Now that they have daughter and a son, Gail Taylor talks through issues of parentage with her children. They have conversations exploring what a father is, what a father does, and what it means when a household is without one. "It's been an interesting year," says Gail, whose daughter was about to turn eight when we talked. She said there had been a "big jump" in her understanding of reproduction. "We had a precocious play-date at our house, older than Chloe, this one knows a lot more about birthing and biology, and she said, 'Where'd you come from, if you have two moms?'" They talk about the role that fathers play in her friends' households, and why there is no person like that in their own. Gail tries to explain to her children that what a father is, really, is a role. There is nothing a father does that a woman can't do. In their household, two women divide that role, and divide, as well, the role of mother. "We say okay: 'Who fixed your bike seat? Mama Trish did. Who takes out the rat that the cat brought in? Okay, that's Mama Gail's job.'" Seen this way, being a father is not a right you claim by biological fiat; it's a right you earn, by doing what fathers do—taking out rats, fixing bike seats—and those are things that anybody can do, including a woman.

And if anybody can be a father, it follows that anybody can be a mother. It follows that a mother is not something you automatically are, but something you do. And anybody—including a man—can do it.

CHAPTER SIX

TWO MEN AND TWO BABIES:
GAY FATHERHOOD THROUGH SURROGACY

"Here Are My Deal Breakers"

"We know more about the female reproductive anatomy than a lot of women we know," Doug Okun points out, not defensively but just reflecting on how much he and his partner, Eric Ethington, have learned, conceiving and now caring for two daughters. They know about fertility drugs; about multiple pregnancies and their risks; about vaginal ultrasound probes; about yeast infections, which one of their twins has a touch of. "How about if I grab the diapers and pajamas and we change them in here?" Doug suggests, disappearing into a room and emerging with fresh diapers and nightclothes for their sixteen-month-old daughters, Sophia and Elizabeth. Doug gently lays them down and changes them; armed with a handful of tubes and creams, he expertly applies salve to one girl's affected region. "It's the same thing women get," he says authoritatively, capping the tube.

It is seven thirty on a weekday evening: dinnertime for the men and nearly bedtime for Sophia and Elizabeth, who, pajama-clad and gregarious, resume motoring around the house and playing. Both men have swapped work clothes for jeans and dark T-shirts. Both are good-looking and in their thirties, with gym-sculpted bodies enhanced by what they call "baby biceps." "I've actually gained muscle in my chest from having kids," remarks Eric, standing at the counter that communicates with the dining room, mixing a drink in a cocktail shaker and unwrapping ingredients for dinner. "Are you hiding back there?" Doug is saying to Sophia, who is standing in a sliver of space between a breakfront and the dining-room wall, giggling at her father. Eric starts cooking sausages and pasta, listening as Doug begins the story of how their family came to be.

"I'm assuming that it's pretty easy to understand why people want to have kids," is how Doug begins.

You could say the Okun-Ethington household got started in a village in South Korea. There for the proselytizing mission every Mormon must complete, Eric Ethington, a Brigham Young University graduate who grew up on a dairy farm in southern Idaho, had an epiphany. It consisted of two things: (1) he no longer believed in Mormonism, and (2) he was gay. Eric found his way to San Francisco, where he took the underground BART train to Castro and Market, then—as now—the gayest corner in the gayest neighborhood in the gayest city in the world. It was 1992, less than a decade after the beginning of the AIDS epidemic, and the devastation was visible everywhere. Emerging from underground, Eric, 24, saw "these two old men who were probably thirty-five, arm in arm, one holding the oxygen bottle that fed them both. They had two prongs into the oxygen and they were moving at this incredibly slow measured pace, like they were eighty years old, and I was so freaked out by the thought of getting AIDS and dying of AIDS that I did not go back to the Castro for another year. I just assumed for years that I would die of AIDS, because everybody did." To "confront that fear," he volunteered as an HIV counselor at a health center, where he came to realize that "AIDS was a very hard disease to contract, and you needed to do very specific things to get it."

While working through this fear, Eric attended Hastings law school, where he had another epiphany. He wanted to have a happy life, which for him meant that he didn't want to be a lawyer. He went into finance, and took a job at Charles Schwab, where a co-worker introduced him to Doug Okun, a Schwab employee who'd grown up in a suburb of Boston, attended Columbia University, and gone to Stanford for his MBA. The men looked at each other, knew that "it was going to be hard to keep this professional," and started going out. On their second date Eric told Doug: "You don't have to say anything now, but here are my deal breakers. The first one is I want to have kids. The second one is, I want to have a wedding ceremony."

Both conditions suited Doug. A wedding ceremony wasn't important to him, but a commitment was. "I'd had a very, very typical gay experience of going to tons of weddings, and often being the only gay guy there and feeling happy for my friends and a little resentful that I couldn't have the same thing," he says now. Doug, like Eric, had always wanted children. "I don't know how much I really considered the logistical challenges or whether I would or would not have them as a gay man, but I remember, in my twenties, I would see a man with kids, and wish that was me." The men agreed that they would like to have their own biological children, something that was beginning to seem possible. They considered adoption, but worried that as homosexual men they would be placed at the end of the queue of would-be parents. In 1997, Doug and Eric moved in together. In 1999, they bought a house together. In 2000, they had a commitment ceremony. In 2001, Doug did a Web search on "gay" and "surrogacy." The first hit was Growing Generations, founded in 1996 to serve would-be gay dads.

Growing Generations is based in Los Angeles, in a swank office suite on Wilshire Boulevard's Miracle Mile. It's no accident that ground zero for gay parenting can be found along this chic and famous stretch of real estate. While Bay Area lesbians pioneered the idea that same-sex partners can be good parents, L.A. has become the world capital of technology-enabled gay parenting. One reason is the state's legislative environment, which has evolved to be surrogacy-friendly. Another is the fact that this is where Gail Taylor, founder of Growing Generations, happened to live. Another is money: For two men, having a family through ART is vastly more expensive than it is for women. A vial of sperm costs a couple of hundred dollars, but parenting through surrogacy costs between $100,000 and $150,000, including payments to a surrogate, an egg donor, one or even two brokering agencies, a fertility clinic, lawyers, and assorted facilitators. Los Angeles offers a critical mass of affluent gay men (though Growing Generations also serves a large out-of-town clientele); a boundless pool of fit, sculpted women willing to act as egg donors; and sturdy family-friendly women from the nearby valleys, willing to gestate the pregnancy.

Perhaps most important, Los Angeles also offers an open-minded medical climate. On the East Coast, clinics can be hidebound in insisting that infertility patients actually be infertile. In Los Angeles, there are doctors who specialize in family-making for gay men. Venturing into the high-end offices of Pacific Fertility Center—a doctor there, Vicken Sahakian, says gay men are 25 percent of his practice—Doug and Eric were struck by how so many California cultures converge, and how intimately.

"If you look at Los Angeles writ large it's got Orange County and San Bernardino, so conservative, butting up against Malibu and Hollywood, some of the most reliably liberal places in the country," Eric points out. As a result, what you have in L.A. fertility clinics are "the gays of Hollywood and the West Side of Los Angeles meeting the good Christian girls of San Bernardino and Orange County, who fully believe in the divinity of life, and think and say that the highest act of love is to be a parent, God is love, and they cannot imagine anybody going through life without a child," even a gay man. In the offices of Sahakian, Doug adds, "you're giving your sperm samples to these little immigrant Mexican Catholic girls, who are the medical assistants, and it's all beautiful. It's the beautiful part of California."

So avid is the demand—and so willing are the good Christian girls, or some of them, anyway—that the practice of surrogacy has been transformed. Not long ago, surrogacy, a controversial but stubbornly resilient niche of assisted reproduction, was a service that a small group of women performed for a small group of women. When it goes well—and often it does—surrogacy is, at least potentially, an example of what feminist bioethicist Barbara Katz Rothman calls "reproductive communism," in which women help each other through all stages of childrearing depending on their desires and abilities. Many doctors believe surrogacy is a crucial way in which women lucky enough to have easy pregnancies can assist women afflicted with cancer or uterine malformation. More recently, though, what surrogacy has become is a service that a small group of women are not only willing but delighted to perform for a growing number of gay men. The U.S. Centers for Disease Control and Prevention report about a thousand births from surrogacy every year. An increasing percentage of these are for gay men, as the *New York Times* pointed out in 2005. That year, according to Gail Taylor, Growing Generations facilitated seventy-eight births, almost 8 percent of all surrogacy births. Other agencies report the same trend. "A lot of the girls won't work with heterosexual couples [anymore]; they will only work with gay couples, because they don't have that other woman hanging around being jealous," says Laura Fretwell, a surrogate who runs an online agency called Angel Matchers from her home in Florida. The surrogates she works with often live in rural states like Arkansas and Florida, and even in these areas, Fretwell says, gay men are "the most sought-out couples for girls to work with. They don't pressure you as much, and they are just easier to work with. They treat their surrogates different."

In many cases, they do so because they are well coached by people like Gail Taylor, who describes surrogacy work as "my first love." Taylor got her start managing woman-to-woman arrangements for the Center for Surrogate Parenting, one of the first surrogacy agencies, whose coordinator, Fay Johnson, is herself an example of the classic reason why surrogacy came to be. Johnson, a gracious woman who lives in Southern California's Simi Valley, is a DES daughter. Her mother, while pregnant, was given diethylstilbestrol, or DES, to prevent miscarriage. As she grew, Johnson suffered some of the characteristic malformations of DES babies: bone overgrowth, in her case, and uterine malformations that prevented her from carrying a pregnancy. The coming to maturity of many DES daughters happened to coincide with the introduction of surrogacy, which burst into the news in 1987 with the infamous case of Baby M.

Baby M involved what is now known as "traditional surrogacy," a term that shows how quickly a revolutionary arrangement can come to seem obsolete. In this case a woman, Mary Beth Whitehead, was inseminated with sperm from William Stern, who lived in New Jersey and was married to a woman with multiple sclerosis. Whitehead became pregnant with a child who was her own genetic offspring. The idea was that she would sign the child over for adoption, whereupon she would receive $10,000. But after delivering, she refused to give up the child. Courts intervened; the intended parents got custody and Whitehead got visitation rights. The case became a tabloid sensation as well as a source of some ideological confusion for feminists. Some sided with Whitehead, arguing that she had been exploited and deserved the baby she did not want to surrender. In *The Nation*, Katha Pollitt wrote a scathing condemnation of surrogacy, calling it a "bucks-for-baby" deal that exploited working-class women—nothing less than "Reproductive Reaganomics." Her argument was eloquent and persuasive; the details of the custody battle make for a dispiriting study of the class dimensions of parenting, as when a prominent child psychologist criticized Whitehead for dyeing her hair, and for buying the child stuffed animals instead of educational toys. Still, some feminists suggested that in the interest of reproductive freedom, would-be surrogates should not be forbidden from entering into contracts like the one Whitehead agreed to.

For Fay Johnson, cultural politics were beside the point. Baby M was her introduction to the fact that surrogacy existed. She and her husband contracted twice with traditional surrogates, resulting in a beloved daughter and a beloved son, both now in their teens. "I was just thrilled to know my husband's children would exist," says Johnson. "One of the reasons I'd wanted to marry him was that I wanted to be the mother of his children, and I got to be." Her husband later died of cancer, she says, and "the last words he heard were from them. What more does any person need—than the last words to be that of their children, saying, 'I love you, Dad'?"

The average surrogate—perhaps this goes without saying—is an unusual person. She tends, according to Johnson and others, to have children of her own; to value family deeply; to have begun her own family at a young age; to enjoy easy pregnancies; to live in a conservative, heartland-type community; and yet to be a bit of a nonconformist, an individualist who enjoys tilting at the status quo. It's true that some intended parents view their surrogate as one more household servant, and one thing agencies do is educate them out of this assumption. "Many of the couples are so arrogant, they think she's a piece of meat waiting; they think they're hiring hired help, they think it's like picking a cleaning lady," says Johnson, who acknowledges that surrogacy is "far, far more complex emotionally than people realize."

Traditional surrogacy, in which the surrogate contributes both womb and egg, is fraught, to say the least. A surrogate who conceives through artificial insemination has a genetic connection to the baby, and nine months' worth of bonding, something that makes everybody a little nervous. As IVF became more successful, more parents began to opt for what the profession now likes to call a "gestational carrier," a surrogate who gestates an embryo created through IVF, using an egg from the intended mother or from a donor. The carrier has no genetic connection to the child, something that negates any legal claim and, it is hoped, also minimizes bonding. The practice has increased the pool of surrogates, who often are women who could never imagine giving away something as precious as an egg.

And for the gay population, Gail Taylor says, all these changes are a good thing. It may be more expensive to involve two women in the surrogacy process, she argues, but ultimately it will get you a better baby and a healthier pregnancy. "It's a lot easier to find all the characteristics you're looking for in this unique process with two different categories of women," Taylor told me matter-of-factly during an interview, sitting on a sofa in her office, wearing dark pants and sandals and a striped black-and-white T-shirt, her tone both heartfelt and businesslike. To better serve their clientele, Growing Generations founded its own egg-donation agency, Fertility Futures, so that gay clients could locate both women they need in a handy one-stop-shopping venue. I found her comment disconcerting, given that feminism was once—wasn't it?—about getting away from the idea that there are different "categories" of women: the bad girl and the good girl, the whore and the virgin, the brunette and the blonde, the woman you had sex with, the woman you married. But for Taylor, the fact that there *are* two categories is important. Egg donors should, she argues, be selected based on looks, youth, brains, health, and psychological soundness, whereas surrogates should be selected according to how well they gestate babies, and how well they work with others.

"In a gestational surrogate you're looking at someone who has healthy, uncomplicated pregnancies; that's compliant, agreeable to all of the circumstances that are unfolding; that's a good communicator, and you're like-minded with all the contractual perspectives: what to do about multiple pregnancies, selective reduction, abortion," she told me. "And then from the genetic part, the egg-donor route, you can have any number of things: you can look at educational level, physical characteristics, ethnic background and history."

"It's a lot easier," she pointed out, "to divide those two bodies."

It could be that much of this is spin, to soften up gay couples who are about to be told just how much enlisting two categories of women is going to cost them; surrogates get about $20,000, egg donors anywhere from $5,000 to $50,000. Yet it's an argument I never got used to. When I mentioned it to Richard Paulson, the egg-donor pioneer who practices not far away at USC, he laughed and said, "Moo." But a number of doctors, most of them at lucrative private clinics, agreed with Taylor. "If you're looking at beauty or physical features you're not going to find that in the surrogate pool," said Vicken Sahakian. "It's a fact. Most surrogates I come across are not typical donor caliber as far as looks, physical features, or education. Most egg donors are smart young girls doing it for the money to pay for college. Most surrogates are—you know, they need the money; they're at home, with four kids—of a lower socio-economic class."

"There's two separate qualities you want," agreed Michael Feinman, a doctor with Huntington Reproductive Center in California, which has one of the largest surrogacy practices in the world.

This, then, was the world Eric and Doug were entering: a world that combines profound parental love with cold-blooded business truths, a world where children are desired, loved, celebrated, wanted, and obtained in a relentlessly commercialized process. Surrogacy is like adoption in that information on intended parents is submitted to the surrogate, who weeds parents out based on what she reads. Starting out, Doug and Eric knew they would be competing with "the L.A. crowd—people who are going to spare no expense, who are in creative fields." So they put together what they called a "marketing profile": a computer-produced mini-newspaper with articles about their interests and personalities, spiced up with photographs that, they hoped, would make them look like good potential fathers. "We made a point of showing us on the beach playing with our nieces and nephews," says Doug.

The men also decided they were willing to work with a surrogate outside California. As with most aspects of family law, each U.S. state sets its own surrogacy laws. Some permit it, others criminalize it; some permit payment, others don't. California is very tolerant of surrogacy: babies are one of the state's unrecognized exports. As a result, there was a long waiting list for California surrogates, and Doug and Eric didn't want to wait for months and months. When they said they were willing to go out of state, Taylor sent their profile to Ann Nelson, a surrogate who lives in Wheeling, West Virginia. Asked what she liked about them when she read their profile, Ann said, "Everything."

"I Could Help These Poor People"

Ann Nelson is a graduate of Michigan State, where she took classes in bioethics and wrote a thesis on reproduction that covered everything from abortion to breast-feeding. Her research included surrogacy, which was in the news at the time because of Baby M. Like Fay Johnson, Ann Nelson was intrigued rather than appalled. "I decided then that I wanted to do it. Because I could. That's not something most surrogates will tell you. Most surrogates will tell you they wanted to help somebody. I wanted to do it because I knew I *could* do it. I thought: 'What that woman is doing is wrong. I could do this, and do it right.'" For Ann Nelson, who is also an environmental activist and a counselor of pregnant teens, surrogacy was part of what she calls "a grassroots thing. I thought, 'I could help these poor people.'"

Ann married a fellow Michigan State graduate, David Nelson, who would go on to be a paramedic and firefighter. They moved to West Virginia, where David Nelson had grown up, bought a house, and had four children. Ann told her husband that she intended someday to be a surrogate, and "he said go ahead, if that's what you want to do." It was no small concession. Surrogacy can be as hard on a husband as it is on a wife, exacting costs both emotional and physical, including, at times, abstinence from marital sex. While the surrogate is being squired around a mall by intended parents buying her maternity clothes and showering her with gratitude, the husband is often at home feeding peanut butter and jelly sandwiches to their own children. That's why husbands are—or should be—included in all negotiations; there have been surrogacy situations that have foundered, rather spectacularly, as a result of neglected and unhappy husbands. In the case of the Nelsons, a man who enters burning buildings to save strangers is probably unusually receptive to the idea that there is moral value in putting your body on the line for the sake of somebody you don't know. Or maybe he didn't believe she would go through with it. "There may have been a part of him that thought this wouldn't happen: another one of Ann's radical ideas," Ann Nelson allows.

Once they felt their own family was big enough, though, Ann Nelson, now twenty-nine, posted an ad on a fertility website's "classifieds" board, advertising her willingness to serve as a surrogate. She was deluged with replies. "There were tons of people writing and wanting all sorts of things." One couple wanted to save money by cutting out lawyers. Another wanted her to agree to tell no one what she was doing, including—she could never figure out how they expected this to work—her own children. In another family she negotiated with, the would-be "dad had disconnected and the mom was just kind of nuts." It was one train wreck after another: surrogacy, when it doesn't go well, can expose and exacerbate hostility between infertile women and fertile ones. "People who have gone through infertility are emotionally and financially bankrupt," says Ann. "Infertile couples—surrogacy is their last choice. To them, every single part of this is just another hurdle to overcome. With a gay couple, it's their first choice. It's the way they get to have biological children, and they're thrilled."

Recognizing that it was time to enlist an agency, she signed with Growing Generations. She had no problem working with gay men: she carries a gay freedom ring on her purse. She was flown out to Los

Angeles, where she went through medical and psychological screening, met the Growing Generations staff, and was accepted as a surrogate. Not long afterward she received the profile of a couple who lived in New Jersey. "We sat at that match meeting, and one of the men said, 'I've known all my life that I wanted to be a dad, and I've known I was gay, and I can't believe I'm sitting here looking at you and that you're willing to help me do this.'"

Everybody who works with surrogates, and the surrogates themselves, agree that the motivation is not just altruism. It is also important to have your altruism celebrated and acknowledged. Part of the motivation is to be thanked, early and often; to feel *appreciated*; to be, as many surrogates put it, "treated like a queen." Many surrogates also want to build an intense bond with another family: to expand their own kinship circle; to achieve the old-fashioned extended-family feeling that's missing in many modern households. "I really wanted to feel as though the people I was doing this with were my friends," explains Ann. "That's how I wouldn't feel used."

And the New Jersey couple came through. "They definitely gave not only me the princess treatment but my whole family," Ann says. "They were so appreciative; they would send surprise packages to my kids, saying, 'Thank you for sharing your mom.'" When they realized that legally it would be advantageous for Ann to give birth in California, they flew her and her family to live in an apartment in the L.A. area. "We had a monthlong vacation," Ann remembers. "I never in a million years would have been able to do that." She bore the men a daughter. When they asked her to gestate a sibling, she thought about it for a while, waited until she felt physically recovered, and agreed. This time, though, when she went for her transfer, the doctor saw a fluid buildup in her uterus, so he cancelled the cycle and prepared to try again. Instead, the couple decided to try a new surrogate. "It was abrupt," Ann says. "They didn't mean to hurt me."

Growing Generations sent her more profiles, among them a single gay man planning to raise a child with the help of nannies. Ann's husband nixed that one. "He said, 'I didn't get into this to help someone have a nanny. I want someone to be a dad, like me.'" Ann was impressed by her husband's reasoning. "I was, like, 'You were paying attention!'"

And then she received Eric and Doug's profile, which she read and, she says, "felt connected." Her profile was sent to Doug and Eric, and they thought, reading it, "This sounds great."

After a surrogate expresses interest, the next step is a face-to-face meeting. Ann and her husband flew out to Los Angeles, and so did Doug and Eric. Doug and Eric liked Ann, but had what is a common anxiety: her weight. This is another place where rural and urban sensibilities sometimes clash. One day, I watched Gail Taylor meeting with Washington-area gay men who were considering parenting through surrogacy or were already doing so. One man broached the question of "a surrogate's physical characteristics," and, after hemming and hawing, made it clear he was asking how much a surrogate could weigh. What followed was a clinical discussion of Body Mass Index, in which Taylor assured the men that a woman can be heavy and have a healthy pregnancy. "Most surrogates have a bit of baby weight," she said. "And most have a disconnect from their physical well-being. Most of them end up eating better during their pregnancy than they did with their own. But I always tell people, if you're looking for a bad habit, it's fast food and soda." Doug and Eric had some worries along these lines, feeling that Ann was overweight. ("Yep. Was then, still am," says Ann.) Taylor assured them their concerns were baseless.

Other than that, the meeting was a revelation. "Here we are hanging out with this fireman paramedic from Wheeling, West Virginia, who grew up there, and his wife, who's from Michigan, and they just like *so* don't have issues around gay people," Doug marvels. When Doug asked Ann why she carried a gay freedom ring, she replied, "Gays are the last people that it's okay to discriminate against, and that's wrong." They still feel in awe of Ann's solidarity. "She's living *out there*," says Doug admiringly. "She's more out there than we are. We're in our little bubble."

After the meeting, the two men took the Nelsons to lunch, and "then we went back to the hotel, and ended up spending hours sitting in their hotel room just talking." Ann had brought them an antique silver spoon as a gift. The next day everybody gave the thumbs-up. They then began to negotiate the many issues that must be hammered out: how many embryos would be transferred; what they would do in the case of multiples. Gail Taylor gave them two valuable pieces of advice: never talk finances with your surrogate—surrogacy situations have also collapsed over money—and never see your surrogate without giving her a gift. "You can never show your appreciation too much," they were told. "They're doing this amazing thing, and you have to make sure they feel appreciated."

And then you have to choose an egg donor.

Jumping into the Gene Pool

"If we're going to jump into the gene pool, let's jump into the deep end," Doug and Eric reasoned. "It's our first egg donor, so of course she would have to be fabulous, your basic Ivy League supermodel." They are saying this ironically, but they also mean it. According to the doctors who work with them, gay men often want good-looking egg donors. "They ask me, 'Would you go out with her?'" says Vicken Sahakian. "They say, 'Give us your honest opinion, if you would basically date this woman.'"

Doug and Eric wanted looks, but they wanted other qualities as well. They wanted brains and a top-notch family health history. They wanted height, so their child would have, as Eric puts it, "a chance of being tall." They bypassed Growing Generations' partner egg agency—they felt Fertility Futures was too Hollywood, too actressy—and went for the Lexus of the egg-donation world: A Perfect Match, another brokerage in Southern California. A Perfect Match specializes in good-looking high-SAT-type blue-chip Ivy-League-or-the-equivalent egg donors, whose oocytes run $10,000 and considerably higher. And when you are willing to pay that kind of money, there are lots to choose from. Scrolling through the list of Perfect Match donor profiles, Eric and Doug decided they were going to have to make a decision tree.

First of all, they did *not* want to embody their love for each other. "Our two criteria were that she could not be like either of us," says Eric, who has sandy good looks: light skin, light hair, the kind of features exported long ago by immigrants or conquerors from the sun-starved climate of northern Europe. "I'm Waspy Scandinavian British, and Doug is Russian-German-Jewish, and we were like: someone who is *not* Jewish and is *not* Scandinavian," explains Eric. The reason had to do with genetics and health risks. Eric's family is vulnerable to northern European afflictions, chief among them skin cancer. "My sister, my uncle, they all have melanoma diagnosed." Meanwhile, "Doug is 100 percent Ashkenazi Jewish," an ethnic heritage that carries a risk of a number of genetic diseases when two carriers procreate together. The two men wanted a woman whose ancestry lay outside those two small gene pools: someone who was Caucasian, but with family roots in central Europe, or Spain, or Italy, or the Mediterranean. "That Levant sort of thing," Eric says, joking that "if you can't be an Italian yourself, maybe your kids can be."

The donor they selected fit the bill perfectly. According to her profile, her heritage was Turkish-Bulgarian. "She was gorgeous, five feet, ten inches, went to MIT," Eric remembers. The problem was, she was eighteen. From a fertility standpoint this was excellent, but from a logistical standpoint it was a problem. The agency required that she get permission from her parents, and when she got around to doing so, the men learned, her parents "put the kibosh on it."

So they found a donor who was Irish and Native American. They contracted for her, and the donor began "cycling" for them, taking drugs to stimulate her egg production. Since she lived on the East Coast, at the outset she was under the care of a New York doctor, even though the retrieval would be done in California. But the New York doctor was premature, they think, in giving her the trigger shot to release the oocytes; though she produced eighteen eggs when she flew to California, only two fertilized when inseminated with sperm from the men. The other oocytes, the men suspect, were immature. But at least they had two developing embryos. The next day they picked up Ann Nelson at the airport and went to Growing Generations. When they walked into the office, however, they "knew something was weird."

The two embryos—they had named them Fred and Ethel—had stopped dividing. Apparently they had died on the laboratory table. The Growing Generations staff was deeply apologetic, and rallied round impressively: "A fair amount of their clients are high-maintenance obnoxious queens," Doug says. "They like us because we're easy-going . . . and they knew we couldn't afford to keep going and going." Growing Generations ensured that they would incur very little expense on another cycle. The men were sitting around talking with agency staff, when their caseworker left to take a phone call, then came back with an odd look on her face. "I said, what, are you going to tell us they're alive?" says Doug. "And she says, 'Yeah, that was the lab, the embryos are cleaving again.'" Their doctor pointed out that Ann's uterus was ready, and they might as well try. "And I'm thinking these are little Frankenstein embryos," says Eric, "and I don't know if I want them to survive."

The embryos were transferred. "Miracles do happen," Vicken Sahakian told them. But in this case neither Fred nor Ethel survived, so egg-donor-wise they were back to square one, looking for a new donor.

Scrolling. Clicking. E-mailing each other saying, "Check out this one!" Getting excited about a donor until they learned that all four grandparents had died of heart-related ailments. There were so many options; their difficulty deciding confirms the work of the Swarthmore

College psychologist Barry Schwartz, who in his 2004 book, *The Paradox of Choice*, argues that too many consumer choices can be a source of anxiety rather than pleasure. "You found yourself saying, 'Mary has a 1550 and is studying at MIT in mathematics, but Jane has a 1390 and she's a political science doctorate and she also plays the jazz piano,'" says Eric. "What are you going to do, get someone with a 1550, or are you going to cheat your child and get them a mom with a 1210?"

They refined their criteria. IQ was important, but so was EQ—emotional quotient. They wanted someone who was sociable and outgoing. Brainy, but not introverted. Like college admission officers, they looked at extracurriculars. "The thing that became very, very important to me, ironically, was music," Eric remembers. "Music and sports, as an indicator of well-roundedness. And sports got further broken down into 'Okay, are these solitary sports like tennis, which would indicate a loner, or are these team sports like basketball or volleyball or badminton or rugby?' Team sports were definitely ranked higher than solitary sports, and piano was ranked higher than violin." They devised an "unofficial algorithm, health plus education times looks, add back social sports."

It was, they said, like hiring someone sight unseen: if you don't get to meet a job candidate, you have nothing to go on except a résumé. "If you fall in love with someone, there are all kinds of things that you don't know at the beginning that you don't care about," Doug points out. "These people we didn't know, so you do look at all of those things."

Again, they found a donor they liked. This time, it was the donor's boyfriend who put the kibosh on, once he knew she would be donating for gay men. It was at this point, Eric says, that "I realized I was approaching this too much as a transaction. I had forgotten about that gut check."

And then the men found a donor they had overlooked. Unlike many other Perfect Match donors, she didn't have a glossy professional photograph. She had just a few photos posted: snapshots, taken in what looked like somebody's garden. She looked real. She looked human. She looked nice. And that, in the end, became the defining attribute. "I looked at her, and I remember coming back to her profile, and thinking: I wonder if this is the one," Eric remembers. "And as I thought about it, it's like: I really like her."

The two men do not give a lot of detail about this donor; people who use gamete donation usually don't, because it would be easy for another person to go online and find that donor. But Eric will say this: "Contrary to Doug's point that you don't fall in love with these people, I did fall in love with her." Their liking intensified when the agency got in touch with the donor. Egg donors are permitted to bring someone along for the retrieval, and this donor wanted to know if the men would pay to fly her mother along. "This is the woman," they agreed. A woman so close to her own mother was the woman whose children they wanted to bear—well, to have Ann Nelson bear.

Dreaming of Twins

There was something else they wanted, too. They wanted to do a "mixed transfer." A mixed transfer works like this: Once the eggs are retrieved from the donor, the lab divides the batch of eggs in half. Half the eggs are fertilized with sperm from the first man, half with sperm from the second. Embryos from each man are transferred into the surrogate.

The goal was reproductive equity. Like many lesbian couples, Doug and Eric wanted to use reproductive science to come as close as possible to having children together. "We always wanted twins," Doug says. "Something we dreamed of was twins who had the same biological mother, and two biological fathers." So important was this that any surrogate they contracted with had to be willing to bear twins. According to Gail Taylor, this is not something Growing Generations encourages. "We typically don't endorse a conversation that somebody wants to get twins." That said, she allows that many clients do end up with multiples, which is not surprising since egg-donor oocytes tend to be so robustly fertile that even transferring two embryos carries a high possibility of twins; according to the Centers for Disease Control and Prevention, 43 percent of egg-donor pregnancies result in a multiple birth. Moreover, men, having paid $100,000 for a surrogacy cycle, are unlikely to want to transfer just one embryo. Eric and Doug not only declared their preference for two children, but tried to maximize their odds.

"We had many discussions about how do we hope to achieve that," Doug recalls. "How do we optimize the situation. It's absolutely a statistical issue. We're people who—I sit down, I do little matrices and tables, saying this is what our probability is. Basically, the doctor told us that if you put this many in, you have this percent chance. I'm taking notes through everything, and we had discussed with [Ann] and she was comfortable with it, that we would put three embryos in. There was a very good chance of the pregnancy taking. A reasonable chance of twins, and a very, very small chance of triplets."

This time, the donor delivered. Half the eggs were fertilized with Doug's sperm, and half with Eric's. But when Ann was flown out for the transfer, the doctor again wasn't satisfied with her uterus. He decided to cancel that cycle, freeze the embryos, give Ann a new round of drugs, and try again. In the process, the embryos were graded. The men were told that the Eric-fertilized embryos were much higher quality than the Doug-fertilized embryos. Whereupon, Eric acknowledges, he got kind of "goalposty."

"I'm in this very kind of visceral male reaction," Eric admits.

"Eric got kind of competitive and male," Doug says.

So Ann went through another cycle. This time she checked out fine, and the embryos were thawed for transfer. But the men had another scare. They were driving to Los Angeles, and the embryologist called to confirm that he had thawed their embryos. "He's like, 'Great, I have two for you, Eric, and two for Dennis,'" Eric remembers. And then they lost the cell phone signal.

Dennis?

"I'm like, I've got to pull over," Eric says. "And Doug is like, 'Those motherfuckers.' We drive like fifteen miles down to get into cell range, we're looking for a tower. I get back on the phone, and I'm like, 'Let's go over this again.' I'm trying to be professional. I'm like, 'I'm Eric, and my partner is Doug. You said you did two for Eric and two for Dennis. Who's Dennis?' There was this pause, and he's just like, 'Let me check.' He's like, 'Oh, it's Doug.' And I'm like, 'Don't give me that shit, I want to know what Dennis looks like, so when this redheaded child comes out, I can go find him.' And of course, the next nine months I'm thinking about Dennis. And we get there, and they have a laugh, and they're like, 'Wasn't it funny,' and I'm like, 'No, it's not funny.'"

There had been another miscommunication, which the men suspect was their fault. The men wanted three embryos transferred: one from Eric, one from Doug, and a third one chosen by Sahakian, whom they had instructed to "pick one that you think looks good." ("Play God" is how Eric puts it.) They assumed the embryos would be frozen separately, but instead they were frozen in pairs. In order to use three embryos, they would have to thaw four. They would then have to transfer all four, or transfer three and discard one. They agonized. They hated the idea of Ann getting pregnant with triplets or quadruplets, and the decisions that would set in motion. They also hated the idea of discarding a human embryo. "For me, personally, I feel like I'm playing God enough as it is, and I do believe that life—I don't believe that life starts at conception, but I do believe that it is precious, and it would just seem to me the ultimate act of this consumer society, to create it and throw it away," says Eric. Doug was torn. The more embryos they transferred, the greater the chances of the pregnancy taking. And at this point, they had already spent about $60,000. "It's this horrible intersection," they point out, "of morals and money."

What made the decision easier was the report from the embryology lab. After freezing and thawing, Doug's two embryos were in good shape—"little troupers," Eric says—while Eric's, despite their initial high grade, seemed to be "collapsing." They figured it was safe to put in all four, since it seemed unlikely all, or even three, would take. "That was a big decision," Doug admits. They two men talked about it at length with Ann Nelson. "It's a big deal," Doug says. "It's her body. We're very cognizant and respectful of that. It's a delicate balance, it's a delicate dance, because you have this—you enter into this legal relationship, and this transactional relationship, about something which is incredibly personal that has real impact on people's lives and their health, and then also has this magical spiritual component, too. It's a very bizarre thing," Doug said. "It's incredibly intentional. Everything you do is so intentional."

Ann agreed to have four embryos transferred. "We took a big risk," she says now, "and I heard about that when I got home. We had gone through a cycle with dead embryos, and a cycle where my lining was not where it was supposed to be, and we were starting to feel desperate. You have this desperate feeling. Eric was much less comfortable with the idea of reduction. Doug was saying, 'Let's do what we have to do,' and Eric was saying, 'I'm scared here,' and at that point I was saying, 'I just want to do this with them and for them.'" When she got home, Ann says, her husband was not happy about the decision to transfer four. She had a blood test and the numbers came back high enough to suggest all four embryos had taken. "I didn't even know how to react," Ann says. "My little girls were with me, and I came home and didn't say anything. I thought: 'How am I going to handle this, how am I going to tell my kids?'" But the first test was incorrect; when a second test came back, the number was so low that, Ann says, "at that point I was positive it wasn't twins."

Early in the pregnancy, the men traveled to West Virginia for an ultrasound. They flew into Pittsburgh and drove to Wheeling, had dinner with Ann and her husband, and the next day crossed the river to Martin's Ferry, Ohio, to the office of the one doctor in the area Ann had been able to find who was willing to treat a surrogate bearing children for gay men. While she was working with Growing Generations, she says, her regular ob-gyn had retired and she had had a hard time finding someone else who would take her.

The men, braced for a confrontation with red-state America, were pleasantly surprised. "It was great, it was really great," Eric remembers. "We're in small-town America, this little town of Martin's Ferry, about five thousand people, right next to Wheeling, the Presbyterian-church-next-to-the-Methodist-church-picture-perfect-America. We're walking into the hospital, these San Francisco gay guys, and the surrogate with her fireman husband. We go in for this ultrasound—we call it the wand of knowledge—they lube up this wand and put this condom on the wand, and I'm just like, 'Oh my goodness gracious.' He puts it up there, and boing, there is one baby, and it's this—of course with the wand of knowledge it's the size of rice, they have told us that if you see a beating heart that's extraordinary. And we go in there, and in the upper-right-hand area of her womb is this vibrant big I-don't-need-to-be-told-that's-a-human-being. So okay we definitely have one, and we're like, one, that's great, and then it was nearly at the end, and he was kind of digging around in one corner, and as far away as possible from Elizabeth is just this faint little beating and this little, tiny two-thirds-the-size-of-Elizabeth, and there were two hearts and they were both beating, and they were like: 'Wow.'"

Twins.

They flew back to the West Coast, and stayed in touch with Ann through calls and e-mails. Later they flew back for a high-resolution ultrasound performed by a specialist, where they learned that both were girls. Eric assumed that both were, biologically, Doug's, given the low quality of his thawed embryos. But there was no way to know. They didn't want to do any invasive testing for fear of dislodging the twin pregnancy. Their anxiety intensified when the specialist mentioned that because Ann's prior pregnancies had been delivered by cesarean section, and because she was now carrying twins, she was at increased risk for uterine rupture, a crisis situation in which the uterus, well, ruptures. The men are still surprised that this risk was not flagged. "It is somewhat shocking to me that nobody, none of the doctors up to that point, nor the agency, identified the pregnancy as particularly risky," says Doug, who believes that one flaw in the process is that every professional—embryologist, broker, fertility doctor, obstetrician—is paying attention to his or her discrete stage of the process. Doug adds that as much as they like Ann and as deeply grateful as they are to her, had they known about the risk "we probably would not have worked with her," because it seemed an unreasonably dangerous situation to put another person in. "If you don't get those babies out in three minutes, wow, they're dead and you're dead. That's where the doctor said: 'As soon as you get to thirty-six weeks, get those babies out.'"

Medical views on the risks of prior C-sections differ, and have evolved: When Ann was starting out as a surrogate, she says, she specifically asked her own ob-gyn, who was independent from any agency, whether she could carry multiples. Her doctor cleared her to carry up to triplets. She says it was her understanding that the risk of uterine rupture was only slightly higher with twins, and she felt it was a risk she could manage. "The important thing was for me not to" go into labor, she says. She figured she would be safe if she took it easy. She was working for Growing Generations, recruiting and screening other surrogates; she could work at home, and she made sure to sit with her feet up on a stool.

And then she went into labor.

"This Remarkable Man"

By then, the twins had almost made it to the thirty-six-week mark: they were at thirty-five weeks and a half. The phone rang early on a Friday morning in the men's townhouse. Eric was in the shower. Doug was getting ready for work. Not recognizing the number that showed on caller ID, Doug didn't answer. The phone rang again. This time he picked it up. It was Ann, using a calling card, which was why the number had been unfamiliar. "This is it," she said. "We're going to the hospital."

Doug ran into the bathroom, calling for Eric. One of the men started calling airlines looking for a flight, and the other left messages at work letting colleagues know they'd be away. They threw some clothes into suitcases; grabbed two Snuglis, the front-pack carriers they'd use to bring the twins home; rustled up car seats; and three hours later were at the airport waiting to board a cross-country flight. Just before they boarded, Doug's cell phone rang. It was David Nelson, Ann's husband.

"They were born," he said. The girls were fine. "I'm really glad we knew that before we got on the plane," Doug says now.

What David didn't say—what nobody knew yet—was that Ann wasn't fine. She was hemorrhaging. The problem was not a uterine rupture, but something else that had developed during the pregnancy. She had placenta accreta, another dangerous condition in which the placenta, the organ that grows, along with the pregnancy, to nourish the fetus, burrows deep into the uterus and is not completely expelled upon delivery. Placenta accreta becomes more likely in the wake of past C-sections, possibly because the placenta is able to burrow more securely into scar tissue. At first, Ann says, nobody suspected anything was wrong. The C-section went well; the girls were beautiful and healthy; Ann seemed to be fine. When she was wheeled back to her room, however, she remembers hearing a nurse saying, "There's a pool of blood under her," and thinking to herself, "That didn't sound good."

Ann called Growing Generations to let them know the girls had been born, and began to nurse the babies. And then, she says, "I just felt kind of out of it. It was because I was bleeding, and bleeding, and bleeding." The blood kept pooling; the nurses changed the sheets on her bed; the blood kept pooling. "They called the o-b, and he said, 'The one placenta disintegrated in my hand, so I'm going to reach up in there and see if any more pieces are in there.'"

"I didn't know," Ann says now, "that that level of pain existed."

The doctor told Ann her uterus should clamp down now, and that the bleeding should subside. But the bleeding didn't subside. Now the staff was calling for a blood transfusion. "I couldn't hear—that's what happens when you lose so much blood," Ann remembers. The doctor did a dilation and curettage, scraping the uterus to remove any remnants of placenta. When that didn't seem to be working, to save her life he performed a hysterectomy, surgery to remove her uterus altogether. "My o-b sat and waited for me to wake up; he told me what they had done, and took me to the ICU," Ann says. "The next time I woke up, I was alone."

David, her husband, was with the babies. Doug and Eric were in the air; landing in Pittsburgh; trying to find out what had happened to their car seats, which had been lost; renting a car; and driving to Ohio. They were directed to a room, dashed in, and saw David Nelson compiling a digital scrapbook of delivery photographs. "David, God bless him, was in the delivery room, the operating room with his digital camera," Doug says. "When we got there, he was in the room with one baby in one hand and his other hand on his computer putting together a photo album. This remarkable man—this remarkable family, really. I don't think he, nor we, realized how bad it was."

The next day Ann was brought out to see them. She was in a wheelchair. Her complexion was gray. Her childbearing days were over. "I just lost it right there," Doug says. "I just started crying as soon as I saw her. The realization what this woman did for us, what she went through, and what could have happened—it's like, how do you express that? I can't."

Twins, and Half Sisters

Even now Doug and Eric cannot grasp surrogacy and why women do it. It's easy, they think, to understand why somebody—anybody—would want children. What is harder to get one's mind around is why someone would want to have children for someone else. "It's really hard to fully appreciate," says Doug. "We are so incredibly grateful. And yet we really don't understand it."

And the risks turned out to be so much greater than anybody had anticipated. (Growing Generations now accepts only surrogates who have had three C-sections or fewer. Gail Taylor says this is not a direct result of what happened to Ann; the agency routinely refines medical criteria.) Ann says now that she and her husband have talked a lot about what happened. "It was hard, even though we were done with our kids and I had said that was my last surrogacy, that's a really abrupt and huge thing to have happen in your life. There's definitely a sadness, a loss, even though I didn't plan to use that uterus again. I would never ever say that I wouldn't have done it. But if someone had told me before I went into it that there's a good chance your children will wind up without a mother, then yes, I probably would have stopped, and not done that. All things considered, they went home and I went home, so that's okay." She says, "Most surrogates would be just like I was, and not think that would happen to them. I never knew a surrogate who had something that serious happen to them. All in all, it was a pretty—it would be a great Lifetime Movie of the Week kind of thing: it's got a happy ending but lots of drama."

Ann maintains—and, as an employee of Growing Generations, it is clearly something she believes—that surrogacy is empowering for

women, including surrogacy performed for men. Not surprisingly, there
are those who disagree. Gay parenting through surrogacy is one of those
junctures where conservatism and liberalism can come together in
mutual disapproval. "Any man with a checkbook can buy a baby,"
objected the bioethicist Barbara Katz Rothman, speaking at a 2003 conference convened by Planned Parenthood. "The pieces are all for sale,"
Rothman said, adding that "when you look who's pushing what babies
down the streets of Park Slope, in Brooklyn, you can see who's going to
be donating what in the market that we're developing." Surrogacy, she
argues, is not an expression of choice. "A certain class of people has
choice, menus, rights," she said. "Another class has far less choice. This
is about the rights of rich people to buy what they want. It's not about
choice. It's about the right to buy." The reproductive communism she
once envisioned, she argued, has given way to "reproductive capitalism."

"This *is* about families of choice," counters Ann Nelson. The reproductive rights umbrella, she argues, should include gay men, and her
own reproductive rights should include the right to help gay men build
the families they long for. "Why shouldn't a gay man have the same
choice as a woman, to have a child? I've heard that—that we're being
prostituted out, we're being used. Again, I made this choice a long time
ago. This is kind of like a little kid who wanted to be an astronaut. I
decided I wanted to do this, and so I did, and the compensation that
comes from that is really more for my family for what they went through.
To me, that money was about making up to my family that they were
helping me indulge my passion." As for the men: she will, she hopes,
always have a connection to them and their family. "It felt like a big
important project that we worked on together, and we became friends
through. That doesn't mean they owe me gratitude for the rest of my life.
It doesn't mean they should have a shrine to me. They don't owe me forever for this."

A Twenty-First Century Pedigree

After the girls were born, the hospital was at first obliged to treat Ann
and David Nelson as the twins' presumptive mother and father. Doug
and Eric had affidavits and other documents clarifying the parenting
relationships, but because of the premature delivery, the documents had
not yet arrived. Because of medical privacy laws, Doug and Eric could

not be alone with the babies until the documents got there. After the delivery they also had paternity testing done. Unlike in California, Ohio does not permit two men to be listed together as parents on a birth certificate. But after paternity testing, it was possible for each man to be listed as the father on one birth certificate. That's because the paternity tests showed that they didn't have to worry about anybody named Dennis. They had gotten everything they wanted. One of the girls was biologically Doug's. The other was biologically Eric's. They do not say which is which, explaining that they don't want anyone to treat one girl differently because of who her biological father is. On the birth certificate, the field for "mother" was left blank.

Consider, then, the uniquely twenty-first-century pedigree of Sophia Rose and Elizabeth Ruby Okun Ethington, two exquisite girls, one with curly hair and a rambunctious temperament, the other with straight hair, calmer, more placid. The girls' first putative father was David Nelson, who has no biological or social relationship to either of them but who was the first man other than the doctor to lay eyes on the girls, and who in ceding parentage to Doug and Eric became—well, there is no term for what David Nelson became. Their nonbiological birth father, let us say. The girls are twins, since they gestated together, but half siblings, since they have different genetic fathers. They each have a different social father. They have the same mother and yet no mother, unless you count Ann Nelson, who bore them, or the egg donor the men selected, whose full identity none of them knows.

Just an ordinary two-parent family, enjoying an evening at home. "I'll put them down," says Doug when dinner is ready. Doug scoops them up and pops them into a bedroom where two cribs stand side by side, then reappears almost instantly. I marvel at how quickly he was able to get two toddlers down: no crying, no lingering, no negotiating, no head-patting, no being summoned, no daddy-one-more-story, no relenting and taking them out of the crib for fifteen minutes. He seems surprised that it would be any other way. What's hard, he wonders, about putting a child to sleep? Or even putting two children to sleep? Why prolong it more than necessary? "We give them baths when they need baths, sometimes at night, sometimes during the day," he explains. "We read to them all the time. We don't say prayers anyway, and we don't need to wind them down, because we don't have kids who need to be wound down." Listening, you'd almost think that parenting was easy.

Well, it is, say the men. At least, parenting is no harder for them than for anyone else. Sometimes they run into prejudice about their parenting ability, though they encountered none in a small midwestern hospital. "We were amazed, wonderfully amazed, at the nurses that were there, such an incredible staff that were just wonderful to us and treated us like princes." But even in San Francisco they get inquiries that question their qualifications, as men, to parent daughters. Or to parent, period. "People assume that men need help," Doug says. "People offer you all kinds of unwelcome advice. They're shocked that you do everything on your own."

"The people that have given me the most grief are women," adds Eric, saying that women like to give him helpful instructions, and even, from time to time, astonished compliments. "Actually, I don't need your advice," he says. "I don't need you to tell me what I'm doing right or not. I don't need your pats on the back. The oddest one that stuck in my head, one woman, her question was: 'Who gets up in the middle of the night to feed the children?' It was beyond her comprehension that of course, we do."

It seemed clear to me, sitting in the house, that two men are fully capable of caring for two girls and nurturing them. But that doesn't dispense with the question of what it will be like for two girls, or any child, growing up without the presence or the possibility of a mother. Does it matter? Does it not? Is a mother a special creature? Must a mother be a woman? What would Freud say about a girl having two fathers? Two Electra complexes? None? One? "It's a fair question," Doug says, "but it's not something we're real worried about. We think it is important for kids to grow up with relationships and role models from both genders, and our kids will always have women in their lives, there are women we've already talked to about" being available to the girls as they go through puberty. One of the men's niece helps with child care; the girls have aunts; the men have plenty of female friends. Other than that, they say, "we feel like, not in all, but in most instances we can provide what needs to be provided."

They also believe that the question reveals biases against men as parents. "When that question is asked, toward men—what will you do in the absence of a mother?—on the one hand it's a very fair question, since you are raising human beings that are of a different gender than you," Eric allows. "But I also think that question can mean: Guys are not really the nurturing loving parents that women are, so where are your daughters going to get the love and the nurturing that they need to become good women?"

"It's not about having a mother," Doug argues. "It's about having women you can share certain things with, and learn certain things from." Among these women is Ann Nelson, whom they see at least once a year, at an annual retreat Growing Generations holds for gay men and their surrogates. And maybe, someday, their egg donor as well. Like Kendra Vanderipe, this donor expressed willingness to meet any children her donation engendered. "I expect that someday we'll meet her," Doug says. "I expect that our kids will want to know who their biological mother is. We also think that our kids are terrific, and we have to assume that she had something to do with it."

"Dude, I'm Gay, Too"

After the girls were born, Eric was taking them for a walk in the stroller. The Okun-Ethingtons live in San Francisco, and Eric likes to take the girls for walks in all neighborhoods of the city. This day he was strolling them through the Castro. "We ran into friends and their twin stroller, and with our twin stroller it pretty much blocked the entire sidewalk, and as we were chatting up each other, one of our friends came up with their stroller, and there were five babies in front of this gay bar on Market Street."

The presence of children, they believe, brings with it profound change both to the Castro and to the gay community as a whole. In Washington, D.C., the well-known Whitman-Walker clinic, founded to serve the health needs of the gay and lesbian community, has broadened its mission to the point where it offers parenthood seminars for gay men and lesbians. Groups like ACT UP, created to promote research into AIDS, have paved the way for groups like Family Pride, created to promote acceptance of gay families. Having children is, the men believe, a political act in and of itself, fully as radical as the Stonewall uprising that started the gay civil rights movement, or the ACT-UP demonstrations of the 1980s and '90s. Now the battleground is for assimilation and domestic normalcy. Doug and Eric were among those gay San Franciscans who hastened to marry after San Francisco mayor Gavin Newsom declared gay marriages legal. They were photographed coming out of City Hall, joyous newlyweds carrying their beautiful girls, who were for the first time the fully legitimate daughters of legally married parents. The girls were legitimate until a few months later, when a court declared Newsom had exceeded his authority and the licenses were revoked. Like a number of San Franciscans I interviewed, the men jokingly refer to their girls as "the little bastards." It is a wry, affectionate term and yet, at the same time, tinged with bitterness and the recognition that they can provide their girls with so many things, but not with old-fashioned legitimacy.

But the ceremony had a lasting impact. The photo of Doug and Eric and the girls went out over the Internet and ended up in *Newsweek* and the Italian *Vanity Fair*. Later, the men got a call from a friend who works for the city's department of public health, wanting to use the photo in a PowerPoint presentation about AIDS outreach. The friend hoped that a photo of two men carrying two babies would have a positive impact on the generation of gay men who came to San Francisco after Doug and Eric, younger men who—lulled by the presence of antiviral drugs—are less careful about protecting themselves against AIDS. As Eric puts it, the photo sends a message of "hope and life and optimism, instead of pounding fear, fear, fear." They hope their example can encourage younger men to take their lives seriously; to know that there are other ways of living and that "maybe it's also time you have the possibility of creating a family."

"When I moved here, everybody was dying," Doug remembers. "When we were in our twenties, the guys in their thirties were dying. The conversations were all about AIDS. Now they're about same-sex marriage. We have a whole handful of new friends who are younger gay men in their twenties or maybe their early thirties, who, when the babies were born, these guys were around all the time, calling up, asking can we come over. All these young gay guys wanted to come and hang out and spend time with the babies, and babysit. Whether they want a family or not, it's this presence in their lives, it's this possibility they see. You end up being kind of an ambassador."

At the same time, they wonder whether parenthood could someday make the gay community less distinctive, less out there, less queer. "Queer has always been the other, the different," reflects Eric, "like you've got to be a freak to be an authentic gay. I don't think that's the future of the gay community. There will always be the queer part of the gay community, as there has always been the queer part of the straight community, but I think that more and more, as gays and lesbians become integrated into society, that we will start to mirror in many ways the straight community."

"As there is greater acceptance, there is greater diversity with any population," Doug adds. "There is an absolute need for progressives and true radicals, it plays such an important role. But the very mainstream act of living in a longtime committed relationship and having kids, and holding down a job, and paying taxes, is also something of high impact, and I think it takes all those things to further the movement of, broadly speaking, civil rights."

They find that they are members now of two groups—groups which, like circles in a Venn diagram, are distinct and yet overlap. One is the gay community. The other is the parent community. "The thing about becoming a parent—and this is what people who are not parents don't realize—is that regardless of your sexuality, there is this thing called the parents' club, and all parents are in it," Eric says. "I'm a dad, and all dads are the same."

"Here in our bubble," Doug says.

"No," Eric disagrees. "Not just here in our bubble. I travel deep into the red states," he says, and when he meets business associates there, "we now have something to talk about. We talk about our children. There's a commonality there that transcends sexuality, and I think that's great. I love being a parent not because it helps my career—goodness sakes, no—but I think it actually does."

Sometimes, though, opposition can come from unexpected quarters. Eric remembers another time he was pushing the girls in their stroller through the Castro. He parked the station wagon and hoisted the twins out, loaded them into the stroller, and made his way down the sidewalk. "I've got six feet of stroller and babies in front of me, and I can remember this one time, right at Castro and 18th, this old couple looking at me—gay bears, gay men looking at me. And this particular man saying, 'Fucking breeders are taking over the neighborhood.' And I just wanted to say: 'Dude, I'm gay too.'"

CHAPTER SEVEN

SINGLE MOTHERS BY CHOICE, AND THE MAGAZINE ARTICLE THAT MADE THEM

Mall-Shopping for Sperm

In 1986, *Newsweek* published a now infamous cover article asserting that a single, college-educated white woman over forty was "more likely to be killed by a terrorist" than marry. It was the high-water mark of a tide of anxiety that had been washing shoreward for more than two decades: part of what Susan Faludi called a "backlash" against feminism and especially against women pursuing success in the workplace. Exactly twenty years later *Newsweek* apologized: Checking back, the magazine found—oops!—that most of the women featured in the 1986 article did marry, and the statistics the article relied on had been misinterpreted. Turns out, women over forty have a good chance of marrying—more than 40 percent, according to 1996 numbers, and probably, the magazine estimated, higher now—and they are more likely to marry the more educated they are. It is true that women are marrying later, but the original article, and others like it, interpreted "later" to mean "never." It also emerged that the "terrorist" remark started out as a joking aside in an internal memo. The editors thought readers would understand it to be hyperbole.

Sorry!

The apology came a little bit late. By then the conversation had been initiated and would run well into the next millennium, a vexed, vexing, tiresome, essential, ongoing conversation about women, work, family, and what may or may not be the domestic consequences of throwing yourself too enthusiastically, and for too long, into your chosen profession, or even just the job you need to live. The same era would give us Glenn Close as the homicidal, family-hungry working woman in *Fatal Attraction*. It would give us Sigourney Weaver (these interesting actresses, so drearily cast) as the boss in *Working Girl,* fretting about her biological clock and yelling at her secretary. The beginning of the twenty-first century would give us Sylvia Ann Hewlett, who in her 2002 book *Creating a Life: Professional Women and the Quest for Children* pointed out that what women who work too long and single-mindedly are really going to miss is a child. As hard as some feminists worked to allay it, "baby panic," as Joan Walsh called the new anxiety in a terrific essay on Hewlett in *Salon,* kept reappearing. An unkillable apparition, it rose up terrifyingly like—well, like Glenn Close surfacing out of the bathtub in *Fatal Attraction.*

It's hardly surprising, then, that the same era gave us Debra Cope, a smart, successful, pleasant-looking woman who in 1998 found herself sitting in a Nordstrom coffee shop in a mall in White Plains, New York, "hoping no one would notice what I was doing."

What she was doing was shopping for mail-order sperm.

Cope had an assortment of donor profiles she had pulled off the website of Fairfax Cryobank, one of the largest sperm banks in the United States and therefore in the world. Fairfax Cryobank, based in the suburbs of Northern Virginia, sits in close proximity to Washington, D.C., the U.S. city that boasts the highest percentage of residents with Ph.D.s and other advanced degrees. It is the opposite, in a way, of Los Angeles, with its picture-perfect body-beautiful egg donors. What you have in Washington is a plethora of lawyers and policy analysts, men with a wallful of degrees from Georgetown and Johns Hopkins. So overeducated are the region's inhabitants that Fairfax Cryobank even offers a pricier category of "doctorate" sperm. It also offers—everything costs extra, of course—personal essays, academic scores, military records, facial features, personality assessments, baby photos, and audiotapes of the donor's voice. In a few cases, Fairfax Cryobank even offers adult photos, something that is rare in the sperm-banking business. In the still rather sexist culture that is gamete banking, it is considered essential to offer glossy adult photographs of egg donors, but uncommon to offer adult images of sperm donors, who, according to one sperm-bank director, would not feel "comfortable" with a photo showing how they look. The premise seems to be that looks issue from the mother, and intelligence from the father, as in the Porgy and Bess lyric, "Your daddy's rich and your mama's good looking." Gail Taylor says that it's the way in which, for years, the industry has been able to get away with selling the sperm of ugly men.

But the truth is, Debra Cope wasn't overconcerned with things like eye color or nose shape or even college major. Cope didn't care whether her donor's hair was coarse or fine, whether his eyebrows slanted up or down, whether his degree was in foreign relations or astrophysics. She did want a healthy donor, and one who didn't have trouble with his weight, since weight was something she had struggled with, and she'd just as soon any child of hers didn't face the same issues. And she was looking for a donor with blond hair, the better to fit in with her own family's appearance. Beyond this, what she was looking for was a man as unlike her former boyfriends as possible. A man who was not complicated. Not tortured. Not—deep.

"A balanced, decent person" is what Debra Cope was after. "I was just looking for a regular, nice guy who seemed to have a pretty good brain. I think a lot of my friends, when they talk about it, they're looking for a date. But I was just looking for some solid guy that I could speak kindly of to my child. Intelligent, but not too complicated."

In more ways than one, Cope personifies the growing number of single women who, like lesbian couples, are using donor insemination to achieve the families they've always wanted, but in their case minus one crucial detail: the partner they formerly hoped they might have. Back in 1998, Cope was nearing forty. She had paid close attention to fertility statistics. She was self-sufficient and then some, earning a six-figure salary working as a financial journalist in New York. What also made her typical was that she had wanted children all her life, but the men in her life had not. Well, maybe they did. And maybe they didn't. They weren't sure. Not yet, certainly. It was just so hard to contemplate, a life so, you know, encumbered! Children were such a commitment! The diapers, the crying! Getting up in the morning! Needing to be put to bed at night! Having to find a babysitter every time you wanted to see a movie! And there was plenty of time to hash all this out later!

It has always seemed to me ironic that assisted reproduction is associated in the popular mind with female procrastination about childbearing. While ambivalent, procrastinatory women clearly exist, their numbers surely are equaled by ambivalent, procrastinatory men, who are one major reason why tens of thousands of women seek refuge each year in the procreative willingness of anonymous sperm donors. "I don't know why women are not marching in the streets," says UCSF's Robert Nachtigall, who has seen all too many female patients who endured too many years of uncertainty from too many noncommittal males. "If anything, guys are encouraged to be more infantile and immature than before. Nobody gets married anymore. The whole thing is—it's nuts!" Reluctant single men: where are the cover stories agonizing about the threat they pose to the traditional American family?

"The child issue was always a divider," says Cope, who was forty-six when I interviewed her. "I had dated lots of guys, a number of guys, including some very long-term stuff. I was with one guy for four years; he knew from the outset that marriage and children and family were important to me, and it was just a big hurdle for him. He had a traumatic family history. You meet somebody, you really love them, and it can take a long time to come to the realization—as it did for me—that as much as I love this person, and as much as I want to spend my life with him, he can't do it. I may have just made poor choices in this regard—I did manage to pick a whole series of men whose readiness to have children, it was clear from the outset, that this was not going to happen. I liked really intelligent, really complicated men, and it goes with the territory. They're so complicated that the simple desire to have a child was—I don't know, too hard. I spent a lot of time feeling guys out on this issue. Even somebody I was very close to, and very involved with for a long time; we had a mutual parting of the ways, and this is why I was able to walk away without feeling any pain. It was so clear to me that this was something that was not going to happen with him. I wanted a child. It's totally conventional. I just wanted to have a kid." And what she realized was that to get one, "I [had] to do it outside all social norms."

Well, not outside all social norms, though it might have seemed that way at the time. Between 1999 and 2003, the number of births to unmarried mothers would grow by 17 percent, to the point where one-third of all births in 2003 were to single mothers. During that same period, Jane Mattes, founder of Single Mothers by Choice, has seen her membership grow steadily. In 2005, Mattes calculates, she welcomed twice as many new members as she did in 1995. Mattes's organization has members in Israel, Australia, Canada, England, France, Belgium, and Germany, and it also has spinoffs and competitors. Single Mothers by Choice is not to be confused with the National Organization of Single

Mothers, for example, another single-mom group that has a website, chat rooms, and a mailing list. Nor is her book, *Single Mothers by Choice,* to be confused with *The Single Mother's Survival Guide*; or *The Complete Single Mother;* or *The Courage to Be a Single Mother;* or *The Single Mother's Book: A Practical Guide to Managing Your Children, Career, Home, Finances, and Everything Else.* In England, there is a single-mothers' unit of the Donor Conception Network; according to founder Olivia Montuschi, "there has been an enormous increase in choice mums over the last few years."

In 1980, SMC founder Mattes really was unusual. Back then, Mattes recalls, it was assumed that a middle-class single woman, were she to become accidentally pregnant, must either marry the father or terminate the pregnancy. There was no plan C, at least not when Mattes became pregnant at thirty-six and decided to keep the child. But surely, she thought, a few plan C–type women must exist. So Mattes, who lived in Manhattan, "put out the word—basic networking, I told everybody I knew to let anybody in a similar situation know that I'm going to have coffee in my house on a certain afternoon. I was looking for mature women who had decided to become mothers while single, where it was a choice. Seven women showed up, and we decided to keep meeting." All the attendees either were adopting or had decided to continue an accidental pregnancy. Soon the meetings were so large that they had to move out of Mattes's living room.

"Women were coming to realize that they could do things that were previously not open to them," says Mattes, a psychotherapist who is now in her sixties; her son is an adult. "Their earning power was greater. The idea of a woman heading a family didn't seem far-fetched." And it seemed better than the expedient marriage and the miserable aftermath: being a single mom from the get-go seemed preferable to marrying, splitting up, and then, after all that grief, ending up a single mom anyway. "Most of us shared the experience of seeing divorces going on also, and we felt like rather than marry someone in order to have a child, if that person wasn't someone we would have married for other reasons, we didn't want to do it. We thought a lot of divorces came out of this pressure to marry." The composition of Single Mothers by Choice has changed as norms have shifted. Of late, Mattes has noticed women coming into the group who are both older and younger than they used to be. Once, the average SMC—a term that's now broadly understood to mean any single mother who has started a family voluntarily—was in her thirties, but now she may be in her twenties, or in her forties. More women are also considering two children, something, she says, that "'most of us [at the outset] would not have considered."

And 75 percent of SMC's membership now chooses to parent by means of donor sperm. Like many gay and lesbian couples, single mothers sometimes fear they will not fare well in the competition for adoptive children, and sometimes they are right. Plus, adoption can be expensive, and sperm donation is startlingly cheap. In many states, if a woman uses a sperm donor she knows, the donor will retain custody and parenting rights. In comparison anonymous donors seem safe, quick, and refreshingly low-maintenance.

Debra Cope started thinking about donor insemination as early as the late 1980s, when, just thirty, she went on a sympathy mission to a friend recovering from a devastating breakup. "I told my mom, if I'm not married by thirty-five, I'm going to get a donor and have one alone," her friend blurted out. Hearing this, Cope says, "I was blown away." It struck her that she could do the same thing, a thought that recurred as she worked her way through the series of child-averse boyfriends. When she herself was nearing forty, her sister had a second child, and around the same time, Cope went to visit a college friend who had also given birth. "Holding those babies and looking at how happy the mothers were, I thought, 'Oh God, I'm doing this.' There was no question. I didn't agonize. I had turned it over in my mind for years: Am I up to the challenge emotionally? How will I tell people? Can I afford it? All of these questions—maybe somebody will come along—the same process that lots of people go through." So she printed off the sperm-donor profiles and went to the White Plains mall. She selected a donor who was blond, and lean; who had a family of his own; who was in the military; who seemed sound, and stable. Presto! Done! She ordered his sperm as well as some supplementary material, including the audiotape, so as to have something to show a child later. She got pregnant more easily than she had expected, had a daughter, Cassie, and found that being a mother was wonderful. And, that being a single mother was something she could manage.

She did have to make career sacrifices. Cope had considered making the jump to a major daily newspaper, but worried that starting a new job and a new family would be too much. After some upheaval in her workplace she moved to Washington as publisher of a trade magazine, satisfying work that demands about 80 percent of her former hours. The job does require travel, so she decided to buy a house with her own parents. This way, everybody could take care of everybody else. "I'm classic sandwich generation," Cope says, so she decided to make it work in her favor.

In doing so, she was joining what the psychologist Clare Murray calls a "distinct group of single parents": women who have actively chosen their route to parenthood, and whose children seem likely to fare better than those from many other one-parent homes. Up until recently, most studies of single mothers have looked at mothers who are single owing to circumstance rather than choice: women who are divorced, or whose partner never stepped up to help parent the baby. Their children were found to do less well in a number of areas. But it was often unclear whether it was the lack of a father that created problems for the child, the lack of a second income, the unhappiness and displacement that follow divorce, or—as seems likely—all of the above. This new round of mothers did have to confront these anxiety-provoking statistics, as well as a new cultural prejudice: What was wrong with them, anyway? Why weren't they married? Prejudice against single women who become mothers by choice is legion: As one woman recounted in a chat room posting, she was sitting at a table with a group of acquaintances, when one of the men began talking about a woman he knew who had conceived using donor sperm. "So how ugly is she?" another man asked, to which the first replied, "Not too bad, but obviously she must have been desperate."

"I was," the mother wrote, "just frozen in horror."

"There was lots of speculation that these women were screwed up; they're all men-hating, et cetera," says Murray, a psychologist at City University in London who has worked with Susan Golombok. Murray came into her office for an interview one morning in 2004 to talk about their work, which had recently produced a study of donor-insemination single mothers. Since the outset, she said, not only society but fertility medicine have been skeptical of single women: When regulations were being written in Britain to govern reproductive science, there was a clause inserted into the 1990 Human Fertilisation and Embryology Act, requiring doctors performing fertility treatment to consider the "welfare of any child who might be born as a result . . . including the need of that child for a father." Gradually, lesbian couples were able to persuade doctors to quietly ignore that rule. Now, Clare Murray pointed out, "Clinics treat lesbian couples at the drop of a hat, but still won't treat single women. They're the pariahs of the assisted reproduction field." This too may be changing: Suzi Leather, until recently the chair of Britain's Human Fertilisation and Embryology Authority, the board set up to administer the regulations, has declared the "need for a father" clause to be "nonsense," and as of late 2006, a movement was under way to replace "need for a father" with "need for a family."

Still, old concerns die hard. In the United States, doctors can treat whomever they like, and many, though not all, will now treat lesbian women. But some still harbor qualms about single mothers. "I think it is so hard to raise a child," I was told by Marcelle Cedars of the University of California, San Francisco. "I wouldn't say single women couldn't do it, because there are lots of single men and women out doing it because of divorce, but I think it is so incredibly difficult to raise a child, even in a family. I am less concerned about my gay couples than my single women. If I worry about family structure, I worry about a single professional woman raising a child." In the United States, it's getting risky to discriminate against gay patients. In 1999, a California practice refused to inseminate a lesbian woman named Guadalupe Benitez with donor sperm. In a test case of reproductive freedom that may well end up in the U.S. Supreme Court, Benitez sued, charging discrimination. But single mothers do not enjoy the same protection as a class, which may be why, after losing the first court case, Benitez's doctor explained that the problem wasn't that she was lesbian, but that she was unmarried. The doctor's clinic won that round of appeals; as of this writing the case was slated to move to the California State Supreme Court.

People worry about single mothers, but they often do so, Murray argues, based on research that does not apply. It's true: children in what might be called the old-fashioned single-mother household are more likely to live in poverty, more likely to exhibit behavior problems, the whole grim laundry list. This research is one reason why single mothers used to be pressured to relinquish their children, and in some cases still are. But later studies have found that it is often the lack of money that creates the adverse outcomes. The lower income associated with divorce is the "most influential factor in the later underachievement of children raised in these families," Murray and Golombok noted in their paper. They point out that there are now so many kinds of single-mother families that it's impossible to generalize about the outcome of children living in them. "There is a significant school of thought that it is family process—parents' state of mind, quality of parent-child relations—that is more important than family structure in determining how a child does," the paper notes. Even so, "whether or not to allow single women access to donor insemination is one of the most controversial issues in the field of reproduction, and in many countries, such as Sweden, it remains illegal."

And the challenges shouldn't be minimized. "On average, two parents do better than one, other things being equal," the sociologist Judith Stacey, in her work on alternative families, has concluded. It's easy to see why this would be the case. Easy to see that child-rearing is more pleasurable and less draining when it can be shared by two parents who get along, two parents who can spell each other and who offer complementary parenting styles. Easy to see that it's wonderful to parent with somebody who loves your child as much as you do, and who would be there if you died or were incapacitated. Another adult to discuss child-rearing decisions with, to offer company, solidarity, mutual bemusement over your children's behavior—all these are enormous advantages, if you can have them.

But sometimes the choice is between parenting alone and not parenting at all. And that, for many women, is the choice today. In fact— while the panicky tone and overall thrust of the *Newsweek* article were wrong—it is true that there are women now who missed out on the chance to marry *in time* to conceive with a husband; that this probably does have much to do with the era we live in; and that in an earlier era, these women might have married early enough to conceive naturally.

As with their lesbian studies, Murray and Golombok were interested in the impact on children of being raised in a household that lacked parents of both sexes. But, Murray told me, when it came to single women they also "wanted to get at the bottom of their motivation: why they had chosen to do it, why they had gone to a clinic." What they found confirmed the reluctant-male theory. "Almost all of the women said they would have preferred to have a child within the context of a loving relationship. The majority want to have a relationship in the future, but time is running out. They'd always had a lifelong dream of having a child, and found themselves in their late thirties, with a string of relationships with men who didn't want to commit. And they didn't want to miss the opportunity to have a child."

As for why they had pursued donor sperm, in most cases it had to do with an unwillingness to dupe a male partner. "They'd chosen to go to a clinic, because they couldn't contemplate the idea of casual sex—a one-night stand wasn't something they could consider. And they didn't want to deceive a man, although most knew women who had. Most had thought about the impact of the child from a financial point of view, and made sure they were financially secure," Murray said.

Comparing them to a cohort of married mothers, Murray and Golombok found that single donor-insemination mothers were slightly older than married mothers. There was no difference in social class, nor in the proportion of mothers who returned to work, nor in the proportion who placed their children in day care. They found no difference in warmth, enjoyment of parenting, or feelings about their role. Solo moms did show a slightly lower level of mother-infant interaction and attention to infant needs; they rated their infants higher on the fussy scale, possibly because there was no partner to help when babies did fuss. But they felt no greater anxiety or stress, and no difference in frequency of contact with friends. They got more help from their extended families than married moms did. All in all, they were highly functional—and deliriously happy. "I haven't ever met one single mother by choice who regrets having a baby," says one SMC I interviewed, Denise Feinsod. "Basically there's a lot of happy, happy women who can't believe they waited."

Murray and Golombok also found that these households were economically sound. Ninety-six percent of the women were supporting themselves. Seventy percent said they were experiencing no financial hardship; 26 percent said they were experiencing some, but doing okay; only 4 percent said they were having a hard time getting by. Ninety-two percent said that there was at least one man in their child's life. Fifteen percent were in a relationship, and 87 percent said they would like to become involved with a man in the future.

The important thing, Murray believes, is that these were women who had *chosen* single motherhood rather than being forced into it. Single motherhood wasn't something that had happened to them. It was a route they had planned, chosen, mapped out carefully, and deliberately taken.

"I'm Thirty-eight and I'm Having Sex with a College Student"

What fascinated me is the way women chose their donors. The widespread prejudice against parents using donor gametes is that they are hell-bent on building a better baby; "yuppie eugenics," the process has been called. This may be true, but when it comes to single mothers, "better" is a matter of taste, as with so many things one shops for. In selecting a donor, many women want to give their child a chance to be free of their own flaws, real or perceived. To take one oddly recurrent example: sperm donation allows women to play out the age-old fantasy: what if I had been born with better hair? "I have curly hair, and I wanted straight hair," said a woman named Ruthellen, who lives in St. Louis.

"It was fun; that's how I looked at it; like a big blind date and you never have the date," said Lisa Schmidman, an elementary school-teacher who decided in her mid-thirties to give a sibling to her daughter, who had been conceived during a marriage that ended in divorce. Schmidman wanted a donor who matched herself and her family: tall, and blue-eyed, like her. She acquired fifty profiles; paid for audiotapes from her favorites; asked friends for advice; chose a donor who seemed smart, funny, and was almost twenty years younger than she was. "You just start thinking: 'Oh my God, I'm thirty-eight and I'm having sex with a college student. It's very funny. My dad's like—'Uh, I don't want to hear this.'"

"I agonized about acne," said Denise Feinsod, who is unusual in that she didn't wait to get started. While still in her twenties, Feinsod, who was working for a Bay Area alternative newspaper, turned to The Sperm Bank of California. "My criteria were medical issues; there's heart disease in my family, and it would be nice to find someone who didn't have it. I found as a whole that it was a very bizarre process. I started out by eliminating donors that wildly didn't look like my family. I eliminated very tall donors. I eliminated other racial people: why bring into the world intentionally someone who isn't going to have that half of their racial heritage available to them? Then eventually, I got to see their questionnaires, filled out by hand, and I could see if they could spell. And then there is this page that is very freestyle: like, what would you say if you could talk to these kids? There was one who had all this very, very, very intense New Age stuff in there. I'm not prejudiced against that, but I couldn't look at it with a straight face. I wanted to be able to show this to a child, and say here's this incredible person who gave us a gift; we don't know who he is and we never will, but he's an incredible person. I didn't know if I could do it with a New Age guy, so I didn't pick him. I ended up picking a Jewish donor. We joke about the tribe. My kids definitely look like the tribe."

While Denise Feinsod related her story to me, she was cooking a frittata in her Craftsman bungalow in Berkeley, wearing bedroom slippers and drawstring pants. It was a Saturday morning. While we talked, two cats lolled on a sofa in a sunny window. Also listening were her two donor-conceived children: Alix, a girl, ten at the time, and six-year-old Cody, a boy. Alix was wearing pink leggings and a T-shirt, sitting on a footstool and knitting a red scarf. Denise said that when she chose a donor she didn't really care whether he had agreed to be identified to the children. "That's a lot to decide when you're twenty, or any age, and I felt like I could really, really understand saying no to that." But once Alix was born, Feinsod realized that this was not exclusively her story. "At that time, I finally experienced Alix as her own autonomous being," Denise said. "And I felt that I had made a mistake." She asked TSBC to contact the donor and ask him to reconsider; the donor replied that "he was so clear on what his role was," and it didn't include being known to off-spring. "He said all the reasons I chose him—'I hope she's happy and I hope the child is happy, and I was just the donor, I feel like that's all.' "

"I don't feel gut-wrenchingly regretful or despairingly regretful," Denise continued, taking the frittata out of the oven and cutting it into slices. "I feel that I really thought about it a lot, and made the best, most loving decision I could, but I do feel like I took a choice away from Alix. She right now says that she doesn't need it and that's fine, but if she ever does get curious, I will wish she had that option." Later, Denise had to decide whether to use the same donor to conceive a sibling, or whether to select a donor who had agreed to be known. Biology won: she decided that she would prefer for her children to have one other person in this world to whom they are both fully related. So what both children have is a manila folder with several pages filled out by some unknown student; neat, modest, slanted cursive words inscribed by a man who to his knowledge had never gotten anybody pregnant; who (Denise asked that identifying details not be revealed; she also asked Alix, first, whether she minded my seeing the profile) had abstained from pernicious sub-stances. He had tried marijuana, of course; this is Berkeley. "I would worry about him if he didn't," Denise joked.

When she asked her daughter if it bothers her that she will never meet her donor, Alix said categorically that it does not. "I have a daddy," she says firmly.

And she does. In her thirties, after bearing both children, Denise Feinsod met Ely Newman. He was in his late forties, with no children of his own. When they were starting to date, Denise thought it best to get to know him by herself, but he argued that he wanted—needed—to meet the ensemble. "He was like: 'If I can't meet your kids, I'm missing a whole part of you.' So the first time he met them he came over for dinner to my house, and we got Chinese food, and Cody was two, and Ely was going to walk to the Chinese food place, and he was like, 'Does anybody want to come?' and Cody was like, 'I do!' and they're holding hands. Cody loved men always," recalled Denise, who watched the two walking down the street and thought how sweet it was to see her son's little hand in the hand of this man she hardly knew; and then thought, with panic: I'm letting my son walk down the street holding the hand of a man I hardly know!

Ely, Denise said, felt "it was a really cool prospect to be part of a family." He married her and adopted the children, and the funny thing is, he looks exactly as though he could be the children's natural father. When you're shopping for a donor whose characteristics strike you as attractive, they may be the same characteristics that strike you as attractive in a man you someday meet. "I married a man who looks just like my cousin from New York, and now we all look just like each other, like a good short Jewish family from New York," Denise pointed out, displaying a photograph of the four of them.

What remains a mystery to her is how couples navigate all the inevitable compromises while enduring the fatigue and distraction of raising small children. How, she wonders, does any couple survive having children *together*?

"When I was thinking about being pregnant, I got to the point that I couldn't believe anybody had had a baby with somebody," she says now. "I still think that would have been really, really hard: to have little teeny babies and try to maintain a relationship."

Feinsod was an early adopter. In deciding that she did not need a relationship with her childrens' genetic father, she was responding to the same feeling that Lori Gottlieb, writing in 2005 for the *Atlantic Monthly,* would later articulate. "I didn't want to date under that kind of pressure," Gottlieb wrote in a first-person essay about choosing to become a single mother, describing how she found herself assessing every date as a potential father. In this day and age, sometimes it really is hard to find the necessary family-making attributes in one person. Motherhood by sperm donation is, it seems to me, a kind of cultural release valve. It is a rational response to the fact that it can be hard for a woman to find, in a relatively narrow biological window, a man she wants to settle down with who is himself ready to settle down. "We're all wildly independent," one single mother said to me. It seems instead that they are wildly organized: the kind of pragmatic women who don't get accidentally pregnant, but who don't want to get accidentally infertile, either.

"I found myself dating to interview fathers, and it seemed silly to me," says Julie Sabala, who lives in Portland, Oregon, and who was thirty-four when she decided to stop father-shopping. Sabala's story is a little different: she did get married, at thirty-one, to the love of her life, a man who wanted children and was prepared to start right away. Tragically, her husband developed cancer and died within a year. After he died, Sabala says, "I gave myself a little time—three years" to grieve. She began dating again, but realized that "finding someone like him was going to be impossible" and she didn't have years in which to look. Her husband had banked sperm with the hope that he would survive and they could conceive together. After a lot of thinking, Sabala decided to use her late husband's frozen sperm to conceive. She consulted her own family, who were supportive. Then she broke the news to her late husband's parents. "My father-in-law was wearing his sunglasses, and had gone completely quiet, and I said, 'How do you feel, are you okay?' He said, 'I'm just trying to figure out a way to move to Portland.'"

Once she was pregnant, with twins, a male friend told her that she was ruining her chances of marrying again. Who would want a woman saddled with two kids?

"I look at it as an asset," Sabala says. "Who's to say that because of them I don't find someone? Why does that limit me? You know, maybe the love of my life will be their soccer coach."

Really, Really, Really Big Love

Among single women who become mothers through sperm donation, there is no consensus on whether to use an anonymous donor or one who is willing to be known. "I hear completely different issues," I was told by Amy, who is bisexual and who at the time was attending one trying-to-get-pregnant support group for single women and a different trying-to-get-pregnant support group for lesbians. "Most of the lesbians are very heavily leaning toward using known donors—gay men—and they really are cautious about shopping for an anonymous donor," she reported. "Whereas in the SMC arena, the fear of a known donor is so intense, the issues around, 'He'll want my child later, I don't want to risk anybody taking my child away.' There's a strong fear in the SMC community around that. For me to go between those is totally wacky."

There is also no consensus over what to call the donor. In 2005, a colleague of mine at the *Washington Post*, Michael Leahy, profiled a single mother by choice named Raechel McGhee. Much like Denise Feinsod, after giving birth to her two children McGhee regretted having selected an unknown donor. As her children grew and expressed curiosity about their donor, she began to wish she could provide them with his identity. In a highly unusual move, she managed to track down the donor, a man named Mike Rubino who—this is also unusual—actually wanted to be contacted. Raechel flew her children to Los Angeles for a week, during which she encouraged them to call him "Daddy." It was okay with Mike, who has established contact with several more offspring, and seems prepared to welcome them all into the family fold, though he later said on the *CBS Early Show* that "I'm a little concerned if any others come forward, only because I don't know how much time I could spend with everyone." The article created an enormous chat-room furor: on a number of sites there were bitter arguments over whether a donor should be considered a "dad." The majority of the women, it seemed clear, did not want their children thinking of the donor as a father; instead they say things like "donor," "biodad," and "the nice man who helped us." Some of the exchanges were so acrimonious that posters departed in a huff.

Still, if they can't give their children a father, many single mothers can give them a different relation, one that is less threatening: siblings.

In 2000, a resourceful Colorado single mother, Wendy Kramer, decided to start a website to bring sperm-donor children together. Kramer had conceived her son, Ryan, by anonymous sperm donation while she was married. The marriage ended when Ryan was still young, and Wendy Kramer raised her son alone. As he grew, Ryan yearned to know his donor's identity, and wondered whether half siblings might have been born into other families. So Kramer decided to harness the Internet to see if she could give Ryan one or both of the things he so deeply wanted. She created the Donor Sibling Registry, a website where parents and offspring can post the name of their sperm bank and the number given to their sperm (and, now, egg) donor and look for "matches": people who also conceived by the same donor, or maybe even the donor as well. Kramer posted Ryan's donor information and their e-mail address, and waited to see if anybody replied.

She also set up a chat room where people could post queries and discuss issues of parenting and identity. Though Kramer works hard to keep things civil, discussions on the DSR board are wide-ranging and often furious: In one of the more curious recent exchanges, a sperm donor (there are a few who post) was inveighing against single mothers. "The losers in the SMC situation are the children who grow up father-less," he wrote. "While there may be a few moms out there who can teach their sons how to hit a curve ball, stand up to a bully, or use a table saw, there are many who cannot." Needless to say, this donor got pounded for his position—by the very women who may have purchased his sperm.

Wendy Kramer's website marked the beginning of another move-ment, another kind of family made possible by the coming together of both reproductive and online communication technologies. The majority of DSR posters are single mothers who want to make contact with their donor, but who are even more interested in contacting their children's half siblings. "The kids are very interested in siblings," Alice Ruby, cur-rent director of The Sperm Bank of California, has found. These mothers want to use the Web to give their children the ultimate birthday present: a sibling. A genetic teammate. They want to use the Internet to expand their kinship circle, to create a unique extended family: self-sufficient women raising children who are far-flung and yet intimately related. In an astonishingly short period of time the number of sibling matches made through the DSR has exploded. As of October 2006, the DSR had almost 7,000 members, with more than 2,500 half sibling matches. Now, there are regular entries on the DSR discussion group from ecstatic SMCs say-ing things like "My daughter Kayla has wanted a sibling for so long and now she has one," and "My daughter and I picked up my daughter's first sister that she has met 4 days ago at the airport. I immediately started cry-ing when I met her sister face-to-face. Her and my daughter look so much alike." An article about a DSR sibling reunion that ran in the *Columbus (Ohio) Dispatch* profiled two young women, Elizabeth Reynolds and Kelli Dail, who learned in their twenties that they shared the same donor. After they met on *Good Morning America*, Elizabeth Reynolds said, "It was like we always knew each other."

The DSR was not unprecedented. The Sperm Bank of California has a sibling registry; so do Single Mothers by Choice and, in England, the Donor Conception Network. But Kramer's all-inclusive Internet registry was a cultural tipping point. In 2003, Oprah did a segment on donor-conceived families. In 2005 and 2006, everybody wanted a piece of the action: a spate of newspapers and television shows featured single women who have organized groups around their anonymous sperm donor.

Call it "really, really, really big love." In 2006, the women of Fairfax

Cryobank Donor 401 appeared on the *Today* show to celebrate their federation, more than a dozen single mothers, with fourteen children among them, united by their taste in donors. But even their ranks are dwarfed by the women who have conceived with Fairfax Donor 1476, an anonymous financial planner with a B.A. in economics who is blond, blue-eyed, 6'2", 215 pounds, with O-positive blood, curly hair, a long-lived set of Germanic ancestors, and a talent for rugby, and who at last count was the donor father of at least thirty-five offspring. So many women have selected Fairfax Donor 1476—and so many have posted their contact information on the Donor Sibling Registry—that the 1476 group created its own website, which is to open to anyone to read, if they have the stomach to listen, over and over, to the melody "Baby Mine" from the Disney movie *Dumbo*. The group plans a mass vacation in 2007. According to its website, which is updated regularly, twenty-three girls and fifteen boys have been born thus far to Donor 1476, with four more on the way. A guest book contains messages from parents living in states as far-flung as Wisconsin and Hawaii, as well as women who are considering using this popular donor. "1476 has made it to our final 3 of swim team choices!" one wrote in the guest book, revealing an unintended use of sites like this. On the 1476 website as well as the Donor Sibling Registry, sperm-bank shoppers can find a donor's entry and view photographs of his existing offspring, to see how his kids turn out before they make a final decision. It's a kind of a *Consumer Reports* for the babymaking world.

On the DSR discussion group one day, there was even a posting from a mother who in selecting a donor had felt torn between two men. She had chosen one and had three "wonderful children," but also confessed to this secret desire to look at photos of kids from the donor she didn't use, maybe even—she had stopped herself—contact that mother. It was the twenty-first century version of wondering *what would have happened if I'd married my college boyfriend?* "I just have this little curiosity about them," she wrote. "Has anybody else experienced this?"

The groups also share information about viruses, ailments, and doctor's visits. Many women who conceive by sperm donor worry about the lack of an ongoing paternal medical history, which remains a key tool for medical diagnosis. One recurring topic on the DSR site is the desire of many parents to force the sperm-and-egg-banking industry to provide updates on a donor's health. In the interim, groups like the one formed by Donor 1476 can assemble a kind of composite sketch. One 1476 mother, a woman named Susanna from California who did not want her last name used, told me in an interview that she noticed that her son was coughing a lot at night, and wondered what was wrong. She had chosen Donor 1476 because he reported being free of allergies. But she had noticed from the online exchanges that a number of 1476 kids seem to have asthma. "I was able to go to a doctor and say, 'We used a donor [and] a few of his half siblings have problems with asthma,' so we're trying asthma inhalers. It's really useful information." In fact, she said, her curiosity about the donor had lessened as she had come to know his offspring. The group, she felt, was building a group portrait of the man, by comparing notes on his collective progeny.

Susanna also appreciates the access the Donor 1476 group has given her to a large group of interesting other moms. "We're going to be linked for the lives of our children. Not only our lives, but beyond. I mean, love it, hate it, whatever." The other mothers, she says, are a stimulating mix. "I don't know anybody who isn't a professional. There's a writer down in Los Angeles, and a couple lawyers, a 911 dispatcher. I am a teacher. It's an incredible network. I feel like I'm part of a sorority. It's like the secret society of 1476 moms." Susanna had recently met one 1476 mom who lived not far away. "It was really interesting to see her sons, because genetically they are closer than my children are to their cousins, but it's kind of like meeting a cousin. We intend to treat it like: here's your friend, and then later, when they understand everything, they can understand why they're so important to each other." For this other mother, she felt, the group offered solidarity: "She feels pressure to have a second one so they're not alone; now that you know there are thirty others, does that change things?"

Debra Cope also decided to look for half siblings. Several years ago, Cope signed on to the DSR, scrolled to her donor number, and saw that another single mother had posted. She contacted her, and they began e-mailing. Soon there was a match from lesbian partners, another from a heterosexual couple. They started sending baby pictures back and forth, comparing traits. "My daughter looks like a clone of me at the same age, but now that I've seen photos of other children, I do see areas of resemblance. They have the same chin and dimples. While Cassie is a brunette and the others I've seen are all blonds, she looks quite a bit like the gay couple's little girl!" Cope wrote me in an e-mail. "As for personality, there might be some similarities, but let's face it—we're mostly just cheerfully boasting about how amazing, cute, precocious, and wonderful our kids are!"

Cope added, "It's a funny little dance, connecting with these other parents. I've initiated two of the three contacts by e-mailing them, letting them know I have a wonderful, healthy child and am grateful to the donor, and that while I'm open to contact, I would prefer to keep it low-key. With the first mom, I've exchanged thoughts on child safety (she is very worried that the donor will somehow track down and snatch her child; I don't share that concern), about how to tell your child about his or her origins, etc. We are in agreement that if and when our kids ask about half siblings, we'd like to have the information to share. I think the gay couple is open to meeting, but I haven't had a chance to feel them out on that, and I'm not sure what I would want." Cope is also interested in keeping tabs on medical issues: "I'm not going to look for the donor, and chances are he's not going to look for us, either! So my best way of gaining insights into any health issues, mental or physical, that may be genetic on the paternal side rests in getting to know other donor families."

One sticking point, however, is the awkwardness Cope articulated: once you make contact—then what? What if you don't like the other parents? What if they don't like you? What if your households are different? What if one parent is more tightly wound? Unnervingly lax? As every parent knows, each household is its own tender ecosystem; it can be hard for any parent to find like-minded, playdate-safe people who share your views on snacks, booster seats, and PG-13 movies. The same is true of donor-conceived households, only more so. Many parents who chose the same donor do not agree on the level of contact they want, or even whether to tell their children that they have half siblings. In one of life's crueler ironies, Ryan Kramer, the donor-conceived son whose yearning inspired the Donor Sibling Registry, has never heard from his donor. After he appeared, now a young adult, on a television show—his mother, Wendy, wrote about this in a posting on the DSR discussion board—they were contacted by a woman who felt sure her daughters must be Ryan's half sisters. The problem was, she and her husband had not told the girls they were donor-conceived, and, Wendy wrote to the DSR group, the couple "had no plans of ever telling them. This was pretty hard for Ryan and still to this day he knows these girls are out there, yet sadly, also knows he will probably never get to meet them and they will never know that he exists." Not everybody wants to be on television. Not everybody wants to be part of an unprecedented extended family.

"Every day I hear the most amazing stories from DSR members

about finding one another," Wendy Kramer, this capable, unsinkable woman, wrote in a rare moment of sadness and regret. "Because of secrecy and shame, this is not so for my own son."

And then there is the million-dollar question: How many offspring might any one donor have? Dozens? Scores? Hundreds? What if half siblings meet and date, even marry? In the United States, there is no central donor registry, unlike in Britain, where a registry is maintained and a single donor's contributions are restricted to ten families. Nobody knows how many children might be engendered by a popular donor. Most banks say they limit donors to a certain number of reported offspring, but just scanning the DSR, it's easy to see that some donors have at least twenty identified offspring. As for 1476—who knows how many future blond rugby-playing financial planners are in preschool just now? "At first we said, there's one, that's cool," said Susanna, who has watched the number of 1476 matches grow, and grow, and grow. Some parents are now beginning to feel uneasy: is it really fair to impose upon a child scores of half siblings? How many is too many? Thirty-five? Forty-five? Fifty-five? Parents in DSR chat rooms are beginning to share concerns about the number of offspring produced by a single donor; and as Wendy Kramer observed in a posting, several donors eventually deleted their contact information from the registry, "because they could not handle more donor kids in their lives."

"I've evolved with it, but I think it would be overwhelming to somebody to come in and see that there are thirty half siblings," Susanna said. "All of a sudden they've got this huge family. For us, we're like, okay, how many more are out there? And on the other hand, when is it going to stop?"

"My Donor Was Fairfax Cryobank Blah-Blah-Blah"

"If you're single and you're going to dabble in reproductive technology, you've got to be prepared. It's more than I thought," says Hally Mahler, whose story illustrates how unpredictable the answer to that last question can be.

Just a few years ago Hally Mahler was an unpartnered lesbian, merrily single, living in the Dupont Circle neighborhood of Northwest Washington, D.C., traveling the world for an NGO-funded company that promotes reproductive health. Mahler had, she says, "different partners in lots of different ports," but no life partner. Arriving in her mid-thirties, she began attending a thinking-about-pregnancy support group for lesbians, and after two years decided to go ahead on her own. For a donor, Mahler turned to Fairfax Cryobank. She was looking for someone Jewish, a criterion that turned out to be harder to meet than she thought. There were only a dozen or so Jewish donors, she says, and some were "Jewish by their grandmothers—a little bit of a scam."

But among that handful was a man she felt very drawn to. "I saw this guy and I just knew he was the right one for me; he could have been my good friend. He was interested in public health, and traveled in Africa, and had hiked in the Himalayas, and wanted to join Doctors Without Borders." She bought some of his sperm, but had no luck. Getting more aggressive, her doctor combined insemination with fertility drugs: first Clomid, then the more powerful injectibles. Mahler grew discouraged; calling from Dulles Airport to get the results of a blood test before she left on a flight for the Ukraine, "I've never been so certain I wasn't pregnant."

She was.

With twins.

Mahler hadn't been planning on being the single mother of two babies. "As a single person, I had an image of what my baby would be like—a daughter, we'd be traveling the world, she'd be attached to my breast as we went off to Senegal; that's what I thought parenting was going to be, the two of us against the world. Having a third person in the picture changed for me, completely, what parenting was going to be. It wasn't this beautiful image, it was an image of being bogged down, not being able to go anywhere, not being able to afford anything."

"I got," she says now, "very depressed."

Nevertheless, as her pregnancy advanced—and the hormone surge subsided, and she felt, gradually, more upbeat—Mahler made a series of personal-life alterations. She realized that she would have to move out of her one-bedroom apartment. She hired a nanny, several months early. By the third trimester Mahler had become so unwieldy, carrying twins, that she needed the nanny herself. "I couldn't take care of myself, could barely get behind the wheel of a car, couldn't cook, couldn't do laundry. When you don't have a partner to root for you, you have to hire one." She contracted gestational diabetes, but other than that the pregnancy went relatively well. "It turned out I had an iron cervix," says Mahler, who kept the twins in utero until the day of their scheduled C-section. Her mother was in the delivery room with her, and outside, waiting, were her father and her best friend.

After the birth of a son and daughter, she lined up as much help as she could. Her parents stayed with her for three weeks, and then she had one friend per week for the first three months. After that, there was the nanny during the day, and then, in the evenings, it was her, alone, with the twins. "I was scared shitless, to tell you the truth," Mahler says. "It was terrible to be alone with them. Actually, if I have to be honest with you, I finally felt comfortable being alone with them for long periods of time when they were about one. As a single person with twins, you're constantly not meeting the needs of one of your children. I look at my friends who held their babies and had this intensive bonding, to be able to hold them and sleep with them, and I could never do that." As the children got older, she felt more competent: "I realized I am a much better mom of toddlers than infants." Even so, it was an enormous transition, going from being a self-sufficient professional to being a single mother of twins. "I did have times when it seemed too much," Mahler says. "I still feel that way from time to time. There is a difference between loving your kids to death—it's not that you don't want your kids, you just want freedom for a little while." And yet she coped: devising, among many other adjustments, three systems for handling travel. If she had to be gone for one night, the nanny would take the twins. Three or four nights, she would pay some friends to watch them. More than a week, they went to her parents in New York.

As she became more comfortable, Mahler became, she says, a "poster child for single people with twins." People started calling her saying they knew this other single woman pregnant with two babies, and would Mahler talk to her? It's not uncommon for a woman using sperm donation to get some help from fertility drugs, and when you get help from fertility drugs, it's not uncommon to conceive two babies. Some friends connected her with another about-to-be-single-mother-of-two who lived across the street from Mahler's new apartment. Mahler invited her over. "She was about five and a half months pregnant. We hung out and had a long talk, and bonded. We are both Jewish, both working in roughly the same field. And at some point I asked if it was important if her donor was Jewish, and she said, 'Yes, my donor was Fairfax blah-blah-blah.'"

Fairfax blah-blah-blah: the woman had revealed her donor number. Mahler was astonished. In sperm-donation support groups, one's donor number is, Mahler says, "a totally taboo topic." Women feel possessive about their donor, and they don't want to be competing with other women interested in buying his sperm.

But the really shocking thing was, the number the other women had just given was the same as Mahler's.

"I was just so shocked, that she said any number, let alone mine, that I could not respond," Mahler remembers. Standing there in Mahler's apartment, the two women realized that they were a family. Or something. What were they? Who knew? Clearly, Mahler's twins and the other woman's twins were half siblings. Other than that—other than that, nothing was clear. "She was very excited about it: even as she left, she said, 'Bye-bye, Auntie!' I was much more hesitant. When we talked again, I said to her, 'Look, this is really wonderful, but it's also a really serious thing.' If we are going to make the decision to tell our children that they have half siblings across the street, we are going to make the decision to be in each other's lives for the rest of their lives. Since then, we've exchanged e-mails, we see each other socially maybe once a month, and now she's more hesitant than I. It's very complicated." The four children could attend the same public school. "Imagine we don't tell them, and they're in the same class, and they find out later, and know that we didn't tell them."

Up to then, Mahler had thought the half sibling issue was one she could deal with when her children were older. But with this revelation, she said, "I became much more serious." She went on the DSR and found several matches—a heterosexual couple who didn't want much contact; a lesbian couple in New Jersey, with whom she corresponded. "The people in New Jersey had a bit of a scare; they were looking at pictures of my daughter, and their daughter walked into the room, and she said, 'That girl looks like me,' and they told her that a lot of people look alike. Their daughter's three, their kids know they are donor kids," but don't know about any half siblings.

Mahler knows of twelve children born from her donor. And those are just the ones listed on the DSR. "I'm fully expecting that this donor probably has sixty children," she says. Increasingly, she says, she feels sorry he is anonymous. "I thought it was the right thing; that giving sperm away wasn't the same as giving a child up for adoption, where there had been a human relationship." Now, she says, "I regret it, a little bit."

CHAPTER EIGHT

ART AND THE RIGHTS
OF THE CHILD

"I Do Think of Him as My Dad"

Eve Andrews, sixteen, noticed one day that her mom had left her computer on, so Eve sat down to check her e-mail. As she did, she saw that her mom's own e-mail account was open. Eve's mother happened to work at Eve's high school, and e-mailed a lot with teachers, and Eve thought it would be interesting to know what teachers at her school were messaging about. She scrolled through the list, looking for topic lines that suggested interesting gossip. A lot of her mom's e-mails seemed to be about "artificial insemination."

Well, that didn't seem totally weird. Eve lived in Texas, and her grandfather raised Charolais cattle, so Eve figured the e-mails had something to do with cattle breeding. "What is Mom doing with Grandpa's stuff?" Eve wondered, scanning the list, which contained, she now saw, hundreds of messages on the same topic. She clicked on one and was struck by the intensity of the discussion. What kind of chat group *was* this? Who *were* these people? "I thought, 'These people really *care* about their cows. These people are frigging crazy!' They're saying, 'Have you done this and have you done that,' and I'm like, 'Wow, people—it's okay!' And then I scrolled down to one where my mom had written one, and it started talking about, like, a daughter, and then I kept reading and it said, 'Yes, I haven't told my daughter yet, but she was born July 28, 1987.' And I was just like—what? That's my birthday."

"It was just kind of a lot for me to take in," said Eve, in an interview a year later.

Eve Andrews is one of perhaps a million children—that's my estimate; no one really knows how many—who were conceived by donor sperm and, like adoptive children a generation before them, grew up unaware of their genetic origins. At that point, of course, Eve didn't know there was anyone else in the world like her. Going on Yahoo, she searched "artificial insemination" and learned that something called human sperm banks exist. Clicking on links, Eve took in the truth. Her father, who had died when she was seven, wasn't her father. At least, he wasn't her biological father. This explained some things. It explained, she thought, why Eve's sister had sensed their father's death as they were driving to visit him in the hospital, while Eve herself had felt nothing. It explained why her sister was tall and model-thin, with long brown hair and olive skin, just like Eve's father—well, like the man she formerly thought of as her father—whereas Eve was shorter, with curly brown-blond hair and fairer coloring and a different physique entirely. It explained why her mom had been asking her, for months, to pick up a copy of *Glamour.* One of the e-mails mentioned an article the magazine had done on sperm donation, and her mom, Eve realized, had been try-ing to find it. "Mom has lied to me my whole life," Eve sat there thinking. She felt angry, and betrayed. But she also loved her mother, and—lying awake all night—she came around to the idea that her mother had done what she'd done because she loved her, too. The next morning she went into her mother's room, where her mother was drying her hair.

"Why have you been wanting *Glamour* magazine?" she asked.

"It's a great magazine," she remembers her mother saying over the noise of the blow-dryer.

"Mom," Eve said. "What's the real reason?"

When her mother didn't reply, she said, "Mom, I know. I know every-thing. I know Dad's not my biological father."

"I'm so glad you're not mad," her mother said.

"I was mad at first," Eve said, "But I'm not *that* mad. But when were you planning on telling me?"

"That's a great question," she recalls her mother answering. Eve's mother explained that she and her husband had meant to tell Eve when it seemed to them she was old enough to understand. But Eve's father had gotten sick when she was still so young, and then he had died, and after that, telling her the man whose death she was grieving wasn't her biological father seemed even harder. Then Eve's mother had gotten married again, to the man who had helped raise Eve ever since. So things got more complicated as life went along, and there were no rules, no guidance. Eve's mother had recently found the Donor Sibling Registry, and canvassed people on the message board, trying to figure out what was the best thing to do for her daughter. She had written the sperm bank to try to get information on the anonymous donor. Eve's mother also told Eve why they used a donor. Her doctor had thought Eve's father was infertile, and recommended donor sperm. His diagnosis would be proved wrong later, when they conceived Eve's younger sister naturally.

They had picked a donor from a bank the doctor recommended, and Eve's mother told Eve what little she knew about him. He lived in California, he was a magazine editor, he was blond-haired and blue-eyed, he had majored in journalism and minored in Islamic studies. Hearing this, Eve was incredulous. It seemed like an unusual choice. It seemed—not like her mother.

"I was like: 'Mom, what were you thinking? *Islamic studies?* That wouldn't have been my pick!' " Eve remembers saying. "She told me that so many of them were doctors and lawyers; she wanted somebody a little bit more spiritual."

And there were qualities of this donor that resonated with Eve. He was into journalism; Eve had always loved reading and writing, whereas her sister was a math-and-science whiz. Best of all, he was—presumably—alive. He was out there, somewhere, carrying around half the DNA that made her. *Her father wasn't dead.* This idea also struck Eve's younger sister, who, when Eve told her, "just started crying. She just took it much harder than I did. I just looked at Mom; what's going on? Mom was like, 'Sweetie, are you okay?' and she was like: 'You have a daddy and I don't.' "

You have a daddy and I don't. The sperm bank had told Eve's mother that when Eve turned eighteen, the bank would contact the donor on her behalf; later she could write the donor a letter and the bank would pass it along. There were no guarantees about what would happen. When Eve and I first spoke, she was still just seventeen, but already she was writing drafts and mulling what to say. She had been e-mailing on the DSR discussion group, asking for advice about how to couch her letter, and—as one of relatively few older donor offspring on the registry—also giving parents advice, when asked, on how to break the news to their own children. "I tell them definitely they need to tell them as soon as possible; the later they wait the harder it is," says Eve, who was trying to brace herself for disappointment, yet unable to contain her optimism as well as her desire to meet a man who would recognize her as his own. His offspring, his progeny, his *daughter.*

"I want to meet the other part that made me," she said. She also wanted to know why this man had donated sperm. What was driving him? Money? The Darwinian urge to perpetuate his genes? "I'm so much like him—I look nothing like anybody in my family. When I was little, I had blond hair, and I have blue eyes. And just as far as personality, I'm more like him. They said that he was very friendly, bubbly, very laid-back, stuff like that, very carefree. I just want to know: Why did he do this? What was he thinking? That's the main question I want answered. To me, that's big. That's just like—giving yourself away, knowing that you potentially have an offspring.

"I want a dad," she said. "I want to feel that connection.

"And," she added, "I do think of him as my dad."

The story of Eve Andrews raises a central and very much unanswered question. What are the rights of children conceived with the assistance of reproductive science? Do children—who after all are the point of all this, the point of everything everybody goes through—have the right to know the truth and the whole truth of their origins? Do children have a right to know the identity of their donor? Are sperm and egg donation analogous to adoption? Do these children, like adoptees, benefit from knowing the person who contributed half their genetic makeup, but who will not be involved in their care or upbringing? Moreover, should doctors rigorously screen patients, the way adoption agencies do, to make sure parents who purchase or otherwise obtain reproductive cells from other people are competent, solvent, stable, and mentally healthy? Should donors be able to direct to whom their donation will go? Should they be allowed to nix households they find objectionable, the way a birth mother can? And if clinics were to screen the parents of children conceived by donor gametes, should they screen all fertility patients, to make sure any child conceived with the help of medical professionals will grow up in a healthy, happy home? What exactly are the rights of the child? What happens when the rights of the child conflict with the rights of the parents?

Whose reproductive freedom is it?

And who is going to make these decisions?

"You need to be able to grapple with the reality that this is the *stuff of life*," Olivia Montuschi said briskly, sitting over tea in the capacious lobby of the Novotel in London. A gray-haired stylish woman with dangly triangular earrings and a steely, determined air, Montuschi, founder of a British organization, Donor Conception Network, is perfectly willing to be one of the deciders, as President George Bush might put it, and her decision—her conviction—is that every donor offspring has a right to know not only the truth, but the donor's identity. "I have certainly heard egg donors talk about giving away what they don't need each month, saying, I don't need [this oocyte], why shouldn't somebody else have it— that it's just like giving blood. Well, actually, it isn't. It's different."

Sitting next to Montuschi was Eric Blyth, a social worker and adoptive father who had traveled to the city to share his views on children's well-being and the need for disclosure. The two are part of a small but extremely influential group of people who have managed to change the way collaborative reproduction is handled in a number of countries, including Britain. Montuschi, Blyth, and others believe that the Eve Andrewses of the world are entitled to know the identity of the man or woman who supplied half the genetic blueprint that built them. So effective has their advocacy been that anonymous donorship has been abruptly abolished in Britain, where donor insemination has been common for more than half a century and where secrecy was once the norm. The change took place in the spring of 2005; now, all donors in the U.K. must agree to have their identity revealed to offspring who have reached the age of eighteen, just like the "yes" donors at The Sperm Bank of California. In Britain there is a centralized registry where any adult conceived after April 2005 can go to find out (1) whether he or she is a product of gamete donation and (2) who that donor is. It is a policy that has also been enacted in Sweden, Austria, Switzerland, the Netherlands, and parts of Australia, where all children may learn identity of their donor upon reaching maturity.

The argument is compelling. For years now, adoptees have been struggling to gain the right to trace their birth parents, an effort so heartfelt, so anguished, and so morally persuasive that it has become customary in domestic American adoptions for birth parents and adoptive parents to meet, and for the child to know the birth mother, and sometimes the birth family. This is not always easy for adoptive parents to deal with; some parents so dread interacting with the birth mother, I was told by one social worker who does adoption screening, that they go abroad, seeking children from foreign orphanages as a way of avoiding the birth mother and her problematic presence. Ironically, once they became secure in their role and believe in their own entitlement to parent their child, some go to the ends of the earth, literally, helping their adoptive children trace their roots. In 1989, the United Nations declared that every child has "as far as possible the right to know . . . his or her parents," as well as the right not to be "deprived of some or all of the elements of his or her identity." Citing human rights laws as well as these assertions—Articles 7 and 8 of the United Nations Convention on the Rights of the Child—in 2000 a woman named Joanna Rose filed suit in Britain's High Court, arguing with a co-plaintiff that she was indeed like an adoptee, and that her human rights had been violated by the fact that she was conceived by anonymous donor sperm. The judge who heard her case was sympathetic; while there was no redress for Rose, who could not be unconceived, or reconceived with a known donor, the human-rights argument has influenced British and European thinking. In Britain, "welfare of the child" increasingly is understood to mean the child's need to know his or her true lineage—to know who his or her genetic parent was, or is.

The movement is being driven by other donor-conceived offspring as well, who like adoptees before them have begun to speak out, in public forums and in Internet groups with names like Donor Misconception and Tangled Webs. There are blogs by donor-conceived children, some of whom describe frustration and confusion, as well as anger at those who dismiss their concerns by suggesting they stop whining and be glad to be alive. Among offspring there is no hard and fast viewpoint; a small portion think donor gametes should be abolished entirely—it's just too hard knowing that your mother and father never loved each other, never physically came together, never even saw each other—while others support reform in the direction of open disclosure. There are also offspring like Eve Andrews who feel that being conceived by sperm donation makes them interesting and special; like plenty of members of the technology-friendly iPod generation, Andrews doesn't resent her means of conception. "I kind of view it as: I'm different. It's exciting, mysterious." Problem is, for parents there is no way to know in advance what camp their child will fall into. On artparenting.org, a website maintained by Harvard Medical School Center for Mental Health and Media, one donor offspring says, "It's not that big of a deal." Another, posting on DSR, feels that being "wanted" is, unfortunately, not enough.

Olivia Montuschi is herself a sperm-donation mother; when she and her husband sought help from a London clinic in the 1980s, the doctor took a photo of her husband and promised to physically match him. They were given no choice of donor, no help explaining anything to their children. That's because they weren't expected to explain. "Our assumption was of course we would tell a child; why would we not?" Montuschi remembers. "Whereas the clinic's assumption was, of course you won't tell a child; why would you?" It was expected that parents would lie; the point of matching the donor to the intended father was to make the lie plausible. And the point of lying was—well, there were many points, but the main point was to conceal why a clinic had been required in the first place.

For years, the shame of male infertility was one chief reason why children were misled about their genetic origins. In a 1995 study, Susan Golombok and colleagues found that of forty-five donor-sperm families queried, not one set of parents planned to tell their children the truth about their parentage. Only 40 percent of the general public was in favor of children being told. In contrast, out of fifty-five families with an adopted child, all but one had told their child. (Interestingly, studies have shown that in adoptive families, infertility is often attributed to the mother, even when it lies with the man.)

In explaining why they did not plan to tell, many mothers by sperm donation said that they would be glad to do so, but the father wouldn't have it. "He really feels he doesn't want her to know, 'cause [what] if she should change her feelings about him," said one woman interviewed. It's testament to the power of children to make you fall in love with them: this father clearly loved the child despite the fact that she was not biologically related to him, but feared the lack of biological relationship would change the way his daughter felt about him. "He is very adamant that she should never be told," another mother said. Another thought her son would be "shattered" by the revelation that he was not genetically related to the father he adored. Still another said she didn't want to tell her daughter, because the donor's anonymity would make it impossible for to trace him and the result could only be frustration. Others didn't know how to explain reproduction to young children, while parents of older children felt they had already waited too long. And let no one underestimate the power of grandparental disapproval: one mother said her mother-in-law would never accept an adoptive baby or a child that was the product of "hifalutin" treatment. "Never expect me to think of them as my grandchildren," the family matriarch had warned.

Montuschi and her husband were the exception. They told their children the truth: starting young, starting basic, starting small. In 1993, Montuschi formed the Donor Conception Network, an organization that helps parents who want to disclose. She became convinced of the need after she appeared on a television show about sperm donation families and was flooded with letters from parents struggling in private with the same issue. "We uncovered a whole group of people who did want to be open," Montuschi says. Meanwhile, other parents as well as adult donor-conceived offspring were speaking out, among them Barry Stevens, a Canadian filmmaker who made a film called *Offspring;* Wendy Kramer, who with her son Ryan has become a passionate voice on behalf of openness and disclosure, and who with the Donor Sibling Registry has done a great deal to encourage civil debate; and an American, Bill Cordray, conceived by donor sperm in the 1940s.

Cordray, an architect, is a thoughtful man who grew up in a household in Utah, assuming the man who raised him was his father, yet wondering why he felt so different from a man he so longed to resemble. "He was strong and athletic and striking, like Anthony Quinn," says Cordray, who did not share many of his father's attributes, a disparity so apparent that "I remember people making comments." As he grew older, "it started to come to me that there were parts of myself that were foreign not only to my dad, but to my mother as well." He sensed he was smarter than his parents (studies have shown donor offspring report feeling mentally very competent, and given the number of med-student donors, no doubt they are) and that he had interests in art and music nobody in his family shared. He became convinced his mother had been unfaithful, something that made him feel estranged from her as well. He also sensed that his mother seemed to enjoy disproportionate authority in the household, a disparity, he says, that is "really common in DI families: the mother has all the power, all the connection. The husband doesn't feel entitled to be the father." Cordray took a long time to come to terms with himself. Then, after his father died, his mother told him the truth.

"My first feeling was disbelief and shock, but primarily, I was elated—it was liberating, totally wonderful to finally have something make total sense," says Cordray, who has embarked on a so-far-unsuccessful search for his donor father. He believes now that disclosure should be mandated, and that families using gamete donation should be screened for fitness. "The main focus should be on the person created. What are his rights:

what does he deserve, what is due to him. And what is due to him is an honest life, and growing up in a family that respects his dignity and autonomy."

Society, he argues, has "an obligation to set up standards."

Barry Stevens, a tall man with glasses and a self-deprecating manner ("Sorry; this is the first PowerPoint presentation I've ever made"), delivered his own story at the 2005 annual meeting of American and Canadian fertility doctors. Stevens was born in 1952 in England, conceived by artificial insemination from, as he put it, an anonymous "provider" at a clinic run by a doctor named Mary Barton. "I don't like the word 'donor' very much," he said, "since they are usually selling a commodity." His sister had been conceived the same way. His father died in an accident when he was in his teens. When he was eighteen, he said, "our mother felt we should know the truth. My father hadn't wanted us to know, but my mother had always wanted to tell us. She had been told by her physician to always, always lie.

"I feel very grateful to my mother for being courageous enough to tell us the truth," Stevens continued, echoing Cordray when he said that "there were lots of things about my father that had seemed unclear, that began to make a lot of sense." He and his sister started on a quest to learn the identity of their donor and to find any half siblings; some of the clinic's donors had been prolific, among them, it's believed, some Jewish donors who were enacting a personal defiance to the ravages of the Holocaust. One half sibling turned out to be a lawyer living in England; they have since become close. His genetic relatives, Stevens reports, now number in the dozens, and he has met many other donor offspring. "Almost all of us," Stevens said, "wish to know the identity of the man or woman who was the origin of half our bodies and a large part of ourselves, and as many or more are very keen to know some of our half siblings. We believe, and I support what the others are saying, that the system of anonymous gamete donation should be replaced with one where identities are available to offspring upon their adulthood. I wouldn't say that we have a right to the information. I would rather phrase it: Nobody has the right to withhold essential and key information about us *from us*.

"That switch for me is important," he added. "If there's some information a government or doctor is holding that is key to you and he won't give it to you, then I don't think that it is right."

Stevens pointed out the importance of knowing one's medical background: how helpful it is for doctors to know what a patient's parents died of or suffered from. "I know of a couple of people who really worried they would inherit their father's medical problems," he said, adding that he also knows of donors, dying of a disease in their forties or fifties, who were unable to contact offspring to warn them. He even knew of one offspring who deliberately did not have children of his own, mistakenly believing that he carried the disease his father had died of.

Stevens also talked movingly about the desire to see your face, your features, your self, mirrored in the face and features of another human being. He talked about the deep-seated human need for recognition, a word that means, literally, to "know again." "Not seeing yourself fully reflected in your family around you—it's bewildering and it's unsatisfactory," he said. "Our culture is full of stories of people who go out in search of their father or their uncle or find a long-lost brother, and these are key moments in the stories we tell ourselves. I think it's also pretty clear in biology that organisms that are simple are capable of recognizing each other, and it's a very strong pull."

He argued, further, that the "climate of secrecy" is destructive to families. This is another philosophical underpinning to the modern adoptive movement: the idea that secrets are lethal, that they create an atmosphere of stress and tension that eats away at healthy functioning. "There may be families where they can lie about the origins and be intact, but mine wasn't" one of those, Stevens concluded. "It created distance where there should have been love, anxiety where there should have been spontaneous joy."

It was a persuasive delivery, and one that many of those listening agreed with. Stevens's audience in the fall of 2005 consisted of counselors and therapists who work for North American fertility clinics. By then, there had been a major shift in industry thinking; in 2004, the American Society for Reproductive Medicine recommended that parents disclose the truth to their donor-conceived children. "The debate is over," said one psychologist.

In truth, the debate is merely evolving: now, it centers on what happens when donor anonymity is abolished. Also at the 2005 conference was Pia Broderick, a psychology professor at Murdoch University in Perth, Western Australia, who demonstrated how laws banning anonymous donation have decimated the ranks of people willing to be sperm and egg donors. Australia, Broderick told the audience, is highly regulated when it comes to reproductive technology. In most states—there are eight states in Australia—IVF is paid for by public health. Three percent of Australian births are accounted for by ART, a rate that grows by 8 percent every year. No payment is permitted to sperm and egg donors—this is also the case in Britain and Canada—so the pool of donors is already limited to people willing to donate for free. In Victoria, and in Western Australia, the state where Broderick lives and practices, there are now laws saying that upon reaching maturity, children can contact a registry and learn the donor's identity.

As a result, Broderick said, "we have very few donors." In all of Western Australia, a state with a population of 1.4 million, there are thirty-five available sperm donors. Egg donors are almost impossible to come by. A fertility clinic in Victoria has thirteen donors. The number of gamete-donation children born in Western Australia fell dramatically after the law was enacted. "It's impossible to get donors," Broderick said. "The supply has completely dried up." In Victoria, one clinic called upon the politicians who enacted the legislation to donate sperm. In 2004, some clinics offered Canadian college students an all-expenses-paid trip to Australia, in return for a small donation. After Broderick delivered her speech, a woman stood to say that her nephew was one of the Canadian students who accepted the offer, and that he greatly enjoyed what turned out to be a fantastic tour of Australia, its clubs and nightlife. "We never had to put our hands in our pockets," she quoted him saying. "We only had to put them down our pants."

Broderick's view is that laws abolishing anonymity are too drastic and represent an unwarranted intrusion by the state into the private domain of parental decision-making. "Legislation doesn't change attitudes and it doesn't change behaviors," she said, arguing that parents "know more about what is best for their children than legislators and counselors." There are many who agree; in this arena of reproductive science, "progressive" is a hard concept to pin down. Advocates of telling children the truth, the whole truth, and nothing but the truth consider Victoria's law among the most progressive in the world. And yet there are those in progressive scientific circles who view these laws with regret and skepticism. They point to the somewhat surprising fact that Susan Golombok's studies of sperm and egg donation found that families who did not disclose the truth to their sperm-donation children did *not* seem to be suffering the stress that is assumed to result from untold family secrets. Instead, like other IVF families, they seemed to be doing very well. In their 1995 report, Golombok's group urged that "the experience of adoption is unlikely to be a useful model for DI, because there are important differences between parental experiences of these two ways of creating a family." The study continued, "It is insufficient to consider only the welfare of the child, which cannot, in any case, be isolated from that of the parents. Thus the primary concern should be for the welfare of the family as a whole."

"We feel that a bit of choice is being taken away from parents," agreed Juliet Tizzard, who at the time of our 2004 interview was director of the Progress Educational Trust, a British group set up to facilitate reproductive science, and the editor in chief of its bioethics newsletter, *Bionews.* Tizzard, who has since taken a staff position with the HFEA, Britain's regulatory body for IVF medicine, described advocates of mandated identity disclosure as "a collection of angry people." Adoption theory is part of what's fueling their movement, she said, but so, ironically, is science, in particular the obsession with the genome, and the idea that every trait in life emanates from our genetic makeup. "The idea has become more popular that genetics is important to your sense of self; that genes form part of your identity. But I'm a bit unconvinced about that. Your genetic relationship to your parents and your social relationship are so hard to separate."

British fertility doctors deplore the change. Ian Craft, director of the London Gynaecological and Fertility Centre and one of the best-known IVF doctors in England—also one of the best paid—was present at the 2005 conference to speak out against mandated disclosure. Craft has been vocal in protesting the new law, and like a number of British doctors who make an excellent living from patients seeking egg donation, he has established ties with a foreign clinic—in his case one in Cyprus—where patients can still go to obtain oocytes from anonymous donors. "Fertility tourism," in which patients visit Cyprus, Ukraine, and Romania to obtain eggs "donated" by women in those cash-poor countries, usually under unregulated medical conditions, is another offshoot of mandatory disclosure laws. Meanwhile, patients who can't afford to travel have no recourse, really. As morally right as it does seem, one ironic upshot of mandated disclosure is that fewer children may be born, thanks to laws directing that these children, were they to be born, would be entitled to know their donor, who, since he or she doesn't want to be known, does not donate, and therefore the child never exists to begin with. When I broached this with Montuschi—the fact that there may be children who will never be born because of laws enacted in their favor—she replied, crisply, "So be it. You can't think about non-existence. I'm not a philosopher, I'm sure."

"I Have This Vision That . . . They'll Be Chasing Me Down the Street"

"The person is thinking of this as a short-term donation," Gail Taylor argues. We are sitting in the Wilshire Boulevard office suite that houses Growing Generations and its partner egg-donation agency, Fertility Futures, talking about children and what they may or may not need to know, and the way this plays out in the United States, where the over-whelming majority of gamete donation is anonymous. Like many Ameri-can brokers, Taylor, whose egg-donation agency serves both gay and heterosexual would-be parents, adamantly does *not* think children need to know the identity of their donors. She does believe they should be told the truth about their conception. But their rights, she believes, do not extend to learning that donor's identity. She, like Pia Broderick, points to the impact on supply. Most egg donors, she argues, would not donate if they had to be identified. Not her egg donors, anyway. They're so young: the cut-off age at Fertility Futures is twenty-six. They're not looking at this as a long-term relationship. They're looking at it as a way to pay their tuition. "The compensation is a significant driving factor," said Taylor.

"If there weren't anonymity, there would be no donation," agreed Sanford Rosenberg, the Richmond, Virginia, fertility doctor who, like Taylor, has to recruit heavily in order to fill the growing demands for oocytes; while Taylor has expanded her recruiting to the academic cen-ters of the East Coast, Rosenberg helped establish an international egg-shipping outfit, Global ART, whereby eggs for American women are obtained from anonymous Eastern European donors.

And anonymity, Taylor believes, is fine. Children, she believes, do not need to track their donors down. What they need is a photograph. Taylor knows about offspring who search for donors; she has read their narra-tives. "The activity they're taking to pursue the identity isn't necessarily to have a relationship; it's to understand that missing link genetically. The kids are at this age where they're starting to wonder who I am, and I think that if the information—specifically a photograph—is available, I don't think there'd be such a quest." Accordingly, Fertility Futures provides a picture of the egg donor. In addition, Taylor says, "We're the only agency in the world that's shooting video." On a laptop, she showed me one of the egg-donor videos. "My professional aspirations are to become a business owner, a proprietor," a beautiful dark-haired woman was saying. "I'd like to save enough money to help other people invest in their ideas and vision. Basically, I'm an adventure capitalist. You get the most reward back from helping other people."

An image like that, Taylor contends, should be enough for any child. Unlike Montuschi, she argues, sperm and egg donation *are* like giving blood. As I listened, I was again struck by the difficulty of reconciling any number of what might be called progressive, in the sense of forward thinking, ideologies. In this instance, the progressive gay and lesbian community—at least, that part of it represented by Growing Generations and its staff—is rather magnificently at odds with the UN's progressive Convention on the Rights of the Child. The former minimizes the importance of biological connection; the latter places biological connection at the heart of parent-child relations. Is it a human right for a gay man or lesbian to have a child with the help of an anonymous donor; or is it a human right for a child to know who that donor was? Growing Generations psychologist Kim Bergman emphasizes that what matters to children is not their genetic heritage; what matters is the narrative the parents create, "the stories you tell them."

In the United States, this debate has not been settled; it has only just begun. Forces like the DSR and The Sperm Bank of California are proving to be agents of gradual change; as parents and children pound at the doors of clinics and share their views in chat rooms, they have persuaded some sperm banks to include a small number of donors who are willing to be identified. And if you search egg-donor websites you can probably find a donor willing to quietly give you her contact information. But what egg brokerages, sperm banks, and fertility clinics themselves will not do is provide ongoing support to parents still trying to figure this out, parents who want to do the right thing by their children.

This is a huge, huge difference between the world of assisted reproduction, which is relentlessly profit-making, and the world of adoption, which is generally driven by organizations with goals other than money. Adoptive parents are supported for their entire lives. IVF patients are cast adrift. As the psychotherapist Jean Benward pointed out in a talk at the 2005 ASRM conference, in adoption the client is "a child who needs a home, a family." Everything that's being done is being done in the service of that child and its needs. "All adoption plans have to be reviewed and formally accepted by a judge; all parents have to undergo home study, fingerprinting, background checks." Adoption is legislated by states, regulated, and socially accepted; moreover, there are resources available to the parents throughout raising the child. In contrast, sperm and egg donation is conducted largely out of sight, with little scrutiny and no provision by anybody for long-term support. At fertility clinics, parents using egg or sperm donation are given, at most, an hour of counseling.

How do donors react to the loss of anonymity? The answer is unclear. Eric Blyth argues that half of all donors would donate even if their identity could be known. But after the law was enacted in Britain, banning anonymity, a poll found that 70 percent of fertility clinics had little or no access to donor sperm. One Scottish clinic was obliged to close. As of April 2006, there was reported to be one active donor in all of Scotland. Olivia Montuschi retorted that clinics had "panicked donors" and scared them off.

On the Donor Sibling Registry, there are postings from a few sperm donors seeking offspring. These, however, represent a minuscule proportion of all donors. Clearly, the majority of donors are not seeking their offspring. But every now and then, a donor relents: Eve Andrews has recently had her tenacity rewarded. Once she was given the go-ahead by the sperm bank, Eve wrote a letter humbly assuring her donor that she was "emotionally and financially secure" and only wanted to know who he was. The bank forwarded the letter to the donor, who agreed to communicate with her. For Eve, their first contact was a revelation: Here was the man who had given her what her mother called her "happy gene!"

"The more we e-mailed and the more I read, the more I figured out where I get my personality," said Eve. "We are so much alike it is uncanny." They spoke on the phone and instantly connected. "My biological father exceeded my expectations," rhapsodized Eve. The donor's wife, she said, "wasn't too thrilled," and his own children did not yet know about her.

While researching for this book, I interviewed a half-dozen egg donors, all of whom were willing to be known. They considered it reasonable that a child had a right to meet them. One of these was a young California woman who preferred not to be named. Contrary to the often negative rap on egg donors and their motivations, she wasn't some six-foot-tall Ivy League supermodel off-loading reproductive cells as a way to pay her tuition and credit card bills. Instead, she had given her $5,000 payment to her parents to thank them for all they had done for her. At twenty, she had finished an associate degree at a fashion institute, and was holding down two jobs, one in public relations and another as a saleswoman at a clothing store. We met at a Coffee Bean on the Santa Monica promenade; as we were going through line to get some coffee, I urged her to get something to eat, and she asked if it was okay if she got a muffin.

This donor had never met the recipients of her egg donation—had never been thanked, even—but she had heard that a twin birth resulted, and felt so gratified that she got teary. She said she wouldn't mind if the children wanted to contact her. "I wouldn't deny the right to that child," she said.

"We're all just one big happy family," agreed Amy Housler, a serial egg donor who has donated eggs six times. Housler is a mother herself; after a hysterectomy made it impossible for her to bear more children, she felt she had excess reproductive capacity—"my family wasn't complete"—and decided to donate that capacity to others. If the families are willing to meet her, she is glad to meet them, though she has no desire to co-opt their children and doesn't think of the children as hers. Some mothers don't want to meet her. For one donation, she communicated exclusively with the husband. But she has close ties to a single mother—"we get together every six months with the kids, the boys will grow up knowing they are related"—and to a gay couple living in Reno.

"I would have no problem doing this non-anonymously," said another donor, a paralegal in Washington who has donated at three D.C.-area clinics, and enjoys the idea of helping the depressed-looking couples she has seen in waiting rooms. "If anybody ever said, 'I'd like to sit down and meet her and talk to her,' I wouldn't mind," she said. Like many egg donors, this woman happens to be, as she puts it, "ridiculously" fertile. But she doesn't want children herself: "I don't have the patience for them." Her fertility is one reason why, in her younger years, she had three unwanted pregnancies, all of which she terminated. In an admission psychologists say is common, this donor allowed that she donated eggs, in part, to make up for the abortions. After her third donation, she felt, "That's a kind of wrong that's been righted. Regardless of how silly that may be. It wasn't like it was ever a devastating experience, but you feel like it's been balanced out."

"It just seems like a very, very easy thing to do to help somebody," added this donor. She does think about the children. Every now and then, she said, a vision rises up in her mind in which the children take the form of the flat wooden dolls that people put in their yards in rural Maryland, where she grew up. "Every now and then I have these horrible visions: I'm like fifty-three, and the children are those faceless dolls they sell in the street down near my parents, and there are a horde of children knocking on the door, saying, 'We found you!' I have this vision that they'll be chasing me down the street. It's just something I laugh about. If they ever did lift anonymity, and all these kids came to find me, I think it would be interesting."

"How Can You Not Just Love Them?"

The thing is, even if donors do agreed to be identified, it doesn't do much good if parents don't tell their children the truth in the first place. As one 2005 study dryly put it: "Legislation imposing identity release donors has little impact as long as the parents concerned do not intend to inform their child about their conception method." And even now, many parents don't. Even in Sweden—where anonymity is banned—half of parents using donor sperm never disclose the fact to their children. And it's not just male infertility that's driving this: recent studies have found that mothers using donated eggs are often equally reluctant to disclose. A 2004 study by researchers Dorothy Greenfeld and Susan Klock found that 40 percent of egg-donor mothers did not plan to tell their children or were unsure what to do. Of those who did plan to tell, only 11 percent had done so. Clare Murray and Susan Golombok did a similar study: in a 2003 paper, they found that just 29 percent of egg-donor couples intended to disclose, and that none had done so. More than half had definitely decided not to tell their children that their genes came from another woman.

What's fascinating are the reasons women gave for not disclosing, reasons that went far beyond embarrassment or shame. In many ways, women are like men: they fear the child will love them less, or will regard the donor as the "real" parent, if told the truth. "I was so intimidated; I didn't want her face taking up my head space," one egg-donor mother told me, recounting a dream in which her anonymous egg donor came to her door, greeted her husband, said, "Oh, look, there's the child," and then left with her husband and the child. Another egg-donation mom, writing in *Elle*, talked about how she fantasized about never having to tell her daughter that they weren't genetically related. It is a hard, harrowing prospect, and there are anxieties, it appears, that are uniquely female. Clare Murray told me that when she and Golombok were beginning their study on egg-donor mothers, they hypothesized that being pregnant would make it easier for a woman to disclose the truth; that prenatal bonding and childbirth itself—hard, physical, and intimate—would make her feel more secure in motherhood. They found the opposite: what pregnancy did, often, was convince the woman that this *was* her biological child.

"The experience of pregnancy and the birth of their child helped them 'forget' that there had ever been a donor," the study noted.

"I don't think of it very often," one mother is quoted as saying. "It's funny, isn't it? I think it's because I actually carried him. I think if I had adopted him, it would be different."

The study also noted the power of domestic competition. Mothers feared their lack of genetic connection would be used against them by other women in the family; they feared it would give female in-laws emotional ammunition. "I didn't want my parents to feel that the baby wasn't sort of theirs," one egg-donor mother said. "I've got a very competitive mother-in-law and, you know, I did feel at the time that there would have been some nastiness there." Still others wanted to protect the child from rejection by family members in general: "You never know how somebody's going to react to it," one mother said.

"We're waiting until my grandmother dies," I was told by one egg-donation mother, who had not yet disclosed to her children. "She has these old sensibilities."

Stigma is a huge, huge barrier to honest disclosure. Adoptive parents face it, but it seems fair to say people conceiving with new technologies face it more acutely, particularly since the portrayal of fertility patients tends to be negative: all these "desperate" women willing to do "anything" to have that baby they "selfishly" covet. In his book *Stigma,* the sociologist Erving Goffman points out that "because of the great rewards in being considered normal, almost all persons who are in a position to pass will do so." Parents who fail to disclose are, as Jean Benward puts it, "reacting to stigma in a predictable way." They are trying to pass. They are trying to seem—to be—normal. That's what families want to be: they want to be normal. And there are so many pitfalls, so many barriers: as the medical anthropologist Gay Becker notes, another harrowing experience is that all-too-familiar situation in which somebody studies a child and then scrutinizes the parents for resemblance; eye color, dimples, cheekbones, hair. Becker, who calls this "resemblance talk," points out that while it may seem harmless, even inane, it serves ancient social purposes. It provides evidence of kinship, not only between child and parents, but within the family. It affirms the child's connection not only to parents but to grandparents; proves the mother was faithful; serves to "reinforce the assumed natural order of things." Regardless of whether they have disclosed to their child, egg-donor moms have to listen to this all day long. "I am Jewish, and work in the community," one egg-donation mother told Becker, "and people always say to me: he doesn't look Jewish."

Becker found that most parents were troubled by these conversations. They worried that the child would be upset, felt even more uncertain about whether to disclose the truth, and were reminded, over and over, of their own infertility and loss. And for some parents, the lack of resemblance is indeed a loss, one that women experience as acutely as men do.

"Women who used a donor egg expressed a great deal of loss, just as men did who used donor insemination," Becker told listeners at the 2005 ASRM conference, presenting her paper on resemblance talk. "I frankly expected to find a greater sense of loss among men who used DI, but that didn't emerge." One woman said, of her donor-egg daughter, "She's beautiful, but there is also tremendous sadness, just this fact that I couldn't see myself in her. That was very powerful, and very hard to get over." Another mother said, "I'll always feel a sense of loss, that I'm not going to see my mother's smile in someone else."

Not all parents felt this way, however. In one of the most moving comments in Becker's presentation, a sperm-donation father said, of his children: "I couldn't imagine loving them any more if I had been their biological father. I have never felt any distance as a result of not being genetically connected. How can you not just love them?"

"He's still my son," said an egg-donor mother. "That doesn't change."

"This is my child," said another mother. "It's like a gift, to have this other combination of genes as part of my family: an incredible miracle."

One thing that studies have consistently shown is that even if donor-gamete parents don't tell the child the truth, they often tell other people, setting up a situation—familiar from the bad old days of adoption—in which outsiders know a secret the child is not privy to. Inevitably, parents regret telling others. Studies also show that all parents crave better advice on how to deal with revelations made necessary by medical science. "I keep trying to prepare myself," said one Rhode Island mother, a self-made businesswoman looking at her gorgeous egg-donor twin daughters, who were sitting on the counter as she baked cookies for me to deliver to David Keefe, the doctor responsible for their conception. "What am I going to say, and when am I going to say it?"

"Do you remember how Daddy and I talked about how hard we had to work to get you, and how much we wanted you?" said one egg-donor mom, sitting in her backyard patio in the Northern Virginia suburbs, trying to broach the topic with her six-year-old son, who kept drawing near to listen and then skittering away. "I just don't know what to say to him," said this mother, who had tried to make a beginning by telling him every night how hard she and her husband had worked to have him.

"I guess I'd like someone else to tell me what to say," an egg-donor mother told Robert Nachtigall. "I'd like to talk to people who have actually gone through this experience with their children and find out what worked and what didn't work," one father told him.

"We're just starved for information," one egg-donor mother told him. "What do people with older ovum-donation children say? How did they react?"

"I'm Glad It's Daddy's Sperm That's Dead, and Not Daddy"

"The majority of youths felt comfortable with their origins and planned to obtain their donor's identity," notes a report conducted by researchers working with The Sperm Bank of California, the tiny but redoubtable nonprofit which, unlike lucrative private clinics and brokerages, has worked to track the outcome of the social experiment it set in motion. With the turn of the millennium, TSBC's first generation of "identity-release" children were about to qualify to meet their biodads, in what was surely the world's first planned release of sperm-donor identities. TSBC found that most of the children were looking forward to meeting their donors. The 2004 study, published in *Human Reproduction,* concluded that most children accepted their means of conception: "They felt very loved by their parents and very wanted, and also (positively) unique," the report noted. About 25 percent referred to their donor as "the donor." Roughly the same number said "biological/birth father," and another 25 percent preferred "father/daddy." One child referred to the donor as "him" or "that guy."

The study found that children from single-mother families tended to feel more positive about their donor than those from lesbian-headed households, perhaps because the nonbiological lesbian partner tended to have cooler feelings toward the donor and the threat he represented. The most common feeling about the donor was curiosity. They wanted to know whether they looked like him, and whether he had changed since he filled out his papers. Half said they felt appreciative of the donor. About one-third were anxious or concerned about meeting him; one wondered "if he is a good person or not." Another one-third felt excited. One felt angry and upset at the donor; another was "resentful that I haven't known him and that I don't know very much about him."

No child expected the donor to pay his or her college tuition.

Eighty-two percent wanted to know what he was like as a person; also high on the list were what he looked like, and whether he had a family.

One youth wanted to know "whether he ever thought of them."

The number one thing offspring wanted, in addition to the donor's identity, was a photo. Several children simply wrote, "I want as much information as possible."

Many thought meeting the donor would "increase their sense of identity."

Parents tended to underestimate the likelihood that their child would want a relationship with their donor. Just 47 percent of single parents thought their child would, when in fact 82 percent of their children said they would like to have a relationship with him. Similarly, 40 percent of parents in heterosexual couples thought their children would want a relationship with the donor, whereas 67 percent of the children did. Most described the ideal donor as a "friend," an unclelike figure "where we do some stuff but not see him all the time." Another said, "Just someone to say, 'hey, that's my dad.'" Seventeen percent replied, reasonably, that the relationship would "depend on what [the donor] was like." The children felt that being told early had made a difference to their well-being. At the same time, TSBC researchers emphasized that this was not an easy situation and that the "families will need additional support" as their children matured.

No heterosexual father, the report noted, looked forward to his child's meeting with the sperm donor.

When I met with Alice Ruby, current TSBC director, in the late winter of 2005, the first wave of identity-release-donor children had recently come to maturity. Of eighty young adults who qualified, only fifteen had thus far come forward to meet their donor. "We kind of expected that," Ruby said. "When we did the research, some of the kids said, 'We don't think we're going to do this when we are eighteen. We're going to be seniors in high school, applying for college. This is going to be a busy, complicated time for us, and we want to have enough time to do it right, emotionally.'" She pointed out that many adoptees become interested in their origins during times of transition, such as their marriage or the birth of a child. "People change and their needs change," Ruby said. "There are some kids for whom it's interesting, but it's not a burning issue. It may be different for different kids at different times." So far, she said, the outcomes had been very positive. Some children had formed relationships with their donors, while others had taken the information and moved on with their lives. One donor had turned out to be emotionally needy, and the offspring had run pretty fast in the other direction.

At the 2005 annual meeting, Jean Benward spoke to counselors about the best way to talk to children. "There is often a seeming lack of interest," she pointed out. "There is a subterranean processing that's going on. Often these questions—'What's a donor? Where did you get her eggs?'—pop out unexpectedly. You are never prepared. They do not submit the question beforehand."

But another thing she has found is that children are remarkably able to empathize with their parents' hardship. Reading comments made by children, she quoted one son saying, to his mother: "I'm so sorry you had to go through this." Another boy said, "Daddy, if you haven't got any sperm, I can give you some of mine." A girl said, "I'm really glad it's daddy's sperm that's dead, and not daddy."

One little girl skipped around the room saying, "I'm a miracle!" and added, "I'm sure glad you and Dad live in modern times, so you could have me!"

But as children get older, they sometimes use the lack of genetic relationship as ammunition in the struggle for autonomy. "I am half-adopted," said one egg-donor daughter factually, as her mother winced. "Mom, technically, you aren't my mother," another girl said. This last remark is of course the comment every egg-donor mother dreads hearing. According to Benward, the child was testing to see if she could speak this semiforbidden thought, and when her mother didn't collapse or get angry, she knew it was okay. Still another girl said, of the egg donor, "It's almost like she is our ex-mother."

And one child said, "I don't think I'm strange at all. I'm just like everybody else. I'm different."

"These relationships involving conception are hugely complex: who is a donor, what do you call them, where do you place them, how to think about it—no one has gotten their mind around this," Benward told colleagues at the 2005 conference. "We don't know whether being born into the family will change the offspring experience from that of the adoptee. People assume it will change the experience, but we don't know exactly how that will work for the offspring. It may work for some, and not for others. We don't know the extent to which parental intent mitigates against genetic continuity. Certainly, parental intent has always been a big part of the stories adoptive parents tell their children: 'We love you, you are so wanted.' That didn't always fly with adoptees. We don't know about that. We don't know about the effect of having large numbers of siblings. We don't know if we have a generation of people coming up who are going to be in crisis because of unintended, unplanned disclosure. We don't know."

She did say, though, that parents should *not* regard a child's interest in meeting the donor as a sign that the parent has failed. "A good relationship with your parent can co-exist with the offspring's interest in the donor."

One particularly difficult situation is when the child knows the donor, or knows of the donor, but is denied access by a parent who does feel threatened. One Saturday in 2005, I went to a conference convened by Family Pride at a Silver Spring, Maryland, middle school, which included a panel of older children of gay and lesbian parents. In the parking lot were cars with rainbows on them, and bumper stickers saying things like "We're a gay family, and we vote." Sitting in the cafeteria were scores of gay and lesbian parents, arms around each other and around their young children, very ordinary, very caring parents listening intently as a panel of teenagers talked about what it was like to be them. Some were the children of parents who had come out as gay and divorced their heterosexual partners. Others were sperm-donor offspring raised from the start in same-sex households. The parents peppered them with questions: What reactions did they get in school? How did they feel about gay marriage? It struck me that the children were in such a complicated position: loving their parents but struggling, sometimes, with the hand they'd been dealt. They spoke honestly about their love, and about their hardship.

After the conference, I met with an adult sperm-donor son named Dakota, who wanted to talk further. He arrived for our lunch dressed in business attire—khakis, argyle socks, striped cotton shirt—and holding a new leather portfolio; he was job-hunting in D.C. Sitting at a café, Dakota, twenty-three, told his story: He grew up in San Francisco, in a lesbian household. His sperm donor had been a friend of his biological mother. The idea, as he understood it, was that the donor "was supposed to be a sort of involved father figure in my life." But the nonbiological mother felt threatened by the donor, who was "basically scared off." The donor still lived in the neighborhood, though, and Dakota grew up seeing the donor's own children, his half siblings. It was, he said, a hard situation to come to terms with.

He deeply loves his mothers (the first nonbiological mother was replaced by another; it is this woman, he said, who helped raise him and whom he thinks of as his second mother). Yet he longed to know the man who helped conceive him. "He's really an awesome guy, and I'm just sad that this person who is half of my genetic makeup was forced out." He also wants to know the rest of his biological family. In high school, he said, he went to the graduation party of a half sister and saw an elderly couple at the party. "I was like, 'Who are these old people?' She was like, 'Oh, these are our grandparents. They don't know about you, though.' It was sort of like a you-shouldn't-tell-them kind of thing. It definitely was, in retrospect, a shaming kind of thing."

His biological father moved to the Pacific Northwest, and had more children; Dakota still sees his San Francisco half siblings. He went back for a baby shower, and saw his biological father "briefly" there. He has his father's business card. I asked if he thought of contacting him: "I think about it every day," he said, but like many sperm-donor offspring, he felt terribly afraid of rejection. And then he said something that seemed to me marvelous, and sad. "Father figures have transformed dramatically," he said, pointing out how these "father figures" can be so much warmer, so much more emotionally accessible, than they once were. The same cultural changes that have altered families have altered fathers, and he wanted one of these, one of this new breed of dads.

"I don't even know my other two half sisters," he added. "I've seen pictures of them. They look like me. We live parallel lives. I'm angry that so much time has passed and every day I lose more time not knowing these people. I'm never going to get it back."

Then he showed that what is true of parents is also true of children: "Family, to me," he said to me, "is one of the most important things."

Children Change You

Research has shown that most egg donors assume that couples receiving their oocytes will be screened to ensure they will be good parents. They assume their lovingly contributed genetic material will find its way to "nice" people. Egg donors could not be more wrong.

At fertility clinics, there is little or no substantive screening of recipient parents. Instead, there is usually a mandatory counseling session whose purpose is to impart information, and to make sure the parents aren't unqualified in some obvious way, such as, say, showing up with axes, looking murderous. As one therapist put it to me, egg-donor sessions are "scratching the surface" compared to the exhaustive home studies of would-be adoptive parents. While the underlying principle of adoption is to do the right thing by an existing child, the underlying principle of fertility treatment is the right of the paying consumer to reproductive freedom.

Reproductive freedom is an important concept in the United States, one that stems not just from *Roe v. Wade* but also from now-repudiated efforts to forcibly sterilize people considered "unfit" to parent. Once upon a time, forced sterilization of the so-called feebleminded was notoriously acceptable; in his opinion in *Buck v. Bell* in 1927, U.S. Supreme Court Justice Oliver Wendell Holmes wrote that "it is better for all the world if . . . society can prevent those who are manifestly unfit from continuing their kind," adding, infamously, that "three generations of imbeciles are enough." Since then, the Supreme Court—like the American public—has, rightly, repudiated this view. "If the right to privacy means anything, it is the right of the individual, married or single, to be free from unwarranted government intrusion into matters so fundamentally affecting a person as the decision to bear or beget a child," the court determined in 1972 in *Eisenstadt v. Baird*. From this can be inferred the right to procreative liberty for all: fit, unfit, and uncertain.

But what about the liberty to procreate with help? Does anybody who wants to procreate, and needs medical assistance doing so, have the right to get it? In some cases the answer is clear: if a person cannot afford treatment and lacks insurance coverage, in the United States the answer is no. As for others: in this country, there are no rules, so doctors decide whom to treat based on their own medical experience and moral code. Some clinics have regular meetings to talk about the hard calls. Sometimes colleagues confer informally by e-mail. The upshot is that some clinics will treat gay men and single mothers, and some won't. A few rigorously screen for drug use and sexual orientation; the rest fall in the middle. According to a 2005 study published in *Fertility and Sterility,* almost 60 percent of doctors agreed that everybody has a right to a child, and 44 percent believe doctors do not have the right to decide who is fit to procreate. Almost half said they were likely to turn away a gay couple; 20 percent would turn away a single woman; 17 percent would decline to help a lesbian couple; 13 percent would turn away a woman with bipolar disorder; 40 percent would decline to assist a couple on welfare paying with social security checks. These numbers suggest that, while there is widespread discrimination based on personal views and biases, any prospective parent can probably find someone to treat him or her. The ASRM guidelines say that programs may withhold services when there are reasonable grounds for thinking patients may not be fit parents, but are not obliged to do so.

And is that so bad? Reproductive medicine has always offered an alternative for would-be parents who might not pass, or fear they won't pass, the not-always-well-defined criteria used to screen adoptive parents. And many times that's a useful service.

"There are many ways of being a good parent, and limitless ways of being a good enough parent," said psychotherapist Joann Galst at another session of the 2005 ASRM convention, pointing out that predicting parental fitness in advance is more difficult than many people realize. This truth has recently been recognized in Britain, where the regulation instructing doctors to assess patients for parental fitness has been scrapped. Galst went on to note something that had never occurred to me before: the really pretty obvious fact that *children change their parents.* "We have all probably been changed as parents based on our interaction with our children," she pointed out. I thought of one couple I interviewed: the mother was in her mid-forties when she conceived, and had never changed a diaper or babysat. She had no idea whether she had a single maternal instinct. "The result is this wonderful mother," her husband marveled. "She is well traveled, well educated, well read, and cleaning up puke, this wonderful amalgam of an amazing mother. Six years ago, I never would have guessed it."

Galst, in her talk, pointed out how few markers there are for parental fitness. Most people who were abused as children do not go on to abuse their own children, and most people who abuse their kids were not abused themselves. Substance abuse, depression: while these are red flags, even they are not fail-safe predictors of unfitness. "A depressed mother who enjoys interacting with her child is different from a depressed mother who doesn't," Galst said, adding that arbitrariness creates inequalities and discrimination. She also pointed out that adoption has a checkered history in this regard. In an experiment performed in the 1970s, a description of intended parents was presented to a number of social workers. There was no consensus on whether the people should be permitted to adopt.

"Denial of the opportunity to parent denies people one of the most meaningful qualities of life," Galst argued, positing that "without agreement on what constitutes good enough parenting and without an adequately reliable way of predicting this in advance of becoming a parent, as professionals we lack adequate grounds on which to breach the individual's reproductive freedom."

Listening to the talk, I had to sympathize. Writing this book, I was struck by how many unprecedented situations present themselves to fertility doctors, and how hard it is to gauge what is okay and what is not. I once interviewed a clinic psychologist who had received a request from a bereaved mother who wanted to use her dead son's sperm, and a traditional surrogate contributing both womb and egg, to create an IVF grandchild. The psychologist had lost her own mother at an early age, and had a horror of "creating orphans," so her clinic said no to the would-be grandmother. Another doctor was approached by an infertile man who, wanting a baby as close to "his," genetically, as possible, asked to use his father as a sperm donor. The doctor called her own father, a minister, to ask what God would think. Her father told her he thought God would approve. The man is now parenting his genetic half-brother. The family, according to the doctor, is very happy. In the absence of clerical advice, some doctors use other standards. The bioethicist Leon Kass invokes the "wisdom of repugnance," while doctors themselves refer to this as the "yuck factor."

Many scenarios doctors face are unprecedented, and in truth, it's hard to see how a consistent set of standards could be drawn, or who would draw them. "Are we going to make everybody like June Lockhart?" asks David Keefe. "There are a lot of psychologists who are very oriented toward normative behavior. I feel really uncomfortable with that. I don't know what's normal. There's a variety of human experiences. A lot of the people you think are the most normal are the most disturbed and do the most damage to their families, and a lot of people you wouldn't give much credit to have done an incredible job in life." If a patient has an untreated mental illness, or a severe disease or disability that prevents her from carrying a pregnancy safely, Keefe would be very unlikely to treat that patient. But he believes it's important to involve the patient in this decision: in some cases, to help a person understand there is some aspect of life he or she needs to work on before considering having a child. And he believes that screening for fitness does raise the problem of eugenics. He reserves the right to treat, say, a schizophrenic who is on medication, a diabetic whose condition is well-managed, or even just someone who is eccentric. Keefe believes in the "ability to see variety as a desirable and good thing in the world."

What makes him much less comfortable, he said, is a different kind of hardship inflicted upon children and families. What can also change a parent—and not for the better—is being the parent of too many children at once.

CHAPTER NINE

BE FRUITFUL AND MULTIPLY:

THE BIG FAMILY, BY

OVERNIGHT DELIVERY

Match, or Coordinate?

Along Route 7, a once scenic, now car-choked corridor that cuts through suburban Falls Church, Virginia, signs have been staked in the grassy median announcing "KIDS STUFF" and "HUGE BABY STUFF SALE TODAY." At 8:30 on Saturday morning the parking lots of George C. Marshall High School are full, front and rear. Inside, a line of shoppers snakes through the front hallway, waiting to be admitted to the annual sale put on by Northern Virginia Parents of Multiples (NVPOM), one of hundreds of regional clubs that have emerged to provide succor to American parents undergoing the increasingly common experience of bringing more than one child forth into the world simultaneously. With fifteen minutes to go before the doors open, shoppers stand edgily in line, straining to get a view of the room's layout. They are dads with newborns in Snuglis; pregnant women clutching cups of decaf Starbucks coffee; grandparents; friends; moms with toddlers on leashes; more pregnant women, some quite pregnant indeed.

Promptly at 9 a.m. the line surges forward. Once through the door, some of the shoppers begin to run.

The determined ones are heading for the strollers section, where some of the choicest items have already been removed by NVPOM volunteers, who in return for helping out have been permitted to shop early and wheel their booty into a "hold" section. But there remain lots of buys for parents facing the challenge of ferrying two or three newborns, or toddlers, between home and playground and store and preschool. There is a used version of the coveted Snap N Go Double, a lightweight rolling frame onto which two car seats, and their human contents, can be placed after being lifted out of a car. There is a Maclaren double stroller; a Graco "duo rider"; a Swan "caravan lite." The cargo strollers are quickly snatched up by shoppers who push them toward other displays, loading smaller items onto the seats and into the underbaskets.

There are bargains everywhere for the resourceful parent of plenty. Hanging from the ceiling are signs denoting the areas where can be found: Feeding Items, Potty Supplies, Bedding, Diapers, Undergarments, Toys, Bottles and Nipples, Car Seat Covers, Baby Changing Supplies. The Safety Gates category is well stocked, as is a "child restraints" section in which can be found every device invented for temporarily enclosing small children while a parent tries to eat a meal, take a shower, or make a telephone call. There are bouncy seats, exersaucers, swings, playpens. There are jumpy seats that can be suspended from a doorway. There are doughnut-shaped lap cushions in which newborns can be propped up, like a sack of sugar, or breast-fed, in tandem.

There are infant baths.

There are breast pumps.

There are lunch boxes.

There are party supplies.

There is a man saying into a cell phone, "Yes, it's got a little pineapple on it. I think . . . yes . . . and this is a white sleeveless shirt . . . and a pair of jeans that's 6X."

There is a boy spinning on his stomach on the floor.

There are many groups of items that attest to the challenges of providing for children who are the same size and have the same needs, often at the same instant. Scattered on tables are a matching pair of Fisher-Price fish mobiles, a matching pair of blue high chairs, a matching pair of "Baby Safari" car seat entertainment toys, a matching pair of 'N Sync "On Tour Collectors Edition" Chris dolls, which seem to have been played with and then stuffed back into their original boxes, where they stiffly lie, peering out of the plastic like little vampires. There are matching two-pair sets of white patent leather Mary Janes, a matching two-pair set of Stride Rite Newport loafers, size 13M, "exceptional condition," each pair carefully taped together, $8 a pair. There are two matching pairs of baby moccasins, brown, size 1, encased in baggies, $4.00 a pair, never worn, perhaps because they were a gift and the parents didn't like them, perhaps because Mom and Dad were too exhausted to fumble with eight tiny suede ties. There are items that attest to the subtly higher level of difficulty in outfitting children who are the same age but not quite the same size: two pairs of sparkly three-strap girls' tennis shoes, for example, one a size 12, the other 13¼. And for the family that prefers to coordinate rather than match their multiples' clothing—and many do—there are two pairs of high-heeled Mary Janes, one in white, and one in a pale gold.

How much of this stuff did parents buy themselves, and how many items were given by well-meaning friends and relatives, who didn't know what would be a good present for parents expecting a brace or even a brood of babies? Would any parent really dress two six-month-olds in hooded terry robes, the kind of robe that grown-ups wear in luxury hotel rooms, and that a baby would promptly crawl out of and/or get tangled up in? Who bought that pair of new-looking leopard-print crib comforters? Were they ever used, and if so, what did that baby room look like? And what about those boxes containing two unassembled "loft beds"? What father came home from a store with a great idea for a way to fit two kids into one room, only to be told by his wife that it was a bad idea to put the twins up really, really, really high?

And what loving parent had the patience and high-order organizational skills to assemble into one sales package two red cotton dresses with white eyelet ruffles at the bottom, exquisite sleeveless dresses hanging together in the little girls' section, pressed and pinned to a wire coat hanger? Somehow it is these two dresses, more than any other item, that evoke the labor and love that is all parenting, but especially the parenting of multiples. The finding and keeping together and washing and drying and pressing and folding and pinning of items for not one child, but two or three or four, all of whom tend to share viruses and snacks as well as socks and underwear. The buttoning, the unbuttoning, the changing, the diapering, the putting away, the taking out, the pushing, the swinging, the vomiting, the cleaning up afterward, the vomiting, the ear infections, the cuddling, the hoisting, the carrying, the putting down, the vomiting, the comforting, the love, the love, the love, entire years spent engaged in tasks that cannot be recalled at the end of the day. And now, the collating and bringing of outgrown stuff to sell, hoping to recoup a little of one's financial outlay and/or make it easier on someone facing the same challenges, and/or raise money for a club that has kept one and one's family sane.

The sale is being run by Northern Virginia Parents of Multiples club members, who this morning happen to be exclusively female: women whose own children presumably are under Dad's care this morning, or Grandma's, or the Public Broadcasting Service's, or all three. The women look liberated and determined. They are everywhere in the big room, overseeing, refolding, reshelving, answering questions, working the cash registers, talking on walkie-talkies. They look as if they have grown used to moving armies. They would make formidable military quartermasters. Still, it seems a busman's holiday: leaving the twins or trips or quads at home so you can go out and sell twins' and trips' and quads' clothing to other parents of twins or trips or quads. They deserve to be at the spa, these moms, getting bikini waxes and facials, having mud packs put on their foreheads and warmed stones nestled along their spines. Not here, tagging matching backpacks. But here they are: firm, stoic, organized, energetic.

Out in the corridor, a dazed-looking new mom is pushing two fat-faced, wide-eyed babies in a brand-new Peg Perego side-by-side stroller with reclining seats. In the ladies' room, parents are saying things like "Oh, these are your kids!" and "Yes, these are John and Leah!" Outside, on the grassy sward in front of the high school, a man in a maroon turban is kicking a soccer ball around with two black-haired twin boys, while nearby more cars are circling the parking lot, looking for a space.

"Who Has Four Kids These Days?"

"Hi there!" says Linda Gulyn as her two boys emerge through the front door, ushered by the nanny, Jill, a sturdy young woman from the Midwest who, one imagines, little knew what she was getting into when Heartland Nannies signed her on with this particular Arlington, Virginia, household.

To be more precise, two of Linda Gulyn's four boys are bursting in the door. That would be the older set of twins, Daniel and Timothy, both four years of age, who have spent the past hour with Jill at a nearby library. The younger twins, Christopher and Jonathan, have been taking their nap in the room they share on the second floor of the three-bedroom brick colonial where the family of six lives. Linda, a professor of psychology at nearby Marymount University, has the boys on a strict schedule. That is one way she and her husband have survived having not one pair of twins but two pairs of twins, two pairs of boys born almost exactly two years apart, four January birthdays all told, all of them IVF babies. At one point Linda and her husband, Peter, were parenting four children under two and a half, a feat unknown to past generations. Even in the heyday of so-called Irish twins, children born less than a year apart, it was virtually impossible for two human beings to conceive four human beings in such a short span of time. Linda learned early on that everybody—and that means everybody—needs to adhere to a schedule that cannot be departed from for anything short of a tornado watch or emergency evacuation. No cozy co-sleeping in this household. No breast-feeding on demand. No potty training whenever the child happens to feel, maybe, ready. None of the other luxuriously unscheduled activities other contemporary households permit themselves. When the first pair of twins was born, Linda—overwhelmed by the demands of twin care, pretty sick herself from a difficult pregnancy, and "very needy," as she puts it—developed a formidable set of systems as a coping mechanism. "I had a diaper system, a formula system, a baby chair system, a bath system," she says now, bemused by the way she became, essentially, the CEO of a household that seemed to be growing as rapidly as Google. Even now, "ten o'clock is juice time. Not ten thirty. There's juice time. Bath time. Everything is literally on the clock."

Even so, a parent cannot control everything. Linda, who went upstairs to wake the younger twins and bring them down, is now explaining what it feels like to be the mother of four little boys who all want to lay claim to her, often simultaneously. "I like to bake, and they're fascinated by that; I literally have them crawling all over me," she has started to say, but now the older twins present one of the younger twins with a book they picked out for him. Whereupon the other younger twin starts to cry. Whereupon—this happens in a nanosecond, so it's hard to know what triggered what—someone throws a puzzle, which skitters across the coffee table and onto the floor, whereupon Linda says, "No throwing," whereupon one of the older boys starts to cry. That sets his twin off as well, perhaps in solidarity. They cry for a while, three boys howling and the fourth standing there deciding whether or not to join his brothers, four boys who very much resemble each other, all their features seeming to match somehow, eyes and hair and skin, everything a lovely light sandy shade. Then, like a squall at sea, the crying stops and the nanny, Jill, goes into the kitchen to get everybody some grapes.

"There would be many moments when all four were screaming," says Linda, a slim, delicate woman in jeans and a flowered blouse. "This one's screaming in his bassinet, and I'm screaming, everyone's crying, really, it was awful."

So petite is Linda, with her alabaster complexion and her soft brown hair, that it is hard to believe that she produced all these boys, in such rapid succession. Hard to believe that during her first pregnancy, which started out with triplets (one of the triplets died in utero and had to be carried to term) she swelled from 120 pounds to an almost unbelievable 205, growing so unwieldy that the only way she could get around was to be pushed in a wheelchair, and loaded in and out of a car. In part it was the horror of the triplet pregnancy and the subsequent nightmarish delivery that made her want to go through IVF again, motivated by the pleasure of motherhood and the love for her children, but also by the wild hope that this pregnancy would be easy; this one would yield a single baby and she could get to see what it's like to have a child under normal circumstances. "I had this idea, if I could just have one, I'll walk around in cute maternity clothes. I could be really cute this time," Linda says, wryly. But after the second IVF treatment the doctor said, apologetically, "Linda, it's twins again." While she emphasized how blessed she feels to have four healthy children, she also remembers, during the second pregnancy, "I was sitting at breakfast with Peter saying: 'Four kids. Who has four kids these days?' Peter was thinking about it, and he said, 'Well, there's this Christian guy at work.' I was like, 'Oh no, what has it come to?'"

Perhaps no change wrought by fertility medicine has been as profound, far-reaching, and underexplored as the explosion in multiple births. It's true, human beings in the past had big families, and some still do, the old-fashioned, time-intensive way, one child followed by one child followed by one child. But having a big family one child at a time is not the same as having a big family—or even a two-child family—all at once, overnight delivery, a gaggle of infants landing together in your living room. Every parent going through it agrees: there is something different about birthing and raising children who arrive as a ready-made ensemble. Among other things, multiples have changed pregnancy and childbirth, which for many women has become another extreme sport: "The kids are coming out and I'm filming back and forth, and then I hear the doctor yelling, 'I want everybody to listen to me!' and he starts barking commands out, you can just tell something's wrong, and I'm thinking oh, shit—what was literally happening, he was having to rip the placenta out; it wasn't coming out and he had to take it out," said one father of twins, recalling the moment when he personally lost it and had to stop filming the delivery of his sons. His wife remembers only whipping around on the gurney, her head lowered to the floor, thinking she hadn't banked enough blood prior to the delivery to replace the blood she was losing now. Multiple births have changed the way parents experience their children's early years; as one mother put it, she and her husband "would lie in bed and just look at each other and say: 'Okay, we have two babies. We have two babies.' That is, when we were actually in bed together." They have altered family dynamics, compelling many parents to adopt a "zone defense" of child-rearing. They have driven up sales of antidepressants. They have changed toy culture and commercial advertisements. More than forty years after Hayley Mills was obliged to play both twin girls in the 1961 movie *The Parent Trap,* matching actresses are a dime a dozen; there is even a theatrical agency devoted to the recruitment of twin and triplet performers. They have altered classroom assignments and created yet another challenge for school principals. Now, in addition to thinking about diversity and gender and ability, who are friends and who are not, principals making up class assignments must take into account the need to keep twins and triplets separate from their siblings—or, maybe, to keep them together. Around the country, state legislatures are wrestling with whether parents of multiples should have a say in classroom placement. "No one knows them better than we do," one father of triplets, who wants to override school policy and keep his children together, insisted in an article in the *New York Times.* His children, he argued, had been together since they were in incubators in the neonatal intensive care unit.

Multiples have changed the demands on parents supervising schoolwork at a time when teachers pile it on for even very young children. A friend of mine, calling to see if she could borrow an old-fashioned hard-topped suitcase in which her fourth-grader was supposed to assemble a diorama filled with items representative of the state of Colorado, found herself musing sympathetically about the triplets mom who, for their school's "state fair" project, was going to have to find three old-fashioned hard-backed suitcases and oversee three dioramas filled with items characteristic of three different American states.

Not that long ago, multiple births were so rare than when naturally conceived quintuplets were born in Ontario, Canada, in 1934, the children were quickly removed from their parents' custody and placed in a theme park known as Quintland, where tourists seeking diversion from the Depression paid admission to watch the world-famous Dionne quints eat and sleep and play behind a one-way screen. Madame Alexander, the famous dollmaker, came out with a line of Quints dolls. Years later, three surviving quints sued, and extracted from the Ontario government a sizable settlement in compensation for being treated as lucrative freaks. At that time, the birth of five babies was truly a marvelous occurrence. Under natural conditions—which is to say, conditions prevailing for human history up to now—twins have occurred in one out of about 80 births. Triplets happen naturally in one out of 8,000. Quadruplets naturally occur in one out of 800,000 births, while quintuplets are one in 40 million. There is a good reason for this, doctors believe: one baby is better for everybody.

"The vital evolutionary importance of avoiding multiple pregnancies in humans cannot be overstated," notes Gillian Lockwood, director of Midland Fertility Services in England, in a paper where she describes the complex surge of hormones through a woman's body every month, awakening hundreds of immature oocytes and quickly suppressing the growth of all but one, which—in a process known as "many are called; few are chosen"—is lovingly coaxed, by accompanying somatic cells, into developmental maturity. Having one child at a time is the norm, for all women, all over the world, and always has been. It is easy to see how singleton babies would have been advantageous for our nomadic forebears. Delivering two babies in the bush without obstetrician or antibiotics, or hauling two newborns through the savanna, or keeping two toddlers out of the way of long-toothed predators, would have been difficult, maybe fatal, for prehistoric mothers and for their children. So nature endeavored to make sure it didn't happen.

"If Nature thought it was appropriate for homo sapiens to have litters, Nature would not have selected against it," argues Alfred Khoury, an obstetrician and gynecologist who works in the high-risk maternity ward at Inova Fairfax hospital, where half the beds at any given time are occupied by women pregnant with twins, triplets, quads, or more. "If twins were something favorable to our survival as a race, it would not be one in eighty. It would be one in five. If triplets were good for us, it would not be one in eight thousand. For us to assume that what nature has selected against we can do, as part of nature, is a very arrogant way of dealing with it."

And yet this is what we apparently do believe, given the rate at which multiple births have skyrocketed. In 1982, there 71,000 twins born in the

United States, a number that had grown to 105,000 by 1997, even though the overall number of births did not rise during that fifteen-year period. In this country, the twinning rate has risen by 300 percent in three decades, to the point where in 2003 there were 129,000 twins born, the highest number ever. Now, *one out of every thirty-three children born in the United States is a twin.* The rate of births involving three or more children has grown even more rapidly. In 1972, just 902 newborns were triplets, quads, or more. By 2003, that number had increased by nearly 1,000 percent, to 8,000 children. About 7,000 were triplets, whose numbers have increased by 700 percent in the past two decades.

These increases are the result of several developments. About a quarter of the increase is due to women conceiving naturally at older ages. As women get older, they sometimes ovulate more than one egg at once, resulting in more spontaneous twin conceptions. Another factor is "ovulation induction": fertility drugs administered alone, often casually, by a fertility specialist or even just an ordinary ob-gyn. Fertility drugs have been around now for more than half a century. In the late 1950s and early 1960s a Swedish doctor treated fifty infertile women with clomiphene citrate, commonly known as Clomid, and of the twenty women who became pregnant, half had twins. With the advent of the more powerful gonadotropins, the truly high-order multiples emerged: the septuplets and octuplets that make front pages and television news broadcasts. Mark Evans, a doctor who helped pioneer selective reduction—a winnowing down of high-order pregnancies through injection of potassium chloride into the hearts of one or more fetuses— recalls seeing the ultrasound of a woman pregnant with *twelve fetuses.* Her doctor had told her about only six of them, unwilling to reveal the extent to which treatment had succeeded, or failed, depending on how you look at it. Evans still has the ultrasound of the uterus of a woman pregnant with eight. He refers to it, with mordant humor, as "my Hollywood squares ultrasound." Ovulation induction creates an extremely uncontrolled situation: once a drug is administered to boost egg production, it is hard for even a trained reproductive endocrinologist to predict how many eggs might be fertilized as a result of artificial insemination or sex. "Estrogen levels and number of follicles do not seem to correlate with the incidence of multiple pregnancy," a number of leading fertility doctors noted in a letter to *Fertility and Sterility* in 2001. This was medical language for "we have no way to tell in advance if a woman who has taken fertility drugs is likely to conceive eight, or six, or four." The letter went on: "Unless we outlaw the use of ovulation induction, we will continue to have this problem."

And then there is IVF itself. Many fertility doctors like to place the blame for multiples on lower-level nonspecialists writing prescriptions willy-nilly for Clomid: "There are a lot of Clomid babies out there," said one fertility doctor defensively, speaking to federal regulators. But the truth is that IVF in this country—and in many others—produces twins and triplets almost half the time. Europe has done better at reducing multiples; some European governments mandate that a single embryo be transferred into a woman under thirty-six, and that older women should be limited to no more than two. As a result, "triplets have virtually disappeared in Europe," a Danish doctor told European colleagues at a 2006 conference, and in Sweden, which has aggressively encouraged single-embryo transfer, the twinning rate has decreased by 5 percent. In the United States, there is no government policy. There is no industry policy, at least not one that's enforced. There are guidelines, which are routinely disregarded. In the United States, more than half of IVF children are born as part of a set. In large part this is because doctors routinely transfer two or three or even four embryos with the vague, fingers-crossed hope that at least one will take.

It's true that more than half of IVF patients, especially the women, actively and sometimes ardently hope for twins. "I was like, multiples: the more the better! Doesn't that sound great! I had no clue!" says Linda Gulyn. According to studies done by Shady Grove Fertility in Maryland, 41 percent of their patients are hoping for twins. "We feel that the most important reason that can be addressed for the high rate of multiple gestations in the United States is that the vast majority of patients consider multiple gestation to be a desirable outcome," another group of doctors wrote to *Fertility and Sterility*. Doctors are often eager to shift responsibility onto patients, against whose demands they portray themselves as helpless. And if patients do want twins, it's easy to see why: many find treatment so stressful and expensive that they want to complete their families as soon as possible. And consumer culture encourages us to believe that multiples are just so darn sweet. In its 2001 Christmas catalog, FAO Schwarz unveiled a line of triplets dolls—Francesca, Alexa, and Olivia—which come nestled in an adorable carrier, with sets of matching clothes. "They are one of our most popular dolls, they really are," says Kim Richmond, executive director of marketing. Richmond adds that twin dolls are by now so common that a company that prides itself on unique toys has to ratchet the number a little higher. "What we really wanted was something that said: 'Oh, wow, where else can you get this?'" The FAO "triplet garden" dolls have been such steady sellers that they now come in Caucasian and African American, and a new set of accessories is added every year.

And of course, multiple births are a staple of newspaper articles, which invariably emphasize the jolly, fun side of having a pack o' babies. "Their joy is fourfold!" was the headline on a 2004 *New York Daily News* article, about a couple whose quadruplets were born nine weeks prematurely and spent a month in the neonatal intensive care unit. "Orlando hospital triples its pleasure three times over," was the headline in the Bradenton, Florida, paper, about three sets of triplets born in Arnold Palmer Hospital in a single day. "Twins for Julia: It's a Boy! It's a Girl! After a difficult pregnancy, the ecstatic star welcomes babies Hazel and Phinnaeus," trumpeted *People* magazine, after Julia Roberts gave birth to twins a month early. A 2005 *National Geographic* story about the glut of triplets in New Jersey—a direct result of the fact that New Jersey is home to a number of major IVF practices—was titled "Triplet Epidemic." The article itself was merrily upbeat, referring to "energetic" children "gumming the furniture" while happy moms looked on fondly.

But every now and then reality intrudes unbidden. In 2002, the *Charlotte (North Carolina) Observer* published an article about a couple expecting quintuplets. The story, which tracked the ups and downs of the pregnancy, was printed early, for publication in a Sunday supplement. Between the day the story went to press and the day it appeared on newsstands and doorsteps, the mother went into premature labor, and all the babies died. There was nothing the paper could do but publish a correction in the regular daily section, which appeared simultaneously with the preprinted feature. "We apologize for any pain the story may cause," wrote the paper's managing editor, "and we offer our sympathy to the family."

The story points to the dark side of multiple births: damaged children suffering long-term effects of severe prematurity, terrible miscarriages, infant mortality, grieving families, parents who are overwhelmed. Over the past three decades, prematurity rates have risen by 33 percent in this country. According to a 2006 report by the Institute of Medicine, one in every eight children born in the United States is now premature. Assisted reproduction is a major reason why. The March of Dimes and the American College of Obstetrics and Gynecology have both issued papers pointing out that the explosion of multiples has resulted in *more* children born prematurely, *more* children with low birth weight and the developmental issues that follow. "When we first started looking at the statistics for perinatal loss, we expected to see that it's low-income women," said one perinatal nurse who assisted the D.C. area March of Dimes in conducting a study of death of children before, during, or after birth, in that time known as the perinatal period. "Instead, it's the corporate, thirty-five and over women." Companies like America Online have hired wellness coaches to educate women about the signs of premature labor, in part because of the high number of mothers of multiples in twenty-first century, high-tech workforces. As one textbook points out, fertility medicine has given the human race an entirely new pregnancy: the iatrogenic multiple pregnancy. "Iatrogenic" refers to a treatment for one medical problem that creates a new medical problem. In this case, treatment for infertility necessitates treatment for a high-risk pregnancy.

"It has all the hallmarks of an epidemic," says Louis Keith, co-founder of the Center for Study of Multiple Birth, and himself a monozygotic, or identical, twin. "So many of them are neurologically and physically damaged."

The full impact of conceiving multiples is rarely understood in advance. A study done by the March of Dimes found that only 35 percent of the American public appreciates the severe risks posed by prematurity. But the statistics are cold and unforgiving. Fifty percent of twins are born prematurely. Ninety percent of triplets are premature, as are virtually all quads and quintuplets. Depending on how premature they are, these children's lungs are often immature and undeveloped, so they cannot breathe unaided. They are hooked up to ventilators, machines that breathe for them, which sometimes scar the lungs so the children will for the rest of their lives be prone to asthma, pneumonia, chronic lung disease, and other respiratory problems. The brain is another vulnerable area: its many folds and creases often have not been fully laid down, and insulating fiber has not formed over developing nerve fibers. As a result, premature babies are susceptible to brain hemorrhages, and to later brain-related developmental problems, including learning disabilities.

"All their organs are premature," explains the perinatal nurse. "You have complications of every organ. There's growth and development issues, feeding issues, developmental delays. A lot of these babies have reflux. Also, there's eye changes, retinopathy; some of these babies will lose peripheral vision. Any organ, when the baby is born preterm, is immature. The brain, lung, GI tract, eyes." Infection is also a constant danger, both after birth and as they grow.

As a result of prematurity and its frequent companion, low birth weight, twins are six times more likely to suffer from cerebral palsy than singletons. Triplets are twenty times more likely to suffer from CP, which in this country is becoming more common even though a traditional major cause of cerebral palsy, jaundice in newborns, has been all but eliminated. "We got rid of one major group of cerebral palsy, just to be afflicted with another," says Carl Gunderson, deputy director of the United Cerebral Palsy Research and Education Foundation.

Then there is death. Fifteen to twenty percent of triplets are lost before viability. Postnatal mortality is also higher for multiple babies: 11 out of 1,000 singleton babies die before their first birthday, but for twins, the number is six times higher, 66 out of 1000. That's 1 in 15, a staggeringly high number. For triplets, 190 out of 1,000 die in the first year, or almost 1 in 5. According to a 2006 study in the journal *Pediatrics,* prematurity is now the leading cause of infant mortality in the United States, accounting for at least one-third of all infant deaths. The risk of having a handicapped child, or a child with Down syndrome, also rises with the number of babies. One study found that at least one handicapped child was produced in 7.4 percent of twin pregnancies, in 21.6 percent of triplet pregnancies, and in 50 percent of quadruplet and quintuplet pregnancies. The average premature infant incurs $42,000 in medical costs in the first year of life alone. The average cost of a triplet birth is more than $500,000.

Then there is—for the high-risk obstetricians who handle these pregnancies and births—the challenge of managing a pregnancy where fetuses are discordant, or growing at different rates. In triplet pregnancies, there is almost always one baby who is smaller, and has different prenatal needs from his or her siblings. The day I interviewed Alfred Khoury, the high-risk pregnancy specialist at Inova Fairfax, he had three cases of discordant twins in his unit. All were situations where one twin fetus was in the 50th growth percentile, and the other was not even in the 5th.

"The one growing poorly, if I don't deliver it, it will die," Khoury said. "What do I do now? I sit down with the mother and give her the options.

You haven't faced it. Why would anybody think about that? If the baby's not growing inside, that baby's not getting oxygen, the uterine environment is hostile. That baby doesn't need to be inside." And yet prematurely delivering one twin necessitates prematurely delivering the other, even though doing so could, and probably will, damage the larger twin. "It's Solomonic," said Khoury, whose pager went off during our interview. "What? What? Talk to me," he said, always alert to a sudden change in a patient's condition. "She has low fluid? Do her that way. Put her on Friday. We'll do her together. I don't know. Find out."

"I've delivered babies as big as your hand, babies that are fourteen ounces," Khoury said after hanging up. "You can't even find a tube to put down their throat that's small enough."

Let it not be forgotten that for most of human history, pregnancy was a life-threatening proposition to the woman going through it. For a woman with more than one baby, the ancient dangers remain. Women pregnant with multiples are more likely to suffer from hyperemesis, or extreme vomiting, "a condition we don't know how to treat," emphasizes Khoury, pointing out that all severe conditions of pregnancy remain mysterious even to the doctors who specialize in them. "We really don't know how to treat anything."

We do know that women pregnant with multiples are at higher risk for any complication you can think of: gestational diabetes, anemia, urinary tract infections. For a woman pregnant with twins, the risk of postpartum hemorrhage doubles, from 4.7 percent to 9 percent. The risk of postpartum infection doubles, to 9 percent. The risk of needing a blood transfusion or hysterectomy doubles. There is a fourfold increase in the risk of a venous thromboembolism, a traveling blood clot that can be fatal. There is an almost fourfold increase in the risk of the mother's dying, and because of the risk of prematurity, women in multiple pregnancies are far more likely to end up on long-term bed rest, which can have lasting consequences, such as loss of calcium and bone density. "They're here forever," says Khoury. "Nobody talks about this." Even something like premature labor—which would appear to affect the babies, mostly—impacts the mother in that it requires giving her drugs, pumping her full of fluids, demineralizing her bones. Then there is the much-feared possibility that the mother's uterus will fail to contract after childbirth, a situation that results in uncontrolled bleeding, and often requires a hysterectomy. Not to mention the "sheer physical discomfort of being basically like a balloon," he says. "You have to get a trapeze for her in the bed, so she can pull herself out of the bed to go to the bathroom."

Perhaps most serious, in a woman pregnant with twins the risk of preeclampsia jumps to 10.3 percent, as compared to 4.4 percent in a singleton pregnancy, and the onset occurs earlier. Preeclampsia is one of the most serious complications of pregnancy, involving high blood pressure, swelling due to fluid retention, and kidney malfunction. It has no cure, except for forced delivery of the babies. "Preeclampsia is a disease only pregnant women get," Khoury says. "To this day we don't understand what causes preeclampsia. If it is severe, you can have seizures, placental abruption, a baby that is very growth-restricted. It's something you really don't want to have. Here we go—a disease that we don't understand and don't know how to treat, and we're creating conditions that double or quadruple the incidence. We're creating conditions that increase the incidence of our most common and most vexing problems that we have no treatments for."

And even when healthy, these children are creating a novel family dynamic, placing unprecedented stress on their parents. "It's a huge imposition on the usual family structure," said Dwight Rouse, an obstetrician and gynecologist, speaking at a 2005 federal conference on fertility medicine and its impact. Rouse also pointed out that the long-term effect of multiples parenting—the social, familial, and psychological impact—has not been adequately studied. Khoury has a PowerPoint presentation that points out that in studies done on mothers of triplets, all mothers reported emotional distress—mostly fatigue and stress—four years after giving birth. They reported difficulties coping, high levels of depression, and psychotropic drug use. The PowerPoint also dryly notes: "Mothers spontaneously expressed regret at having triplets."

Even when it goes well, raising multiples represents a profound change in family life and structure. While working on this book I put out an e-mail request to a local D.C.-area multiple club, asking to interview parents willing to talk about raising twins or triplets or more. I was inundated by e-mails. The request was forwarded from one club to another; the replies came from Lexington, Kentucky; from Oak Bluffs, Massachusetts; from Littleton, Colorado; from Plano, Texas. They came from members of the Bluegrass Parents of Twins and Multiples Club; the Gemini Cricket Parents of Multiples Club; the Denver Mothers of Multiples Club. The e-mails said things like "I am at home on bed rest but will be glad to share my experience with you" and "I live in the Dallas area and am the mother of 3½ year quadruplets" and "My name is Debbie and I have Triplets" and "I had my twins when I was 50" and "I heard you are looking for stories, and boy, do I have stories." I could not respond to every one, and started frantically caching them in a special e-mail folder. Some of them languish there still, unanswered. I felt overwhelmed by technology—in this case, online communication technology—and the quantity of material it could deliver overnight. It was an experience, it seemed to me, not unlike what many of these parents were living. Technology had given them riches beyond their imagining, riches delivered in a massive, sometimes unmanageable dose.

CHAPTER TEN

"IT'S ALWAYS A PARTY WITH TRIPLETS":

THE ADVENT OF HIGH-ORDER MULTIPLES

In November 1997, seven babies were born to an Iowa woman named Bobbi McCaughey, bursting into public attention after their community kept McCaughey's big pregnancy a big secret. The father worked as a billing clerk at a car dealership; the mother was a seamstress. After delivery the McCaugheys were showered with gifts and media coverage, given things like a lifetime supply of diapers and seven years' worth of cable TV. The situation represented the culmination of what had become, in this country and elsewhere, steadily escalating births of "supertwins," a coinage invented to describe extremely high-order multiple pregnancies. So inured was the public, in fact, by this time, that a Washington, D.C., family who gave birth to a mere six babies a few months earlier had been all but ignored, receiving exactly nothing until a *New York Times* article (which also noted that race may have been a factor; the family is African American) pointed out their plight. And the McCaughey seven still occupy a special place in the world's imagination: They hold a place in the *Guinness Book of World Records*, tied with a Saudi Arabian family for the most number of babies born together who survived. Each December the children are dressed in coordinated outfits and posed on the cover of *Ladies' Home Journal*, to satisfy whatever it is in the reading public that enjoys looking at massive quantities of children gestated simultaneously and dressed alike. Looking closely, it is possible to see that the children are not so similar. All are heart-stoppingly beautiful. Two, however, suffer from muscular disabilities related to cerebral palsy, and have a tendency to sit with their legs positioned a little awkwardly, their eyes not entirely focused, smiling gamely.

In so many aspects the McCaughey pregnancy is classic. Now, for a multiple pregnancy to garner real attention, there have to be not just lots of babies, but lots and lots and lots of babies. It's also representative in that some of the children are fine, others are a little less fine, and two are disabled. And, it's representative in that the simultaneous birth of so many children is due in part to socioeconomics. Which is to say, money—the lack thereof. One of the terrible ironies of fertility medicine is that among those who seek treatment for infertility, it is invariably the poor and working-class who end up with the most mouths to feed; the most diapers to buy; the greatest need for a minivan, or even a small school bus; and children with the most expensive and long-lasting medical problems. In the United States, one reason we have supertwins is because of the lack of insurance coverage for IVF. Only fifteen U.S. states mandate insurance coverage for any fertility treatment. Only seven mandate coverage for IVF, which remains the most sophisticated and controlled form of fertility treatment. When insurance companies can get away with denying coverage for IVF, they consistently do. And one reason they can get away with it has to do with the persistent notion that infertility is a self-imposed, "voluntary" affliction that doesn't deserve coverage, whereas things like, say, stomach-stapling, or treatment after a car accident caused by drunk driving, or back surgery to repair damage done in a fall off a ladder, of course, do. The profession itself is complicit: IVF doctors aren't exactly lobbying for more insurance coverage, and some openly say they don't want it, because what happens when insurance covers a procedure is that the cost—and the doctor's salary—drops, often dramatically, as the insurance company exerts its considerable negotiating leverage.

And so less affluent infertile patients—and remember, lower-income adults are more likely than affluent ones to suffer from infertility—pursue the solution they can afford: pharmaceuticals. Almost invariably, the extreme cases of high-order multiples are the product of fertility drugs alone. Bobbi McCaughey conceived her septuplets while taking a drug called Metrodin. Almost invariably, these drugs are taken by patients for one reason: they cannot afford IVF. "The multiples in this country—the fives and sixes that get onto the front pages of the paper—are poor women," says Louis Keith. "They're not Hollywood stars."

Certainly, socioeconomics was a powerful factor for Melissa Middleton and her husband, Brett. Of all the strong and loving parents interviewed for this book—and there were so many—Melissa and her husband struck me as standouts. The story of what she and her husband went through, and are still going through, to conceive and care for their children, is a powerful corrective to the notion that fertility patients are picky people bent on superior offspring, or somehow unequipped to meet parenthood's more serious challenges. It also shows what happens when patients can't afford IVF.

In 2003, Melissa and her husband were in their early twenties, and lived in Texas. Both were graduate students—Melissa was working on a community counseling degree—and both were also working. After several years trying to conceive, Melissa was diagnosed with polycystic ovarian syndrome. They could not afford even a single round of IVF, so Melissa's regular ob-gyn put her on Clomid. When that didn't work, he sent her to a fertility practice, where Melissa told the doctor explicitly that she did not want high-order multiples. "I told him I didn't mind having twins, but I wanted to be sure that we did everything possible not to have more. He told me it was very unlikely that we would have more than that."

A surprising assertion, considering that her doctor put her on a regimen of powerful gonadotropins. The doctor performed IUI, in which her husband's sperm was concentrated and injected directly into her uterus. When she went for her sonogram, on the screen were four gestational sacs. "I was really shocked," she said. The doctor recommended selective reduction of at least two of the fetuses. "We just couldn't do that," says Melissa, who resolved to carry the pregnancy despite the fact that she's a "small person, 5'4", about 120 pounds," and "I didn't have wide girth in my hip."

Just months into her pregnancy, Melissa was feeling so much discomfort from carrying four growing fetuses that she had to go on bed rest. By week sixteen she was experiencing severe contractions. Soon she had to be admitted to the high-risk maternity ward, where she was in a private room for two months, taking magnesium sulfate, an antilabor drug, as well as morphine, "anything to relax my body enough so that I wasn't having these contractions." She spent a month with her feet elevated, trying to enlist gravity in an all-consuming effort to keep the babies in utero. "I had my head on the floor; the bed was tilted all the way back, the position where your feet are all the way up. I couldn't get out of bed."

Despite all this, at twenty-five weeks her cervix began enlarging. At a certain point, when the body decides to expel a pregnancy, it cannot be persuaded to reconsider. Melissa was wheeled into the delivery room for an emergency C-section and the babies were removed within five minutes of one another. "I didn't see them; I only heard them mew," Melissa says. "They were really, really tiny. They weren't breathing. They did emergency ventilation; put me back in recovery, told me that there was a 30 percent chance that we would bring even one of them home."

The next day Melissa was wheeled up to see the babies. She and her husband had them baptized right away, in case they did not survive. As is often the case with parents of multiples, they gave the babies names beginning with the letters—A, B, C, D—by which they had been known in utero. For three weeks, all four babies survived, hooked up to machines. And then one—Alex, the healthiest of her children—contracted an infection around the site of a crucial IV line. Before the doctors could decide what to do, the infection floated to his heart, causing a blood clot and the massive shutting down of organs. Within days Alex was in so much pain, his suffering so intense that Melissa and her husband made the decision to take him off life support. He died in her arms. That same day, her other three children had surgery to repair openings between their heart and their lungs.

After their operations, Benjamin, Callie, and Donovan stayed on oxygen for several more months; they came home at different times. Now it was their home that became an ad hoc NICU unit: Melissa and her husband learned what it's like to care not only for three newborns, but three newborns with severe, unfolding, and difficult-to-diagnose health problems. Though medicine can now keep alive children born at twenty-five weeks, doctors are only beginning to understand the ailments that afflict these children. "Benjamin stopped breathing one day and we had to give him CPR and take him back" to the hospital, Melissa told me. At the hospital it was also discovered that Benjamin had a severe detachment of one retina. They took him to Baylor University hospital, where doctors were obliged to remove the lens from that eye. Benjamin is now blind in one eye, and his vision is impaired in the other, possibly as a result of the severe bilateral cranium bleed he suffered in the hospital, which resulted in hydrocephaly and, they suspect, had a permanent impact on his ability to see spatially. "We take him to the retina institute in Dallas," Melissa said. "He has a vision teacher, he has an orientation mobility instructor, and he also—we have the coordinator for the Center for the Blind, who works with us a lot. He just has a whole slew of doctors for his eye problems. All my kids at one time had early intervention; they were in occupational therapy, physical therapy, speech therapy."

All the babies, when they came home, had severe reflux, or feeding problems; they were throwing up after every meal, making it hard for their very small bodies to grow. Benjamin had what's known as "failure to thrive." For the first year, he lay on the floor, unable to walk or crawl. "We took them to every specialist we could think of," Melissa related. "Gastrointestinal doctors, nutritionists, speech therapists. We took them to a feeding and growing clinic, anything we could think of who could help." Benjamin had a feeding tube inserted, which helped him gather strength.

As for the parents—here is the week of a loving wife and husband whose chief goal in life was to become parents and raise active, healthy children. "My mother comes in on Tuesday," said Melissa. "We have a sitter on Fridays" so Melissa could resume her counseling degree program, which was derailed by the birth of so many children. "Tuesday from 1 to 4:30 we have therapy, and Thursday from 10 to 11. Friday we have Easter Seals. The vision teacher comes Tuesday and in the morning on Thursday, about nine o'clock. Monday, the orientation mobility teacher is here, so really, Wednesday is the only day that we don't have any therapy appointments scheduled, and typically those days a lot of time is allocated for out-of-town doctors' appointments." Meanwhile, Melissa said, "we've turned our house into a big baby therapy practice: we've got the big ball and the swings. Whatever stage they've moved into, we've bought all the toys. We have really, really pushed that."

In 2005, when we first spoke, there were typically two nights a week when Melissa and her husband would be up all night. "My little girl had perainfluenza, which was really big around here. It turned into a respiratory infection, which turned into pneumonia, which turned into bronchiolitis." She added: "Pretty much the first year, we didn't take them out of the house, except to doctors' appointments. We were pretty much under house arrest." When I asked her if she regretted having the kind of treatment she did, she replied, "I've thought a lot about that. I love my children dearly, but probably we would have opted for having IVF at a time when we would have been able to afford financially to do that."

More than a year later, I communicated with Melissa again, in e-mail exchanges, and she reported that she had "changed a lot." The family had moved, because of her husband's job. She had completed her counseling degree and had a job helping other families withstand the fear and challenge of the NICU. She had become a volunteer for the March of

Dimes, helping families in similar situations understand that they could not only cope, but prevail. Her three surviving quadruplets were still small for their age, but healthy and growing: "The doctors say they are following their own growth chart." Callie and Donovan both had challenges—one had an abdominal feeding tube; the other had had some recent surgeries—but were in a lively preschool environment, and thriving. Benjamin was still meeting with vision specialists; there was talk about pre-braille skills and cane training. The doctors said there was a 40 percent chance Benjamin would someday be totally blind, but Melissa and her husband took comfort in the fact that there was a 60 percent chance he would keep his vision. Either way, she wrote, she knows Ben will be okay. "He is a three-year-old boy who has never been told he can't." She added, "We are doing really fun things, too." They were getting out of the house: going to the zoo, the pumpkin patch, to museums. "For the kidlets' birthday present," she wrote, she and her husband took them to the Wiggles Live.

Melissa said that her husband is "amazing: I couldn't ask for better."

"I've had people tell me that we deserve it, because we didn't have them naturally," this equally amazing woman also told me, in our first conversation. "Because that's what God did not intend for me to have. They're saying that it's my fault for trying to have them in the first place."

"It's Embarrassing"

Of course, IVF does not exempt families from multiples. In fact, IVF causes multiple births as often as fertility drugs do. It just doesn't cause the birth of so many at one time, at least not usually, at least not anymore. In the early days of IVF, doctors had no idea how many embryos to transfer to achieve a pregnancy. So they often put in as many as they had, and waited to see what would happen. As Geoff Sher puts it, for years the scientific approach to embryo transfer was "throw a bunch of spaghetti against the wall and see what sticks."

In this case, "early days" is of course relatively recently. "I heard of one practice that transferred seventeen embryos," says Cornell embryologist Lucinda Veeck Gosden, describing an episode from the 1990s. "He transferred eight, and didn't ask me about it until I was loaded with Valium," said Janis Elspas, media coordinator for the Triplet Connection, a national multiple-birth networking group, describing her doctor's approach to "informed consent," back when her own triplets were conceived by IVF in 1997. "The doctor said, 'I have all these healthy embryos, and I hate to freeze them.'"

"In the 1980s and early '90s, they used to put four, five, six embryos in there," says Dr. Siva Subramanian, chief of neonatology at Georgetown University Medical Center, where several high-profile sets of extreme multiples have been delivered. Dr. Subramanian's practice has been transformed by multiple births. With each delivery of quints, quads, triplets, or even high-risk twins, he has a team of neonatologists assigned to each baby, standing by the table poised to count the number of heartbeats in six seconds, look for breathing problems and congenital malformations, and instantly gauge the damage that prematurity and crowding have done to children who sometimes emerge weighing just a few pounds.

Even today, there is enormous uncertainty with every transfer of more than one embryo. That's because, for all the ballyhooed advances in the field, doctors and embryologists still have no idea which embryos, if any, have the ability to turn into children. A patient can undergo three IVF attempts and have all of them fail. The fourth time her doctor might decide to transfer five embryos; this time, all five might implant. Some scientists think there is a "cohort effect": if one embryo is good, chances are its littermates are good, too. Here's the important thing: doctors still have no reliable assay, or test, to identify which embryo among many is the embryo with the ability to develop into a fetus. They have no way of sorting good from bad, winner from loser, wheat from chaff. When doctors tell a patient that an embryo is "good," or that it's graded "A," it is an educated guess, to put it mildly.

"The state of the field is to look under a microscope and say: 'That's a good one,'" acknowledged James Trimarchi, a scientist who at the time directed the IVF lab at Women and Infants Hospital in Providence, Rhode Island. Trimarchi was talking to young cell biologists who in the summer of 2005 were taking Frontiers in Reproduction, a six-week course in laboratory reproduction offered at the Marine Biological Laboratory in Woods Hole, Massachusetts. Remember, embryo diagnosis is a basic part of treatment, required in every single IVF procedure. And yet it can't be reliably done. "The cell biologists are going to help us solve the problem," Trimarchi told the scientists listening to him. He was being hopeful. "We can't solve it now."

In his lecture, Trimarchi compared the rich, constantly evolving human embryo to a car. As with a car, an embryo has a number of systems, distinct yet interrelated. It has the nucleus, the centriole, the mitochondria, the microtubules, the cytoplasm. Any of these systems could have a defect, particularly in IVF embryos, which may or may not—nobody knows—be affected by exposure to fertility drugs, to culture media, to other laboratory conditions including (one study suggested) what shelf of the incubator the embryos are placed on. What science needs to do is come up with some gauges: a dashboard that can light up if one embryonic system or another is going wrong.

To show just how unsatisfactory the current tests are: For the class that day, Trimarchi displayed a list of seven embryos his laboratory graded when they had been allowed to develop for three days. At day 3—it is believed—the best embryos will have developed eight cells; they will be symmetrical; they will display little fragmentation. Embryo 1, Trimarchi explained, projecting a bunch of numbers onto a board, fit all these criteria. Rated 8CF4S1, it had eight cells, little fragmentation, and was perfectly symmetrical. Embryo 2 had the same score. Embryo 3 was graded 7CF4S1: just seven cells. Embryo 4, which had nine cells, was rated 9CF4S1. Five was more whacked-out: 9CF3S1. Six was developing slowly: 6CF4S1. Seven was 7CF3S1, a hodgepodge of odd numbers. He asked the class to guess which embryo was most likely to grow and thrive. They chose, naturally, 1 and 2. Trimarchi told them that in this case, 6 was the one that lasted long enough to be transferred. "It's embarrassing," he said. "We're deciding people's fates."

"When you run all the numbers, the truth is, we really don't know anything," Trimarchi told me. At Women and Infants, he said, scientists have put together a database of every IVF case that resulted in a pregnancy and every case that didn't, looking for a pattern in the way the embryos were graded. He took the data to some mathematical modelers, and, "the conclusion was that there are some very complex relationships that we can't even begin to understand."

"There is so much we do not know about the human embryo that we need to," said Trimarchi. Like every scientist in the field, Trimarchi deplores the lack of federal funding to study human embryos. "We've gotten kids from ugly-looking embryos, and we've transferred beautiful embryos into people who are back months later for another cycle. The relationship between the looks of an embryo and whether or not it's going to implant and develop to term is extremely loose. It's like: That one looks good," says Trimarchi, who is always seeking an improved way of gauging embryo quality. In the past decade a new test has emerged: pre-implantation genetic diagnosis, or PGD, in which a single cell is removed from an embryo and sent to a lab for chromosomal testing, to the tune of several thousand dollars. But PGD labs can test only the nuclear DNA and not, say, its cytoplasm, which is the energy pack of the embryo. "You can have a fantastic nucleus and shot cytoplasm," Trimarchi says. And nobody knows whether PGD could damage the embryo. Well, actually, they do. PGD probably does damage the embryo: they just don't know how much and in what way, according to a leading academic scientist, Dagan Wells of Yale University, who also lectured for the biologists at Woods Hole.

Another day, Trimarchi gave me a tour of the research lab that Women and Infants maintains in Woods Hole. Trimarchi and another scientist were working with human eggs. This group of eggs was "surgical discard," meaning that they had not fertilized during IVF, and the patient had given permission for them to be used in research. Trimarchi and his colleagues were exploring in vitro maturation—a lab technique that, if it works, would permit scientists to take eggs that emerged immature and culture them to maturity. The trouble is, the only way to tell whether an egg has been cultured properly is to fertilize it, and what you get when you fertilize an egg is, eventually, a human embryo, and what happens when you get a human embryo is—alarm bells buzzing, sirens going off, federal sanctions, your NIH research funds cut off, your equipment confiscated, and maybe, who knows, you go to jail. And that's why nobody knows what a good embryo looks like. You can't study them—not if you get federal money, and most good scientists do. The long-standing ban on federal funding for human embryo research means that any scientist who receives federal funding for *anything* cannot use any portion of that funding—or the equipment it purchased, or the room whose rent it funds—to conduct research on embryos. What human embryo research goes on in the United States must be done in privately funded labs, or in research facilities that receive their funding from sources other than the federal government.

The impact is huge. There are crucial territories of assisted reproduction that remain a mystery to science. "At a certain point, what you want to be able to do is to fertilize the egg and then study its development to make sure it's okay," Trimarchi told me. "The only alternative, eventually, is just to put it into a uterus and see what happens. This is pretty much what happens with every lab technique for IVF. You fool around in the lab as much as you can; you test it and observe it and inject markers, but what you cannot do is fertilize it to see how the embryo would develop. At a certain point—and the point comes pretty soon—you just have to jam it into the uterus and see what happens."

Jam it into the uterus and see what happens. Jam three into the uterus and see what happens! Jam four! Jam five! The ban on federal funding for embryo research is not only affecting people who suffer from debilitating diseases, who hope that someday science, in the form of embryonic stem-cell research, may deliver a cure. The ban is also affecting, more immediately, infertility patients whose doctors can't recognize a good embryo. The ban on embryo research is affecting, with devastating consequences, the children born as a result of too many embryos being transferred, children born two and three and four at a time, born in some cases to die, or live a life of disability, simply because doctors still know so little about what's going on during embryonic development.

"The current problems with IVF in the United States, and they're substantial, exist because there is no public funding at all, because all the NIH scientists were prevented from doing any work," the Australian embryologist Alan Trounson is quoted as saying in *Merchants of Immortality*, Stephen Hall's fascinating book about the private entrepreneurs who are driving stem-cell research with the hope of reversing human aging. Trounson, one of the world's leading embryologists, points out that the hidden result of this ban is the epidemic of multiple births. "Anyone who was funded by NIH was not permitted to do any work in this area. So currently, research in IVF is driven by private IVF clinics. It's market-driven research, and it results in large numbers of embryos being put back into patients, and a lot of multiple births, and a lot of fetal reduction."

"We Can't Stop It"

Another irony: The very pro-life factions that have managed to thwart embryo research are contributing to a situation where babies die before, during, and after childbirth. Pro-life opposition to embryo research is a crucial reason why the most loving, most ordinary, most bent-on-a-family, most salt-of-the-earth-type couples are enduring the most overwhelming grief imaginable. "There's been years and years of lost capabilities," says Robert Stillman, a doctor at Shady Grove Fertility. "We're so far behind where we could have been."

And if those capabilities had not been stymied, then—who knows?—it might have been possible to avoid what happened to Tammy and Steve LaMantia, a couple living in Northern Virginia who married in their late twenties and started right away trying to build a family. Unfortunately, Tammy turned out to have PCOS and other problems. Fortunately for them, the Fairfax County school system, where Tammy was an elementary school reading teacher, provides IVF coverage under its health insurance plan. The LaMantias found a clinic they liked and a doctor they trusted, and on an initial round of IVF, they ended up with eight embryos. According to Tammy, the doctor told them for a woman her age, normal procedure at that time—1999—would be to transfer three embryos. The doctor added that two of the embryos seemed particularly high-quality, so they could do two if they wanted to. Tammy—who remembers wondering what "you *could* do two" meant, exactly—wanted to be sure the procedure would succeed. Since three was normal procedure, she told the doctor to go ahead and transfer three.

"That was a bad decision," says Tammy now, sitting on a couch in her living room, a lovely woman with shoulder-length brown hair, wearing jeans and a soft blue sweater, the sort of friendly, pretty, relaxed, accessible woman that any child would want for a teacher. Her husband, she says, still feels guilty about it.

Whether it was or was not a bad decision could be argued. At any rate, three embryos were transferred. Several weeks later a blood test confirmed Tammy was pregnant. Early tests suggested she was pregnant with twins. At a subsequent sonogram, however, Steve sat intently watching the ultrasound screen. "How did that one get into the same sac with the other?" he asked. Looking closer, the doctor saw that Tammy was carrying three fetuses. Two were so close together that one fetus had been hiding the other. Their hearts were beating in unison, so the sound of an extra heartbeat had not been picked up on the monitor. What Steve and Tammy had was triplets: two identical twins, and one fraternal sibling. Two embryos had taken, and one of these embryos had split in two, or "twinned," something that happens to IVF embryos more frequently than naturally conceived ones, for reasons nobody understands.

The sonogram yielded further disquieting news. The identical twins were monoamniotic, meaning they were growing in the same sac. The twins also shared the same placenta. This was far from ideal. In the best-case multiple pregnancy, each fetus has a separate placenta and is encased in a separate sac, ensuring that each will have its own source of nutrients. Having twins in the same sac meant, among other things, that Tammy was at risk of developing twin-to-twin transfusion syndrome (TTTS), a serious disorder in which two fetuses are so interdependent, and so inconsistently fed by the placenta, that one twin acts as a parasite on the other, siphoning off much of the oxygen and nutrients. A twin-to-twin pregnancy is a volatile situation to manage; what's good medical care for one twin may be bad for the other, and the existence of a third sibling makes the situation even more problematic, Tammy remembers being told by Alfred Khoury at Inova Fairfax Hospital, where she had gone for her consultation.

"I thought, 'Oh my gosh, why is he saying all this?' " says Tammy, who listened astonished as Khoury laid out the problems that could arise. Tammy was put on bed rest at sixteen weeks, told to stay at home with her feet up. Just before her nineteen-week appointment, she began suffering back pain, which is one sign of premature labor.

At that nineteen-week ultrasound, Tammy's doctors saw that she had in fact developed twin-to-twin transfusion syndrome. Ominously, her cervix was already thinning, suggesting that her body was preparing to go into labor. All of the babies, however, displayed healthy heart rates. The doctors performed an emergency cerclage, stitching Tammy's cervix to hold it closed. They drained off excess amniotic fluid and began consulting with national experts in the management of twin-to-twin transfusion syndrome. Tammy was moved into the high-risk pregnancy wing, where doctors could watch her around the clock as she lay on her back in the hospital bed, a position that, ideally, she would occupy for the next several months.

Tammy's mission was simple: to keep her three babies in utero past the magical twenty-five-week point. It became her one, all-consuming task. Steve began bringing Tammy cases of Ensure, which she drank, can after can after can. Everything was going reasonably well: her doctors were closely monitoring the levels of her amniotic fluid, and she was receiving drugs to stave off contractions.

Then, Tammy recalls, "At twenty-one weeks, six days, I went into labor." She was given a stronger antilabor drug, but it had no effect.

"We can't stop it," Khoury was telling her, unbelievably. A neonatologist was called; he told them that babies so extremely premature would not live longer than twenty-four hours. Their organs were simply too unformed. A vaginal delivery was decided upon. Tammy was given Pitocin to bring on the contractions that previous medications had been administered to prevent. Her husband called Tammy's parents to tell them that labor had begun, and the triplets were not expected to make it. "What do you mean, they're not going to make it?" Tammy's mother said, incredulous.

Tammy was given an epidural to ease the pain of three sequential vaginal deliveries. She wanted the medication; wanted anything to help her pretend that this wasn't real, this wasn't happening, she wasn't giving birth to three not-yet-twenty-two-week-old fetuses, they weren't emerging four months before they should. "I didn't want to feel giving birth to them," she remembers. "I thought: 'They're not going to stay.' I remember saying to Steve at one point, 'This is a bad dream. I'm going to make this a dream. I'm going to make it so that it's not real.'"

But it was real. One baby—the fraternal sibling, a boy—emerged, already stillborn. And now a girl was being born, the larger of the identical twins. About ten minutes later—"I never thought I would forget the time" Tammy says now—the third baby was born, another girl. Both twins were alive, but clearly would not survive. People were rushing around, leaving the room, returning. Tammy lost awareness of exactly what was happening, at least until she was faced with a third dilemma for which nothing in life had prepared her. Here, abruptly, was a nurse, asking: Did she want to hold the babies?

Did she? No, Tammy replied, she didn't want to hold them. "I didn't even want to see them." Maybe if she didn't see them, they would not exist. Maybe if she did not hold them, none of this would have happened and she could just get dressed and go home.

"You're going to want to hold them," the nurse told Tammy, kindly. "I know you don't want to now, but you will wish that you had."

"And then finally it struck me," Tammy says now. "*Of course I want to hold them.*"

And so she held them: she held her dead, and dying, babies. She begged Steve to hold them, and he did. So did Tammy's mother, who, with Tammy's father, had driven down from Lancaster, Pennsylvania. And now another nurse was asking, "What are their names?" Tammy and Steve named the boy—who would not get a birth certificate, being stillborn—Jackson. They named the larger and smaller twins, respectively, Maya and Maddelena. Tammy was wheeled into a recovery room; she would later learn that "there was a butterfly on my door, which is the universal symbol for somebody whose baby didn't live." Several days later she was discharged. A week later, they had to organize a funeral. "So we called—between Steve and me we called maybe twelve of our friends," remembers Tammy. "Everyone came. We didn't invite everybody to the funeral; that was for family. But we had a reception afterward. It was pretty awful, but they came and I'm glad they came." The logistics still seem incredible to her. "I just remember sitting at this table, ordering caskets for three infants, thinking, this isn't happening."

When you experience what is known as "perinatal mortality," the experience doesn't go away. It stays. It burrows in. Tammy was, she realized, in a severe depression. She began seeing a therapist familiar with fertility treatment and its aftermath. "I owe her SO many boxes of Kleenex," says Tammy. Days would go by where she talked to nobody except her family. Eventually she began to realize that it was time to restore some normalcy to her life. So she talked to her principal, who welcomed her back. Tammy wrote the teaching staff an open letter, telling them what had happened and how to approach her. It seemed one small thing that she could do in memory of her babies. She could help educate the public, even a little bit, about how to react to someone who has lost a baby. She didn't want people to think she was fine, because she wasn't. She didn't want them to think they could not ask her about what happened, because they could. She wanted to talk about the children, who after all had existed. They were supposed to be here with her now. She was supposed to be on maternity leave, taking care of them. "I said, 'I don't want you to avoid me. I don't want you to not ask me about things.'"

Tammy didn't feel ready to enter a classroom, so her principal let her work in the book room. During this time, one close friend put Tammy in touch with the Center for Loss in Multiple Birth, or CLIMB, a nonprofit group based in Anchorage, Alaska, with members in the United States, Canada, and around the world, and a website that by now has been translated into Chinese, Russian, French, and Spanish.

CLIMB exists because of fertility treatment and its tendency, when it succeeds, to succeed too well. You can go on the website and read the stories. They are narratives just like Tammy and Steve's, harrowing stories of people who always wanted children; who had to work harder than most to get them; whose pregnancies went horribly wrong; and who, as a result, found themselves losing one, or two, or three, of the children they'd wanted so badly, and living, forever, with that loss. Tammy would get the CLIMB newsletter and read it over and over. "Steve said, 'Why are you doing this to yourself?'" she says, but it helped in what Tammy describes as "that constant search to feel normal. To not feel so isolated." She found that there were other people in her area who also belonged to CLIMB. She and Steve went to a meeting, where they were invited to tell what happened. Tammy couldn't. So Steve told their story. There were six couples at that meeting. One of the women, Tammy saw, was in worse shape than she was.

"Her husband was saying: 'She doesn't talk to anyone. She doesn't even accept phone calls from her mother.'"

She found out, too, about a separate infant loss group that met in her area. That group and the CLIMB meetings were, she says, "my best therapy ever." She learned that Inova Fairfax Hospital, like many hospitals, now has an annual memorial service for parents who have lost children before, during, or after birth. She went to the service, where a nurse read a poem that described, exactly, what Tammy had been through. She thinks it might have been the same nurse who told her she would want to hold her babies.

She finds now—sitting on her couch, trying to describe the transformations that have taken place within her in the past several years—that she is living with several incremental layers of guilt. There is the guilt about the decision to transfer three embryos, rather than two, something that Tammy and Steve, it seem to me, take far more responsibility for than they should. There is the guilt about having visited her elementary school for a brief baby shower the day she was supposed to start bed rest. There is the guilt about how hard she finds it to visit the cemetery where the three babies are buried. "I never like to go there, really," says Tammy. "I go sometimes, but I don't like going there with other people. I dealt with that guilt for a long time. I finally realized: I remember them at other times and in other places. I don't have to be at the cemetery to think about them." A friend gave her three memorial wreaths, which are hung on a wall in an upstairs bedroom.

And there is the guilt that she held the babies for what seems, now, like too short a time.

"Months afterward I was so depressed about it," says Tammy. "I said: 'Why didn't I hold them longer?' They don't know that I loved them. I still feel that way. I didn't hold them for very long. I just wanted it to be over with. I just begged them to let me go home that day. I think they lived for a little less than an hour, the girls. I didn't hold them the whole time. That's my biggest thing I regret—I wish I had held them until they stopped breathing. They were alone when they died."

But how would you know? Who lays plans like that? Who studies, in advance, the lingering psychological impact of not having held your three premature newborns longer than you did?

What map exists for this uncharted territory of the heart?

"A Consumption Specialty"

It's true that patients sometimes pressure doctors to transfer more embryos than might be best; true, too, that lack of government research has severely retarded embryo evaluation. But it's also true that many doctors are implanting too many embryos, even when they know better, as a way of increasing success rates. In 2006, the American Society for Reproductive Medicine issued guidelines recommending that no more than two embryos be transferred into a woman under thirty-five and that doctors should consider transferring only one. However, the ASRM's research and membership arm, the Society for Assisted Reproductive Technology, declines to censure, rebuke, penalize, or criticize doctors who violate these guidelines. In part this may be because to set a strict policy—to set a concrete, professionally recognized standard of care— creates a situation in which violations could attract that most dreaded of outcomes: malpractice suits. Some outspoken doctors think this is going to happen anyway. "I can imagine a scenario where people discover— lawyers in particular—that you can sue the physician for not following the guidelines," David Keefe told colleagues in 2004. Predicting that multiple births are a lawsuit "waiting to happen," Keefe said, "I can see a case coming back to a doctor who put in three when you could see [transferring] two. I can see that happening."

To date, though, many doctors have been loath to transfer just two embryos, much less one. The explosion in multiples also has to do with the competitiveness of the IVF marketplace, where clinics recruit patients by advertising high pregnancy rates. Many clinics with the highest pregnancy rates also have high rates of multiples.

Some think that the current system of government regulation— spotty and inadequate as it is—may be exacerbating the problem. Until the early 1990s, the field of IVF was all but unregulated; in the United States, doctors were permitted to advertise any level of success, with no government verification of data. Concern began to grow that some clinics were inflating their success rates. "People were being taken to the cleaners," says Sam Thatcher, a veteran reproductive endocrinologist who runs a clinic in Johnson City, Tennessee.

Then in the 1990s, one of IVF's periodic, massive, weird scandals erupted. A Fairfax County, Virginia, fertility doctor, Cecil Jacobson, was found to have artificially inseminated, with his own sperm, many unwitting patients. They thought they were getting an anonymous sperm donor, when what they were getting was Jacobson, who refused to apologize or admit wrongdoing. Moreover, Jacobson was found to have committed another form of fraud, in some ways crueler: he had injected some other patients with a hormone that tricked the blood test into reflecting a pregnancy, indicating that his fertility treatment had succeeded. He would then inform the unhappy patients, looking at an ultrasound, that they had miscarried. So egregious were his crimes—they were crimes, and he was jailed—that a national outcry ensued. Congress called, at last, for government regulation of IVF.

The law that passed in 1992, and took effect in 1996, did nothing to prevent doctors from inseminating patients with their own sperm. What it did was direct clinics to provide information about how many IVF cycles they perform each year, how many succeed, and a few other details. Rather than have the monitoring performed by the U.S. Food and Drug Administration, which exists to regulate drugs and medical devices, oversight was assigned to the U.S. Centers for Disease Control and Prevention. It may seem a bureaucratic distinction, but it is beneficial to the doctors: the FDA is a regulatory agency; the CDC is not. It is the job of the CDC to collect data, conduct research, and promote public health. Not to regulate.

As regulations go, the law is almost toothless. Any fertility clinic that doesn't want to participate in the CDC's monitoring program doesn't have to. There is nothing mandatory about the reporting system. Of those clinics that participate—and most do—only 10 percent have traditionally had their data verified by the federal government. "There is so much lying," says Sam Thatcher.

Thatcher, like some others, argues that the regulation has backfired; that the chief upshot has been to fuel more multiple births. The CDC's annual publication of treatment outcomes put pressure on clinics to come up with good results, and one way to come up with good results—that is, a high pregnancy rate—is to put back lots of embryos. "It has done more damage than good," says Thatcher. By now, he argues, IVF techniques are good enough that two embryos should suffice for almost any patient. In Thatcher's clinic, the triplets rate plummeted after he instituted a two-embryo transfer policy for women under thirty-six. "We don't have triplets pregnancies anymore," he says. At his clinic, patients no longer have a choice. Embryo transfer is not their decision. "There need to be mandatory rules," he says. "You cannot put back more than two embryos. The ASRM guidelines don't work."

Around the country, many clinics are failing to follow Thatcher's lead. CDC figures for 2003 show that many well-known New York clinics still had high triplet rates. For example, the program at St. Luke's Roosevelt Hospital had a *14 percent* triplet rate in women under thirty-five, transferring an average of 2.6 embryos. At Cornell, the triplets rate for this group was 10.3 percent, the average number of embryos transferred 2.4. "We transfer less embryos than we used to," said Zev Rosenwaks, predicting that later CDC reports would show improvement. "But the truth is, at the moment we do not have any embryo viability [tests] that are useful. . . . Hopefully we'll be able [someday] to transfer just one, but at the moment, the technology isn't quite there. We have to accept the fact that when we put embryos in, particularly at the third day, we cannot be sure, despite the fact that they've gone from six to eight cells, that they will implant." Figures for 2004, released as this book was going to press, did show some improvement at Cornell. But around the country triplets rates varied wildly, from 1.5 percent at Thatcher's clinic to a whopping 20 percent in women under 35 at the clinic affiliated with Baylor, and the average number of embryos transferred varied greatly as well.

Yet clinics with a commitment to lowering the triplets rate, like Thatcher's, have been able to maintain a high success rate. In the D.C. area, Shady Grove Fertility has made a concerted effort to reduce triplets. In 2004 the average number of embryos transferred in women under thirty-five was 2.1; the triplets rate was 1 percent. Recently, Shady Grove made single embryo transfer the recommended treatment for women under thirty-six, and those using donor eggs. Shady Grove can do this in part because they offer a "money-back guarantee" for patients who, in return for a high upfront fee, receive a certain number of IVF rounds. If one round fails, they get several more. If treatment fails entirely, most of the money is returned. This way, patients can have fewer embryos transferred, knowing they will get more tries. Doctors in hospital-based practices sniffed at this money-back guarantee as too commercial, but it's hard to argue with the results.

Elsewhere, patients are left to make the profoundly significant medical decision of how many embryos to transfer. In some sense, multiple births in the United States are the result of a perfect storm of contributing factors: pro-life opposition to embryo research; the demand for total reproductive liberty—absolute freedom of choice—on the part of infertility patients' advocacy groups; and, not least, doctors who benefit from transferring practical, psychological, and moral responsibility for decision-making to patients. But patients going through IVF are not the medical experts. Plus, they want treatment to succeed. They've paid a lot of money. And rarely are they aware of the full range of risks. "Most people are not rational when they make decisions; that's why I have two sets of twins," says Linda Gulyn, who, as a psychologist, is fascinated by pressures driving patients during the few minutes most have in which to decide how many embryos to transfer. "I remember [the doctor for her second IVF round] saying, 'We could do one or two,' and Peter and I saying, 'Put two in.' It was irrational. I was tired of going through fertility treatments. I didn't want two [twins]. But not so much that I would have said, 'Just put one in.' I wonder what it is, what it is that leads you to making a not-good decision." And the thing is, she says, fertility doctors carefully avoid making that final call. They make a recommendation, Gulyn points out, and then "they throw it back at you."

David Keefe argues that IVF doctors benefit from a phenomenon known as splitting. When something goes wrong in a multiple birth, the patient blames the ob-gyn who delivered the babies, and not the fertility doctor responsible for their conception. "It's not your problem if she leaves with twins and triplets. The obstetrician, he's the one who has all the responsibility for the problems I created. You don't even see me anymore. All you think is: you're grateful. So it's a classic case of what we call splitting, in organizational psychology. All the goodies come to one person, and all the baddies go to another. I'm the good guy, they're the bad guy. [Fertility doctors] make huge amounts of money; they're incentivized to get pregnancies at all costs, which drives the numbers of twins and triplets up, and there's no disincentive not to create them. It's a structural problem, and you cannot rely on moral incentives. Everybody else will do it. It's a culture. It truly is a culture. People think medicine is a science, but it's a human civilization that has mores, practices, behavior, just like a primitive tribe. In medicine, our practices, mores, and behaviors are veiled in a cloak of professionalism, and even worse, there's a kind of a cargo cult of science. Cargo cults were these Melanesian tribes that during World War II got their first exposure—

they were Stone Age tribes that had been completely immune to technology until World War II. Bombers would crash and they thought these things came from heaven; they carried these relics around, like hubcaps from the planes. They would carry these things around like they are totems. That's what happened in science. People go in, and believe that science is a good force, and in fact it's actually being used by some business interest to advance their interests. People have to understand: there's very little science behind what we do. There was science before, but the people that are practicing it are not doing science. They're applying science, and doing it for business."

So what exists now is some doctors behaving responsibly, and others acting much less so. The impact, especially on families that are fragile or troubled to begin with, can be devastating. In 2003, police officers in Towamencin Township, Pennsylvania, arrived at the house of Tamra and James Seymore, summoned by a report of an unresponsive child. Police found the father holding "the limp and lifeless body of his four-year-old son." The child, who was described in a police report as "grossly underweight," was one of four quadruplets born to the family as a result of IVF performed in 1998. The child was dead of "inanition," a type of starvation caused by lack of food and liquid. Walking through the house, the officers encountered animal urine and feces; "tremendous clutter and a maze of plastic baby gates, . . . five small dogs and three cats . . . and three surviving quadruplets . . . asleep in one of the two rear bedrooms . . . enclosed in a pen, the walls of which were composed of various baby gates lashed together with plastic zip ties." The sink was filled with "dirty clothes, bottles, and trash." The three surviving quads and an older sibling were taken into emergency custody. Family members told police that the parents, who were described in the police report as "college-educated individuals," had been normal and accessible until the quadruplets were born, whereupon, overwhelmed, they shut out all offers of help. Both parents were convicted of involuntary manslaughter and sentenced to terms in state prison. Their defense lawyer said they were "unable to cope with the situation."

Every year around Christmas, the *New York Times* runs profiles of impoverished families whose circumstances make them worthy of special charity assistance. In a city where poverty is known to tens of thousands, these unfortunates are singled out as the "Neediest Cases." On November 21, 2004, the newspaper profiled Cristina Gopal, who had given birth to quadruplets after receiving IVF treatment. Gopal, a postal worker who had waited for four years for a promotion to the job she wanted, had to abandon her long-sought position almost immediately. Forced to go on bed rest in her first month of pregnancy, she would be unable to work for a year and a half. Her husband was a stock boy at Macy's. The babies were born at twenty-nine weeks—more than two months premature. "I was given the boys first," Gopal related in a first-person account. Two weeks after she brought the first two home, "the hospital called me to get the two girls. I started crying. I couldn't stop. I said, 'If I can't handle two, what am I going to do with the other two?'" Gopal, who lived with her husband and their children in a walk-up apartment in the Bronx, told the *Times* that "if my friends asked me if I wanted anything, I would say, 'Diapers and wipes.' Forget about the clothes. It was difficult in a walk-up, but we would wash them every day."

In the fall of 2004, an Arizona doctor transferred five embryos into the uterus of Teresa Anderson, a perfectly fertile young woman serving as a gestational surrogate for Luisa Gonzales and Enrique Moreno. "They told me there was a one in 30 chance that one would take," Anderson, who has never had any difficulty getting pregnant, later told the *Wall Street Journal*. The transfer resulted in quintuplets, to the surprise and dismay of both surrogate and parents, who were so overwhelmed by the costs of five premature babies that the surrogate waived compensation.

Their doctor, Jay Nemiro of the Arizona Center for Fertility Studies in Scottsdale, declined to comment to the *Wall Street Journal* about his embryo transfer policy. According to the CDC, in 2003 his clinic had a triplets rate of 17.4 percent in women under thirty-five. The paper conducted an investigation of embryo-transfer practices across the profession, and found that doctors routinely violate ASRM's voluntary guidelines. It did get a comment from Rifaat Salem, director of the Pacific Reproductive Center in Torrance, California, who told the *Journal* that he doesn't hesitate to transfer "10 or 12" embryos in some patients, even if they are under thirty-five, as long as he feels the embryos are of low quality and unlikely to result in a pregnancy. The paper also profiled one of his patients: Heesun Hall, who at age thirty-four, according to the paper, reluctantly permitted Salem to transfer four embryos during her first round of IVF. Hall and her husband felt uneasy about the number transferred. She ended up pregnant with quadruplets, which the couple reduced to two. The remaining twins were extremely premature, and one died. At two years old, the *Wall*

Street Journal noted, the remaining child still required a feeding tube. Dr. Salem told the newspaper that "since then we've learned a lot" and "she's one of the cases we all should remember."

It should of course be pointed out that many doctors do follow the guidelines, and that many speak openly and often about the need to transfer fewer embryos. David Grainger, president of the Society for Assisted Reproductive Technology, the membership group for U.S. clinics, says that SART does monitor clinics who have unacceptably high multiples rates; that such clinics are asked to justify these rates and to explain what they plan to do to bring them down; and that "membership in SART depends on clinics following guidelines." But there are no brakes on clinics that don't belong to SART. And it still seems legitimate to ask whether the professional organization is doing enough to rein in doctors who transfer too many. To date, no clinic has been expelled from SART for high multiple rates. After the *Wall Street Journal* article appeared, SART held a routine meeting at the 2005 ASRM convention. One of the topics was a new online clinic registry that SART plans to maintain itself, publishing data and success rates apart from the CDC. (In part, this may be because in 2004 the CDC, in a rare act of independence, stopped relying on fertility doctors to provide their own data, and assigned the job to an outside consulting firm.) During the presentation a group of doctors lined up at a microphone to make comments. One doctor said, "One of the missions of SART has been to lower multiple rates. Some people are following this, and others aren't. This [website] would be a great place to have a prompt that comes up: if you did not follow the guidelines, why did you not follow the guidelines, and then give an answer."

There was lots of nodding from doctors who presumably felt that egregious cases of extreme multiples were making everybody look bad. The SART official chairing the meeting replied, "I think that's a really interesting idea." However, she noted, "I think we're not quite ready for that."

"My biggest problem with the infertility people is that it has become a consumption specialty. It's no longer medicine," says Alfred Khoury. "There so many of them out there, they compete among each other to see who gets the patients, so they will do anything to maximize the chances of achieving a pregnancy. I just had a patient who came in who had an incompetent cervix, who lost a pregnancy last year at twenty-two weeks. The baby just kind of fell out. She lost it. It died. She's an IVF patient; she goes back to the specialist, who really wants to get her pregnant, he puts two in, she gets pregnant with twins. She will probably not make it with twins, if she didn't make it with a singleton. Now, here is a woman who comes into my office, ecstatic about being pregnant, and she's told, you're probably not going to make it. She's told, your best chance is selective reduction. This woman cries and cries and cries. Why do [patients] not realize that it's not somebody else's problem?"

What drives Khoury crazy, he says, is that patients come into his office expecting that modern medicine can make all problems go away. "Do you understand what mental retardation means? What cerebral palsy means? What deafness, blindness in a baby mean? Do you understand what it means for you as a family?" Khoury tells them. It's his job to discuss all scenarios, lay out all risks. At the outset of these conversations, Khoury says, "Most people don't understand. They're focused. They're on a mission. They say, 'I want this pregnancy, and they tell me you're the best doctor in northern Virginia.' I say to them: 'I was born in Bethlehem, but my initials are not J.C.'"

Khoury—intense, avid, passionate, honest, angry, cautious—is a perinatologist, or maternal-fetal medicine doctor. Both reproductive endocrinology and maternal-fetal medicine are subspecialties of obstetrics and gynecology. The rise of the first, fertility medicine, has led to a flowering of the second—those doctors who, as Khoury puts it, "clean up the mess." Like "stuck twins," the specialties are interdependent and, to a certain extent, mutually hostile.

"Sometimes I feel like our subspecialties are at odds," says Khoury's medical partner, Barbara Nies. "If they did the obstetrical care, maybe they would think twice before they do some of the things that they do."

Khoury himself is blunter.

"All my friends who are in reproductive endocrinology have a special place in hell," he says. "Well, maybe someplace between heaven and hell."

"We're All on Zoloft"

Another question is whether, in leaving the embryo decision up to patients, doctors adequately inform them of the risks. Many fertility clinics ask patients to sign a boilerplate consent form, attesting that they realize that transferring more than one embryo increases the risk of multiples. The form neatly exempts the fertility doctor from responsibility for adverse outcomes. But in the reproduction field, there is a subterranean conversation going on about how much detail to put in the warning. It's one thing to sign a boilerplate form; it is another thing to be truly counseled. In innumerable interviews, patients told me they did not feel warned; the multiples consent form was one in a sheaf of forms they had to sign, and the embryo-transfer decision was discussed, in an ad hoc way, at the last minute. "We had two women deliver quads, and neither woman was told in advance that she had a chance of having quads, or even triplets," said the perinatologist Barbara Nies. "Neither woman was talked to about multifetal reduction. One was IVF, and one was IUI. Both of them were very angry when they came for a consult with us, and we told them the things that could happen. 'Why didn't my IVF doctor tell me this?' they said."

"I see a lot of parents with high-order multiples, and the level of counseling they have had about high-order multiples is minimal," another perinatologist, Ronald Wapner, told government officials at a 2005 federal conference on fertility treatment outcomes. "The triplets [parents] that walk in have no idea of the risk of cerebral palsy, the risk of delivery under twenty-eight weeks, all of those real-life complications that go along with preterm birth. There is counseling and there is counseling, and we need to really understand what level of counseling is necessary."

And then there is the question of when the counseling takes place. Kristina Jorgensen-Harrigan, who lives in Louisville, Kentucky, e-mailed me after seeing my inquiry on her local multiples board, volunteering to chat about life with young triplets. Hers were then twenty-one months. Kristina, a graduate of the University of Louisville with a master's in elementary education, married when she was twenty-eight. She had suffered for years from endometriosis, and after a number of unsuccessful surgeries to clear it up, she and her husband adopted a girl, Megan. When Megan was about two, Kristina got "baby fever" and set out to try to have another child. By then her husband was working for a company whose insurance policy covered 50 percent of IVF treatment. So they went to a hospital-based fertility practice, where Kristina had a doctor she liked. During treatment, however, that doctor moved away, and Kristina was left with a doctor she didn't know. After undergoing egg retrieval, she and her husband ended up, she recalls, with four embryos. Three days later, on the morning of the transfer, she said, "I knew I was going to be a nervous wreck, so they gave me two Valium."

Up to then, she said, there had been no real discussion of multiples and the risks they entail; no discussion of how many embryos to transfer. Now the doctor came in to talk it over. Kristina recalled that he told her one of the embryos had been graded A; the other was a seven-cell embryo, and the two remaining embryos were considered grade C. "He told us that C-quality embryos almost never take," Kristina said. "His suggestion was that we go ahead and place all four back in, because the two were going to die off anyway."

Her husband was reluctant, but Kristina talked him into it. At that point, she said, the doctor brought out a form for her to sign, saying that they understood they were putting all four embryos in, and that this was not the norm. "I'm on Valium," she said. "They could have told me anything."

Two weeks later, Kristina got a call while she was teaching. She was told that she was pregnant, and that her hormone level was extremely high, indicating a multiple pregnancy. After the next blood test, it was even higher. She called her husband, saying, "My God, we could have triplets, we could have quads. We lived in a 1,000-square-foot house. I was already panicking."

She started hemorrhaging; rushed to the doctor, she apparently miscarried one of the fetuses. Her pregnancy, she thinks, started out as quads. Now it was triplets. "Every time we went in to meet the fertility doctor—and I do mean every time—he tried to convince us to reduce." Her husband would then ask the doctor, "Which of your daughters would you have eliminated?"

The triplets were born at thirty-four weeks; not so bad, for triplets. Baby A—the designation for the fetus closest to the cervix—was born first. They named her Abigail. Next came Baby B, whom they named Brigid. Baby C, Caroline, "came two minutes later; she was so wedged in, bless her heart, her face was all smashed on one side." All babies were taken to the neonatal intensive care unit, where they variously received nasal cannulas, feeding tubes, bilirubin lights. All of the triplets had issues, minor and major, which continued to unfold. At nine months Caroline was not rolling over; both she and Brigid at first were considered developmentally delayed and received physical and occupational therapy. Both Caroline and Brigid were also speech-delayed, though most of these issues have now resolved themselves. The smallest, Abigail, made it onto the growth chart at eighteen months.

It is Caroline, little Baby C, about whom Kristina worries most, she told me. "Caroline has some sensory thing going on, some weird sensory thing—her reaction to loud noises and smells. She has this thing where she carries big knitted afghans everywhere she goes, wraps them around her neck and shoulders, which provides a lot of deep pressure and helps calm her down. She's a different kind of kid. She has a hard time coming to midline with her hands. She can't do a pat-a-cake. Her trunk is very stiff; her arms are very weak. When she grabs things, she grabs them with a lobster claw rather than fisting." It is unclear what is wrong, Kristina said. There is something called "sensory integration disorder," and the doctors are keeping an eye on her to make sure she doesn't develop that. For some reason—maybe because of the way she was wedged into the womb—heavy pressure has a calming effect. "When she was first sitting up, she took the plastic credit cards, the [fake promotional] ones that came in the mail, Caroline would take them and smash them on her forehead and her nose and hold it there when she got really stressed out. She loved the remote control because it was long and rectangular and she could put it on her forehead and hide the world. She figured out early on, when she started crawling, that if she pushed her head on somebody that it would help her feel better. She was coming to me every few minutes and doing it. Then she started figuring out that Megan would let her do it, so Megan would say, 'Oh, Mommy, Caroline is doing her deep pressure thing on me.'"

After the birth of the three girls Kristina went back to work, teaching, and her husband switched to working nights. They had to move into a larger house, with larger mortgage payments; they bought a used minivan. After several months Kristina realized that she was suffering from postpartum depression, "which I understand is really common for moms with multiples: on top of working and trying to maintain, and never seeing my husband, I got up one morning and started crying and couldn't stop for three days." The theme of uncontrollable crying entered into many conversations I had with mothers raising high-order multiples. So she went on medical leave.

As a result, she and her husband were rarely able to go out together. Out of all the babysitters they knew, only one would watch all four children. Others required a co-babysitter, which is reasonable, but also twice the hourly expense. "As much as I love my kids, and as much fun as I have all day long with them, our whole lives changed," Kristina said. "We lost friends. We don't have the money to go out and do things anymore." Meanwhile their older daughter, Megan, suffers from the attention given to the triplets: "She'll say don't hold the babies, hold me," Kristina said.

"Our marriage has basically fallen apart," Kristina continued matter-of-factly. "We are going through therapy to try and get it back to normal, but we've basically become two adults that live in the same house, and take care of kids, and that's it. There is always the worry of how are we going to get this done. The house is always a mess. We always have ten loads of laundry to do; we sort of pass each other in the day, hey, did you do this. I have another friend that had triplets and a three-year-old. She and her husband are going through therapy. You become so consumed with trying to take care of your four, you forget about each other. We were talking this morning—[her husband] feels like he's on the bottom of my list. I feel the same way. When you've got three kids who are the same age, it's totally different than having a four-year-old and a two-year-old." When they adopted Megan, Kristina carried her in a sling and lavished her with attention. Now, "I always feel that somebody's getting left out."

Kristina and I first spoke in 2005. When I contacted her a year later, in the fall of 2006, she and her husband had separated, and were planning to divorce. The triplets were doing much better, including Caroline, but her older daughter was in therapy, to help her cope.

"I very often want to just, you know, strangle the fertility doctors for not preparing us for this," Kristina told me in 2005. "I laugh more every day than I ever have in my life, because they are just so much fun, and I've cried more than I ever have. I'm seeing a psychiatrist along with a therapist."

The psychiatrist visits helped, up to a point.

"I'm well medicated, let's just say that," she said. "Like all the other triplets moms I know. We're all on Zoloft. We're all on varying degrees of Zoloft and Prozac. I don't know anybody in our group who isn't on some sort of antidepressant. I don't think you could survive without it."

At the end of our telephone conversation, she urged me to call her back if I had any questions. "We're always home," she said.

"I'm Always, Like, One, Two, Three"

It would be a mistake to imply that people with multiples regret having multiples. All the parents I spoke with are incredibly grateful for the children they have, and are doing everything possible to provide loving care to them. Many felt euphoric when they found out that they were going to have more than one child. "I have a foursome now," said one prospective father and avid golfer, Scott Coleman, whom I interviewed when his wife, Tracey, was early in a triplets pregnancy. Scott, ecstatic that treatment had succeeded, had already purchased a minivan complete with "traction control, antilock breaks, side airbags, and a DVD center in the back." Once the children arrived, they felt their lives immeasurably enriched. Tracey, in an e-mail, reported constant trips to the doctor; life as a blur; speech and physical therapy; reflux issues; and so much happiness she was sorry to see the early months ending.

To epidemiologists and health-care workers, supertwins are an epidemic. To parents they are something more rich and complicated. They are a joy; a blessing. They are every waking hour. This being America, they are also an ideology. A lifestyle. When you become a parent of multiples, you join an identity group.

And so it was that parents of triplets became one more advocacy organization descending on the nation's capital when in July 2005 the Triplet Connection held its annual meeting in Washington, D.C. As always, the meeting featured seminars, meals, speeches, excursions, the traditional opening-night triplets parade, and lawn games—including, of course, a four-legged race. The convention took place at the L'Enfant Plaza, a downtown hotel that in one respect was an awkward choice of venue: the underground parking garage has a very low ceiling, so a warning was posted on the website advising families driving minivans with storage bins on the top—as most triplets families do—to unload before entering the garage. The media were invited to a photo shoot at the Jefferson Memorial. The population of triplets in the United States grows every year, and so, naturally, does the annual Triplet Connection photo: on the steps of the memorial could be seen white triplets, black triplets, red-haired triplets, blond-haired triplets, girl triplets, boy triplets, triplets wearing matching T-shirts that said "property of Jesus," teenage triplet boys wearing shirts that said "Fort Jennings Soccer." There were triplets wearing yarmulkes, triplets in matching blue Hawaiian shirts. There were three dark-haired eight-year-old-ish girls, all wearing orange capes.

"Look at the triplets!" said tourists, inevitably.

"Anybody who wants to be in the picture, sit down now!" said one of the organizers, as news photographers squatted in preparation.

A triplet girl with an American flag shirt fell down on the pavement, and everybody made a sympathetic noise.

"On the count of three!" said one photographer. "Everybody look at

Julio!" There was one photo of everybody; one photo of just the triplets standing. Innumerable strollers were parked to one side of the memorial plaza. The largest and best triplet stroller, an aircraft-carrier-type thing that costs more than $1,000, is known as a "triple decker," and here were triple deckers with plastic toddler seats, triple deckers with infant car seats, triple deckers with George Bush bumper stickers. At one point a security guard came over and wanly pointed out, to nobody in particular, that strollers must not be left unattended in post-9/11 Washington. His warning went unheeded. He eventually went away.

In one sense it was an unlikely place to hold a gathering of parents who are outnumbered by their own young progeny. The Jefferson Memorial rises alongside the Tidal Basin, a picturesque body of water that, in the wake of a heavy rain, was on the verge of flooding. In some places there is a slim railing between sidewalk and water, with plenty of room to wriggle through; in other places there is no barrier at all. After the shoot, children started spilling down the steps toward the water, antsy from holding still so long. The parents started desperately herding them, trying quickly to load them into strollers. "He wants to run," said one woman despairingly to her husband. "Can we go home now?" said one triplet. "No," said his mom. "Not yet."

I sat on the steps chatting with parents; one family had driven up from Lynchburg, Virginia, about three hours away. They were not sure the trip had been such a good idea. They didn't realize how hard it would be to fit everybody in the hotel room. They had a four-year-old boy and three-year-old triplets. And the chaos of the hotel lobby! "They're a little too young," said the mother. "They need enclosed spaces."

It had started to drizzle. A stroller went by carrying three little girls, all in pink raincoats, with umbrellas secured beside each seat. The children were starting to scream. A parent was walking by with a tote bag that said "It's always a party with triplets." Swarming up the steps toward me were three girls, about eighteen months old. Their mother was devoting most of her energy to keeping them away from the water. "I'm always counting," she said. "I'm always like, one, two, three."

After the photo session, everybody made their way back to the hotel for a series of discussions. Parents of quads and higher are also welcome to join the Triplet Connection, and they were there as well, including one man wearing a "quad squad" T-shirt. The hotel lobby was a mass of parents, children, mega–diaper bags; people were trying to wedge strollers into elevators, with varying degrees of success. Amidst the commotion was one man sitting at a lobby table playing cards with three boys, an island of calm. Entertainment was being set up for triplets of all ages: in different meeting rooms were clowns and magicians and women in tutus saying things like "Can you bark like a dog?" A venting session was being held for teenage triplets. On the second floor were parent seminars on financial management and marital relationships and other relevant topics. In the session devoted to education, parents were talking about overseeing homework. They commiserated about trying to help when, for instance, all of your children have different lists of spelling words from different teachers. "The teachers say that all the homework is the same, but it's not," lamented one mother, who gathers her triplets and two older siblings around the kitchen island, and does her best to help everybody study. "It's like running a marathon."

The big concern was classroom placements: whether the triplets should be in class together, and if not, when to split them. One woman who described herself as the mother of "three surviving quadruplets" worried about the breakup of two of her girls. Up to now, she said, they had been in the same class together, but one girl was going to be separated into a class for the deaf and hearing impaired. Another, who was home-schooling, worried about a son who is "very smart but non-ambitious; he quits easily." Another wondered what to do when "one is intellectual and ambitious, and feels he has to do the work not only for himself but for his brothers." They talked about what to do when the school principal, at the assembly at the beginning of the year, asks your triplets to stand up so everybody can clap. "I'm afraid they're going to make a spectacle out of my children," said one mother. Everybody nodded knowingly. Among parents of supertwins, it is accepted that your children will be stared at and discussed, that questions will be asked about whether you did use fertility drugs or didn't; strangers will consider themselves entitled to know how you conceived and how your family gets through the day. Among such parents there has developed, understandably, a sense of aggrievement. Still, I admit I sat there wondering what exactly is the difference between a principal asking a group of triplets to stand—which, apparently, is a wrong thing to do—and a triplets networking organization having a big photo shoot and inviting media to attend. Which, apparently, is not a wrong thing to do. Were the kids being made a spectacle out of by the very organization that caters to their welfare? Or was it simply an opportunity to educate the world about what these families are going through? And anyway, when you are getting triplets together like this, who could resist taking a big photo? It's hard, sometimes, to know how to treat the spectacle the children naturally create, and the parents were trying to figure it out. Dress the triplets alike? Or not? The contemporary thinking seems to be no, but as one woman pointed out, "I sometimes do for logistics, so I can see them in a crowd." One husband, himself a twin, talked about how he and his brother, growing up, would deliberately destroy one outfit whenever they were given two that matched. The humorous, easygoing moderator of the group, Stephen Kennedy, headmaster at a private school in Atlanta and himself a triplet father, told of taking a trip to Montana while his wife was pregnant, and running into two elderly twins out for a walk. They stood chatting about their upbringing; when he asked for advice about raising multiples, "One of them said, 'Our parents dressed us alike and we did everything together, and we've never married, and we've never made any other friends.'" And that, Kennedy said with a laugh, "was very scary."

Mostly, though, in his skilled veteran educator way Kennedy took pains to allay anxieties. He talked about how, as an educator, he had seen a steady escalation in parental worrying and how, for parents of multiples, the worrying is exponentially magnified. "For the most part," he reminded them, "children survive pretty well the situation they're in."

For the most part. But there is such unpredictability. This is true of childbearing in general, but so much more true of high-order multiples. Triplets can work out wonderfully. They can be a tragedy. They can be anything in between. Two of the babies can turn out fine, and the third can be hard of hearing, or afflicted with a touch of scoliosis. The third one can emerge malformed, with the gastrointestinal organs incompletely formed and no bowel opening; the third can die slowly and inexorably, never leaving the neonatal intensive care unit, like the triplet of one woman I interviewed, who with her two daughters releases a balloon every year to remember the child who didn't survive. There is no way to know which category your own pregnancy will fall into: whether you will dodge the bullet of prematurity or loss, or whether it will hit you once, or twice, or three times. Down in the hotel lobby, Janis Elspas, the busy and congenial media director of the group, had put together an Excel list of parents willing to be contacted by media. On Elspas's list was a family whose triplets all had attention deficit disorder. Another family consisted of two lesbian moms, both pregnant from the same sperm donor. One was pregnant with a singleton, the other with triplets.

Elspas and I chatted about the situations, and decisions, that parents of multiples face. "It is possible to have a successful triplets pregnancy!" she emphasized, which is clearly true. Surveying the scene, I could only think what a massive gamble a high-order multiples pregnancy is. And these are your children. This is your life. This is their life. You don't get to do it again. You don't get to go back and say, "I've changed my mind, this is too hard. On second thought, let's transfer a single embryo."

CHAPTER ELEVEN

DELETING FETUSES:

SELECTIVE REDUCTION,
ART'S BEST-KEPT SECRET

What a patient who wants to reduce the risk inherent in a multiple pregnancy can do—if she has the will—is get in touch with one of a select group of doctors who are willing to take three fetuses and, as one puts it, "turn three into two." Some larger IVF practices have a doctor on staff who performs selective reductions, a procedure in which one or more fetuses are eliminated through an injection of potassium chloride, which stops the fetal heart. The other IVF practices invariably know a specialist to whom they can quietly direct parents facing, suddenly, the question of whether to eliminate one or more fetuses so the others may thrive. This, then, is another "choice" facing today's woman and her partner: a termination situation in which the emotions are somewhat different from those surrounding abortion. These are wanted pregnancies, avidly sought. Though abortion itself is not pleasant, it seems fair to say that reduction is even more agonizing, with more complex emotional consequences. One infertility website, Fertile Thoughts, has a chat room devoted to the procedure. The contributors are invariably women, either deciding whether to undergo; about to undergo; or recovering from undergoing a reduction of their pregnancy. The discussion threads have titles like "three-year anniversary" and "five-year anniversary" and "just starting to grieve." It is an anguished conversation, punctuated by computer emoticons, usually the weeping ones, or—when someone is trying to comfort somebody else long-distance—the little emoticons that hug. When I looked at the message board, the postings included one from a woman who had not reduced her triplets and ended up losing all three, and who described herself as "a shell of the person I was." There was one from a woman marking the anniversary of her reduction from quads to twins, a day she remembered with "such clarity, but so, so, so much sadness." Other entries said things like "Is the procedure painful?" and "I feel sorry for anyone having to make this awful decision in the making of their family" and "I wonder all the time if I did the right thing" and "There are too many of us who have been faced with this."

Actually, nobody knows how many people have been faced with this. Multifetal reduction is one of fertility medicine's most unpleasant secrets, and one of the best kept. There is no way to know how many IVF pregnancies start out as triplets or quads and are quietly turned into something more manageable. The annual CDC report does not record the ultimate fate of each multiple pregnancy. And it ignores multiple pregnancies engendered by drugs alone. Nobody tracks the outcome of those. According to one textbook on multiple pregnancies, *Iatrogenic Multiple Pregnancy*, by the year 2000 some 5,000 reductions had been formally reported worldwide, by some well-established IVF centers. The textbook notes, however, just how inadequate these numbers are: "Undoubtedly there are many others that have not been reported," it writes. Undoubtedly: around the world, at least a million IVF cycles have been performed since 2000. There are scores of postings on the Fertile Thoughts website alone, at any given moment.

What's interesting is how difficult it is to obtain remotely reliable numbers. The Centers for Disease Control and Prevention does not publish selective-reduction figures; according to the CDC's annual report on fertility clinic outcomes, any numbers it does have are considered unreliable.

"This is a very sensitive topic," says SART president David Grainger, meaning that it's sensitive, personally, for patients, but also, politically, for doctors. SART maintains its own database of fertility clinic procedures and outcomes, but Grainger said that he would be "reluctant to assign any numbers to" IVF pregnancies that are selectively reduced each year, because any numbers SART has may be underreported. In short, nobody is tracking them, or at least, nobody is making them public. During a lunch interview, a therapist who works with developmentally disabled children remarked to me that she had three colleagues who—knowing all too well what prematurity can do—had reduced their own pregnancies to singletons. One woman, and she herself knew of three reductions.

Certainly, the practice is busy at the office of Mark Evans, one of the few doctors who not only performs reductions but is willing to discuss all qualms, ethics, issues, outcomes. Evans has seen no decline in his own patient load, which finds its way to him from around the world. Over the years Evans has debated on television and in print about the ethics of reduction, often sparring with pro-life opponents. Coming as he does from the "chutzpah school of obstetrics," Evans is often the doctor quoted way down in the newspaper articles about this or that high-order multiple birth, the lone voice pointing out that sextuplets are really not altogether a good thing.

Evans, who describes himself as an obstetrician-geneticist, is a pioneer in fetal therapy, well-known for saving babies in utero. He was one of the first doctors to devise genetic therapies that can be administered to cure fetuses in the womb who are afflicted with disabilities. He developed the first in utero correction of a genetic disease, SCID (so-called bubble babies), using stem-cell transplants. He has also pioneered fetal surgeries, including bladder shunts for fetuses with urological obstructions. The goal for all this is the delivery of a healthy, unafflicted baby. In some cases this can be achieved by treating a fetus in utero. In some cases it is achieved by sacrificing a fetus in utero.

"You could make the argument that we never should have invaded Normandy, because we knew that we were going to lose the troops on the beach, but there was all of Europe at stake," says Evans, who proposes this analogy as a way to think about selective reduction. Seen this way, reduced fetuses are the equivalent of those American GIs who drowned in the surf or were shot and killed making their way to Omaha Beach during the D-day landing. They died, but France was liberated and all of Europe saved, and the tide of the Second World War—and with it, humanity—was turned.

In 2005 I spent two days with Evans. He had just moved from a hospital into a private office in an ob-gyn practice on Manhattan's Upper East Side. It was a boutique practice located discreetly on the ground floor of a townhouse, not far from the gift shops and nail salons of Lexington Avenue. Before patients may be admitted, they must ring a buzzer and wait for the door to open. Inside, the crowded waiting room offers upscale leisure magazines, including *Parenting, Panache, Golf,* and the *New York Times Magazine* special section on high-priced homes.

The day I was there, down the hall in a tiny examining room, a sonographer was saying, "Do you want to see the screen?"

The sonographer, Rachel Greenbaum, was sitting on a high stool next to an ultrasound machine. Near her, flattened unhappily on an examining table, was a patient: pale-skinned, fine-featured, tall, in her thirties. The woman was wearing a hospital gown. Beside the woman was her husband, sitting in a chair, holding his wife's hand. He too was pale, and like his wife he looked miserable. The way the ultrasound screen was turned, Greenbaum could see it, and so, craning, could the husband. "Yes, I'd like to see them," the patient said firmly.

"I'll just take a few pictures and I'll show them to you," said Greenbaum.

"Them" applied to the three fetuses in the woman's belly, where they had been growing for slightly longer than eleven weeks; the pregnancy was nearing the end of the first trimester. Evans has found this to be the best window of time in which to perform a reduction. Waiting that long provides time to see whether the pregnancy might reduce itself naturally through miscarriage. By eleven weeks, moreover, the fetuses are sufficiently developed that genetic testing can be done to see which are chromosomally normal.

While she was waiting for the measurements to be taken, the woman on the table told Greenbaum how she'd ended up here. After undergoing a series of miscarriages, she and her husband had sought treatment, and were diagnosed with male-factor infertility. Like many couples suffering recurrent miscarriage, they had IVF performed, with the hope of using pre-implantation genetic diagnosis to test the embryos for chromosomal problems. They hoped to transfer only chromosomally sound embryos, and reduce the chances of another pregnancy loss. But, she said, they ended up with only three embryos, which was not enough to risk doing the PGD procedure. All three were put back. All three took.

"Triplets," said Greenbaum.

"So they tell me," said the woman, her voice hollow.

And sure enough, on Greenbaum's screen were three little honeycombed chambers with three fetuses growing in them. The fetuses were moving and waving their limbs; even at this point, just shy of three months' gestation, they were clearly human, at that big-headed-could-be-an-alien-but-definitely-not-a-kitten stage of development. Greenbaum periodically magnified one fetus and brought it into focus; she would then freeze the frame, manipulate some controls, and take precise measurements. She was doing two things: measuring the fetuses to assess their growth and see if any one was lagging; and taking a "nuchal translucency," measuring the fluid behind each fetus's neck. An excess of nuchal fluid suggests a possible problem: Down syndrome, for example. The procedures took about ten minutes, during which the patient could only peer upward at the ceiling and wait for it to be over. "What kind of creatures do they look like this time?" she asked, at one point. "Last time they looked like shrimp."

"They don't look like shrimp; they look like babies," said Greenbaum kindly.

"We're not having shrimp," said the husband nervously, affectionately.

"They are all measuring at eleven weeks and six days," said Greenbaum.

"That's right," said the woman wonderingly. "It is twelve weeks tomorrow."

So far there was nothing anomalous about any of the fetuses. Her job done for the moment, Greenbaum turned the screen toward the patient. "That's the little heartbeat," she said, pointing to the area where a tiny organ was clearly pulsing. "And there are the little hands. There's the head. The body."

"Oh my God, I can really see it!" cried the patient. "Oh my God! I can see the fingers!"

"Okay!" she said, abruptly, gesturing for the screen to be turned away. She began sobbing. There were no Kleenex in the room, so her husband gave her a paper towel, which she crumpled to her face. The patient spent the rest of the procedure with her hospital gown over her face, so she would not have see any more of what was happening.

Day 1

What was happening was day 1 of a two-day process, developed and perfected by Evans, a large man with the occasional impatience of a person who all his life has been smarter than most of the people around him. In high school, Evans says, he became interested in genetics, and soon perceived that the most interesting action, genetic-therapy-wise, was going to be prenatal. He became known as an adept practitioner of fetal therapy: somebody with the medical know-how, hand-eye coordination, and the nerve to treat a tiny creature growing deep and invisible to the naked eye, within a vulnerable belly. In the early 1980s, Evans says, he was contacted by an ob-gyn who had a patient pregnant with quadruplets from fertility treatment. The patient, not even five feet tall, was too small to carry four babies to term. The doctor saw no solution but to abort them. The woman, unwilling to sacrifice a pregnancy she had worked so hard for, asked whether it might be possible to do a "half abortion." The way abortion is normally performed, through vacuum suction, this would not be possible, but the doctor called Evans to see if there might be another way. "I don't know if it's possible, but I know how I would do it," Evans replied. The woman was sent to Evans, who, as he puts it, stabbed two of the fetuses with a needle. "Not an elegant technique," he acknowledges. But it worked. "I reduced four to two," he says, "and the two are in college right now."

The technique has been refined. Today, Evans would perform CVS, chorionic villus sampling, on his reduction patients. In CVS a small portion of the placenta, which shares the chromosomes of the fetus, is sucked up into a hollow needle. The tissue is shipped to a lab for an overnight test to determine the makeup of key chromosomes. In particular, the lab will look at chromosomes 13, 18, and 21 (the sources of some of the most common and serious birth defects) to make sure the fetus has the normal complement of two. The same test can also determine the gender of each fetus. Evans today also studies the position of each fetus. The point is to seek a rationale for which one, or ones, to reduce. In the absence of genetic flaws or location criteria, if there is something that Evans feels suspicious about, he may use professional instinct to decide which one to eliminate.

Most of Evans's practice does not involve reduction. Mostly, he does genetic counseling and testing of pregnant women, many of them with high-risk pregnancies, to whom, in the majority of cases, he gets to deliver the blessed news that their baby is normal. "We get to give good news to 95 percent of the people we see," Evans says. The days I spent with him, Evans had three sets of patients coming in for reduction, in addition to his regular patient load. The first was the unhappy couple, who left the office shaken. Another was a woman who had flown in from Puerto Rico and who arrived for her ultrasound alone, clearly terrified and needle-shy. She shrieked throughout the CVS procedure, making the tense and difficult process of aspirating three incipient placentas even more tense and difficult. This work takes a variety of skills. Evans must be able to look at a sonogram, make a razor-sharp 3-D version of it in his mind, then figure out where the things he sees on the ultrasound—three fetuses, three placentas—are located under the skin. When performing CVS, he plunges a needle either up through the vagina or down through the belly; pinpoints the placenta, which at this point is still frondy, like a marsh; and draws a tiny piece into a vial. He can do all this remarkably quickly, but because this patient kept squirming and screaming, he kept losing his position. "The more you relax, the easier this goes," he kept saying, trying to soothe her. Finally the procedure was finished and the patient was sent away to await, overnight, the results of her tests.

His third couple was not a heterosexual pair, or not exactly. They were two women, smart good-humored high-energy women, one of whom early on in the consultation said, almost rhetorically, "So how did two mostly straight women end up together?" She proceeded to answer her own question. This woman was a physician, practitioner of a high-level surgical specialty involving young children with severe birth defects. The other had a job that also involves children in difficulty. Like the other patients seeing Evans for reductions, the women were gracious enough to let me sit in, but asked that identifying details not be revealed. Both women—I'll call them Jane and Emma—were very open and likable. According to Jane, the doctor, over the years she had had several long-term relationships with men, but the relationships didn't survive her career and her intense work schedule, and after a while she couldn't take any more heartbreak. Meanwhile, Emma was showing interest and that, long and short, is how they ended up together. They identify, they guess, as bisexual. The story mildly interested but did not in the least faze Evans, since when you work in an Upper East Side gynecological practice you tend to get a broad view of human love relationships. At one point one of Evans's ob-gyn colleagues popped into his office wanting to know whether a patient who had recently gotten a prenatal test could use the same tissue sample to do a discreet paternity test to see who was the father of her baby, her husband or her boyfriend. (The answer, Evans found after checking, is: yes, she could.)

Emma was the one who was pregnant. During Evans's inquiry about genetic histories, the women said that they had selected a sperm donor of mixed nationality. Emma rolled her eyes, indicating that there was a long back story on donor selection, which they did not go into. It emerged that on their first intrauterine insemination attempt Emma had suffered a miscarriage. But the next time, she got pregnant with quadruplets. "We went from famine to feast," said Jane, explaining that they felt carrying four babies to term would be far too risky. These women knew all too well what serious disability does to a child and to a family. Among the professionals who care for them, severely disabled children are known, with respect and gravity, as "gorked kids." The women's goal was to do everything possible to avoid gorks. Evans outlined what would happen: measurements would be taken, CVS performed, the samples sent off for testing. He reviewed the criteria for deciding which to reduce: chromosomal abnormalities, fetal position, instincts. "And then, if absolutely nothing else matters, and if everything else is equal . . ."

"Selection for sex," finished Jane.

"If there's a gender difference, we'll talk about that," he acknowledged.

"I used to be totally not willing to talk about gender," elaborated Evans, who has constructed his own ethics during more than twenty years of practice. At the outset—in the absence of anything like rules— he worked with a bioethicist to develop guiding principles. For years, he says, the majority of sex-selection requests came from Asian and Indian parents, who tended to want to keep the boys. That he would not do. Increasingly, however, what he sees are "requests from everybody, half and half, boys and girls." What people want, now—and what seems, to him, morally more acceptable—is the Holy Grail of the modern two-child family: balance. What people want now is one boy and one girl.

The two women had been thinking about this. They asked themselves what gender would fit better into a two-woman household. Anecdotal evidence suggests they are not alone, and that increasingly gays and lesbians are a market for fertility techniques that permit sex selection. There are a number of ways to do this. Sex selection of embryos can be obtained from some IVF clinics using techniques that include spinning sperm to separate X-bearing sperm from Y-bearing sperm, as well as PGD testing of an embryo's chromosomes. At the 2005 ASRM convention, one San Francisco doctor took the microphone during a discussion session to ask whether anybody else was getting sex-selection requests from gay and lesbian couples, and what people thought about doing it. "We do seem to get quite a few requests for sex selection," said the doctor, adding that it was her impression that gay men were asking for boys, and lesbians for girls. "This is just one of those down-in-the-trenches issues we all see and struggle with." Another doctor got up to say that he considered gay and lesbian sex-selection to be acceptable under the banner of reproductive freedom. "We do see requests," he said, "and we do honor them."

Now it was time for Evans to do his day 1 examination of Emma.

Everybody proceeded to the tiny examination room, where Emma positioned herself on the table. With her physician's instincts, Jane was having a hard time not seizing the ultrasound wand from Rachel Greenbaum. She contented herself with scrutinizing the four fetuses on the screen. "They are all tucked in really nicely into their little nests," she said, fascinated. "The most I've ever looked at in utero is two."

Greenbaum, whose demeanor is perfect for her job—warm, friendly, yet exquisitely well trained and professional—took the measurements. The growth of all four babies was fine. The nuchal folds were fine. There was no apparent abnormality. Here, roughly, is how the four fetuses were lying:

		B			
MOTHER'S			A	MOTHER'S	MOTHER'S
HEAD	D			CERVIX	FEET
		C			

Studying the positions, Evans quickly made one decision. "I want to test C," he says, "because I'm trying to keep it." Faced with four fetuses, he explained, he normally tests only three; testing all four is "gilding the lily." C was a candidate for keeping because C would be hardest to get at. To reduce C, he would almost certainly have to go through B, cutting off any other options. But in order to justify keeping it, he needed to make sure C was normal.

The woman were planning to reduce two of the fetuses and keep two. They were grateful in a way that they ended up with quadruplets rather than triplets, because having four made the decision to reduce more clear-cut. They were also grateful that when they first contacted Evans, he did not pressure them in one way or another, just cited the statistics; in going from four fetuses to two, their chances of losing the entire pregnancy declined from 25 percent to 7 percent. And if they fell into the 7 percent who lose the remaining twins, that, he said, would be "despite the reduction, not because of it."

For all their joking ("couldn't you reduce all of them?" Jane said once) both women found the procedure agonizing. "The baby is kicking," said Jane, watching A on the ultrasound. "I hate seeing that."

"It's killing me that we're going to do this," she continued, while on the table Emma was undergoing three CVS procedures with stoicism and calm. "I never thought I would feel that. I'm the most pro-choice person. I'm vehemently pro-choice."

"If you want to see D, you've got to look now," said Evans, who was almost done. Sure enough D was there on the screen: big-headed, waving its arms. "We're done," said Evans presently, having tested A, C, and D. He thought B was almost certain to go, being closest to the skin of the belly and therefore the most accessible.

Then Emma got up from the table, and the women left for the day. Tomorrow they would be called, like the others, once their test results were back. After they left, Evans had a scare: the messenger who took the morning's set of samples to a lab outside the city had gotten stuck in traffic, and there was no way he could make it back into Manhattan, pick up this set, and get them back to the lab before closing. Which would mean the two women would not get their test results in time to have the procedure done tomorrow. Evans, whose professional life is dictated by the unforgiving rhythms of the female reproductive cycle, had to switch to a Chicago lab to which he could overnight this last sample. Even if it didn't arrive till morning, Chicago could perform the test within hours. It all involved quite a bit of irritation, and a number of calls on his BlackBerry: to the first lab, the second lab, the messenger agency. "You've got professionals on both ends," he said, "held together by high school dropouts."

"Dos Niñas!"

Midmorning the next day, Evans got the results back for the first two sets of reduction patients. He scribbled them on the back of an appointment sheet. He delivered the results, first to the Puerto Rican woman, who entered his counseling room this time accompanied by her mother—two heavily lipsticked, well-cosmeticized women in tailored suits. Evans told the patient that all three of her fetuses were normal. The patient let out an elated squeal. This, said Evans later, is unusual: often couples are relieved if there is something wrong with one fetus, since this makes the decision to reduce easier to bear. "We can leave you with two," he said.

"Do you know the sex?" asked the mother of the patient.

"All are girls," Evans said, in front of him a sheet on which he had written

A: XX
B: XX
C: XX

"Two girls! That's what she wanted!" said the mother.

"Dos niñas!" said the patient, who ran out of the counseling room and into the waiting room, where she summoned her grandmother and another elderly woman, and suddenly there were four abundantly accessorized Puerto Rican ladies in the room looking ecstatically at Evans, who tried to explain what would happen next. Because there were no abnormalities and no sex differences, he said, he would go to the "next level of subtlety" in determining which to reduce.

"It's up to you," said the mother of the patient.

"It can't be three?" said the patient herself, wistfully.

Evans outlined the risks again, reviewing the loss rates for triplets. The patient's mother was clearly in favor of reduction. They had been over this before, she said, with the IVF doctor in Puerto Rico. "It's too much," she said of triplets. "The risk would be too great. Sometimes you have to do unpleasant things to have a family."

"Very unpleasant," said the patient, who went obediently into the examining room, where she was accompanied by one of the older women, whom she introduced as "the wife of my grandmother's brother." The great-aunt-by-marriage was equable and calm; she took her place in the chair beside the examining table, and held her grandniece's hand while Evans reviewed the measurements and map from yesterday. He opened a pack of instruments; a square napkin with a hole in the center went over the patient's belly, which he swabbed with Betadine. "I'm a coward," said the patient, bracing herself. Beneath the makeup it was apparent how young she was. Evans and Greenbaum, wearing surgical gloves, inspected the ultrasound and evaluated the situation. Evans decided, based on location, to go for baby C, which was the most accessible.

Ironically, selective reduction is quicker than CVS. It takes a smaller needle to inject a chemical down than it does to draw a placenta piece up. Even so, the procedure demanded great skill, dexterity, and resolve from Evans and Greenbaum. Reduction requires three hands: one to hold the ultrasound transducer on the patient's belly; one to inject the needle and maneuver it into a position near the fetal heart; another to draw out the metal rod at the core of the needle, and replace it with the vial of potassium chloride. Evans, who is left-handed, did all these things at various times; he sometimes had two tools so close together, working over the patient's belly, that he seemed to be cutting a steak. At various points, Greenbaum assisted by holding the vial until he needed it; holding the transducer; and coaching him into position, watching on the screen and issuing directions. Evans worked for a while trying to get the needle into the right spot. "I'm not in," he said at one point, tensely. Then he pinned C with the needle, and pushed the plunger to release the chemical. The fetus, which had been undulating and waving, went still. "Let's check the other two," said Evans, and they moved the transducer to see the other two fetuses, still there, still waving, two hearts beating, unaware of what just happened to the sibling they would never have. "When I take the needle out, you may feel a little burning from the chemical on the tip," Evans said. Then he asked the patient, "Do you want to see your twins?"

"I don't want to see the other one," the woman said quickly.

"I chose my words carefully," Evans told her. "Do you want to see your two daughters?"

And so he showed her the two daughters, girls she had gone through God knows what to get, babies so important to this family that three generations of women had made the flight from San Juan to New York to ensure the safety of the pregnancy. Contemplating her two living fetuses the woman's face displayed an explosion of emotion. Her face went suddenly and intensely red. Then Greenbaum did something very kind: she adjusted the machine so that it showed, suddenly, a vivid 3-D image of the twin fetuses, an ultrasound image so rich and detailed that it looked like a digital photograph taken from deep within the belly. There were the two faces, the two fontanelles, the two nasal ridges, the eight limbs, everything. "That's the heartbeat, the hands," said Evans to the patient, who was holding onto her great-aunt's hands and gazing, rapt, at the ultrasound. She still had shoes on her feet. The Betadine was running down her stomach in bright red stripes. And now—somebody, apparently, had summoned them—the two other women were in the room; four women oohing and aahing at the babies on the screen.

"They're beautiful!" said the woman's mother, overcome.

"You can see the arm going up and down," said Greenbaum, pointing to one. "This one is a busy baby!"

"A busy baby!" joyfully repeated the patient's mother.

"It Could Have Been You"

After that procedure, Evans had a few moments catching up at his desk. During such interstices we had a number of conversations: about, for example, whether what he just did was an abortion. This is something

Evans has been asked before, but it is a conversation he is always willing to explore. "Technically this is not an abortion, a procedure that kills the fetus and empties the uterus," he said. "The bottom line is, abortion ends the pregnancy. We don't end the pregnancy. We very specifically don't end the pregnancy." Clearly, though, reduction is a sad and traumatic experience, and we talked about the importance—once the procedure has been finished—of celebrating the twins. "Everything I do has a history," says Evans, who over the years has learned to focus on the end, rather than the means. "We want them thinking *twins*. Here are your *twins*." We talked about the abilities his job requires: spatial sense, hand-eye coordination, knowledge, empathy, the ability to meet a patient and "instantly assess what level you are pitching to." We talked about IVF practices. Evans, who deplores the amount of "schlock" in New York medical practices in general, said that some of his patients were victims of reckless behavior on the part of doctors, but the majority of IVF multiples were, he believed, created by professionals acting as responsibly as they could.

While we chatted the sonographer, Rachel Greenbaum, was eating a salad. Greenbaum is a lovely, warm, black-haired woman of Iranian ancestry; a recent mother herself, she was pumping breast milk between procedures. She had been recently hired by Evans and, just back from maternity leave, was trying to get used to this part of her job. "I don't particularly like doing the reductions," she said. "I find it very stressful. With every patient, I think: 'If it was me, what would I do?' Some of these people tried to get pregnant for the past five years, and prayed to God, and now that they are pregnant they are telling God you gave me too many. I sometimes feel like we are playing God, and find that very emotionally stressful." A skilled sonographer specializing in high-risk pregnancies, she had held jobs with hospital practices, where there was always somebody else to do the reductions. But now that she was working with Evans, she could not avoid them. "I am here doing reductions and I would go home to my baby, and I thought maybe this is not the job for me," continued Greenbaum, who is Jewish and takes seriously her religion's admonition not to take a life. What sustains her, she said, was the knowledge that the reductions she had been involved with were done for sound medical reasons. She would never, she said, work at an abortion clinic. "This is as close as I would get," she said. "Here it is completely different. You are helping people have healthy babies. Have babies, period."

Still, she says: "It's a very hard procedure, because the baby is moving, and you are chasing it. That is what is very emotional—when the baby is moving and you are chasing it. Do you still feel emotional?" she asked Evans.

"I've come to look at it as: the finished product has a much better chance of surviving," replied Evans, who had been following the conversation intently and clearly took it seriously. "Look—you never want to dehumanize it, because then you get cavalier. You have to keep the big picture in mind. We're not losing one. We're saving some."

The first time Evans ever performed a selective reduction, he said, was in 1984. At the outset, about 75 percent of his patients had gotten pregnant using fertility drugs. The number of babies was high, the imperative to reduce clear-cut. Back then, he estimates, about 40 to 45 percent were pregnant with quads or higher; the same percentage with triplets, and 10 percent were twin pregnancies in which one fetus had a serious problem. But now, he says, things have changed. Now, 75 percent of his patients have gotten pregnant through IVF, and the size of the pregnancies has gone down. Now it's 5 to 10 percent very high order multiples, 20 percent quads, 60 percent triplets, and about 10 to 15 percent twins.

"When we started, the first patients I ever did, we had nothing to lose," he reflected. "My second patient had octuplets. Anything we did was better than nothing. And as with everything, when you do something radically new, you start with the life and death cases where you have nothing to lose. . . . [Then] as you begin to develop the safety and efficacy of any procedure, the indications for it liberalize, you move from life and death to quality of life."

All of which has compelled him to rethink his moral calculus regarding twins. When he was working to establish bioethical principles, Evans decided that he would not reduce a normal twin pregnancy. He would take somebody from three to two, but he would not take somebody from two to one. "The rationale we used was: One, every OB knows how to take care of twins, and two, the outcome is not as good as with singletons, but good enough. And number three, all these were fertility patients and if we could get them to twins, that was that much closer to their family ideal. And four, we didn't know what the risk might be of damaging one of the fetuses by the procedure. Because of all of the above, it didn't feel ethical to go ahead and do that."

But Evans's thinking has changed. He is willing now to reduce two to one, and he does so. Not a lot, but the incidence is increasing. There is now data showing that reducing one twin does not affect the physical well-being of the twin who remains. Plus, he pointed out, "We're seeing an aging population." Many of his patients are women in their late thirties and early forties, some married for the second time. Both partners may already have children, and what they want is one child together. "The average age of my patients who reduce to twins is thirty-eight, and the average age who reduce to a singleton is forty-two." For a woman reducing a twin to a singleton, the pregnancy loss rate drops from 8 percent to 4. Also, he points out, "the appreciation that twins are not quite as safe as people assumed they were is relatively new, and I've been making people aware of it." Evans has written articles arguing that it is ethical to reduce a twin pregnancy. After all, he argues, if it's okay to reduce from one to none—that is, if you support abortion rights—then two to one should be okay, too. The idea is still controversial. "When we first started, all but the most ardent right-to-lifers had no trouble with [reducing] quads or more, and nobody, including me, thought we should be doing two to ones. Twenty years ago the ethical debate was with triplets. But now as far as I'm concerned, there is no doubt about triplets, and the ethical debate has moved to twins."

And it is a debate. At the same time that Evans and others have been refining and publicizing the data about the risks of gestating twins, other experts have been showing women how they can take care of themselves to enhance their chances of carrying a multiple pregnancy to term or close to term. Among these is the nutritionist Barbara Luke, who has published books on successful gestation of multiple pregnancies, pointing out among other things that eating lots of protein helps the babies grow. Luke also published a 2004 article in *Fertility and Sterility* showing that babies who started out as high-order multiples and were reduced to twins did not do as well as babies who started out as twins. The implication was that reduction may be bad for a pregnancy. This argument drives Evans—and many perinatologists—nuts; they believe it's not fair to compare reduced twins to twins who started out as twins. "Reduction is not bad for you," Evans says. "Starting out as quints is bad for you." When a pregnancy is lost after reduction, he argues, it is not because of the reduction, but because of the original situation. One pities the parents caught between two medical crosscurrents: people like Luke suggesting, rightly, that women can raise their chances of delivering a healthy multiple pregnancy, and those like Evans, arguing, rightly, that for some women, protein bars and multivitamins won't suffice. In truth, there is no way to know for sure which women can successfully gestate multiples and which cannot. During scores of interviews, I was astonished by the number of women whose counseling amounted to a fertility doctor eyeing her up and down and saying, "Well, you're big enough to carry twins." In fact, it's very hard to identify in advance which woman is going to have an incompetent cervix, or an irritable uterus, or any of the other insultingly named conditions (one woman's boss thought she had an "angry vagina") that can complicate the attempt to carry more than one child at time. In this case, size has little to do with anything.

And, in the middle, are the people who have to make this unlooked-for choice, based on their own priorities and instincts. Among these was the unhappy couple from my first day at Evans's office, who, he said, fit one common pattern. Generally, Evans has found, patients face the procedure with dread and grief, but by the time it's concluded—when they get to see their twins on the sonogram—they are already feeling better emotionally.

The long-term psychological consequences of reduction are still little understood. According to Isaac Blickstein, one of the world's experts on multifetal pregnancies, what studies have been done suggest that the aftereffects of selective reduction are different from those of abortion. In general, parents feel better after the reduction is over. But grief can return, postpartum, as parents contemplate the babies they have, and think about the one, or ones, they lost. ("It is so weird to rejoice and mourn at the same time," as one Fertile Thoughts poster put it.) One study found that one-third of women who underwent selective reduction reported persistent depression a year later. At two years, most were feeling better. But, Blickstein notes, "psychoanalytic interviews with women who underwent [selective reduction] describe severe bereavement reactions including ambivalence, guilt, and a sense of narcissistic injury, all of which increased the complexity of their attachment to the remaining babies."

"Complexity of their attachment to the remaining babies" seems an apt way to put it. I encountered this one day when I was interviewing a mother of twins. She was talking by cell phone, driving from home to work—the only window of talking time available to many parents. This mother happened to mention that she had reduced triplets to twins, and talked about what it felt like now. "In some ways, the selective reduction becomes harder to deal with after the babies are born," she said. "They are babies with personalities, and you become attached to them, and you think: I could have ended up with one of these being gone, and the one that is gone could have been one of these."

"Now that I know my two daughters, looking at them, I think: it could have been you," she continued. "You terminated an embryo that could have turned into a baby you would love as much as the ones that you have, and that's hard."

"I am pro-choice," she said, "probably more now than ever—I know how hard pregnancy and childbirth are. But everything becomes a little more complicated in my mind, and you just can't help but wonder a little bit: what if we had chosen a different embryo? I wouldn't say that I think about it every day, but I do think about it. It's not like it's a chapter that's closed. People will say to me: 'Could you imagine triplets, how hard that would be,' and I can't help but think: I did have triplets. It was a decision between me and my husband. We talk about it, how hard it is. He talks a lot about how sad he feels about it, and he's surprised how sad he feels about it. And he does feel sad."

One of Each

At 3:40 p.m. Evans was eating muffin crumbs from a paper bag and waiting for the Chicago lab to call with the results for Jane and Emma's quadruplets. While sitting, he showed some favorite photos. One, which he keeps on his wall, is of Evans meeting the late Pope John Paul II. The Catholic Church is opposed to IVF, seeing it as an unnatural procedure that usurps the province of God; it goes without saying that the church also opposes selective reduction. Evans was in Rome for a conference, and "John Paul loved to lecture to ob-gyns"; and some of the attendees got to meet him. Evans "used my minute to talk about stem cells." He also keeps on his laptop the dramatic photo of Bobbi McCaughey in a hospital bed, her belly rising massive into the air like Mont Blanc. He said the only reason the seven McCaughey babies stayed in as long as they did was because the one closest to the cervix was a "transverse lie," meaning that it was lying across the opening, stopping the dike, preventing the others from exploding out.

And now here were the two women, Jane and Emma. And here were the results, the sexes stacking up like this:

A: XX
C: XY
D: XX

Two girls, one boy. All normal. And so, for this last reduction procedure of the day, one decision had been made. Evans would reduce B, which had not been genetically tested, because B was most accessible. As for the other one to go: he told the women they could have two girls—that is, he could keep A and D—or they could have a boy and a girl, in which case he would keep C and one of the girls. "If you want one of each, I'll keep one of each," Evans told them.

They wanted one of each. "I have mixed feelings about it, but I think boys and their mother have a very special relationship," said Jane, who several times alluded to what she perceived as the difficulty of raising adolescent girls. And with that, she and Emma had done it: they had selected for sex. They had made their choice not on the old-fashioned, now unacceptable idea that boys are superior to girls but according to the more modern notion that boys and their mothers may have easier, less tempestuous relationships, or maybe just a different relationship, from girls and their mothers. At any rate, the boy would stay. One of the girls would go.

Now Emma was on the table and everybody was looking at four fetuses on the sonogram. The screen had been turned so that even Emma, the pregnant mother, could see it. Evans decided that in addition to B, he would reduce D, because of her position, farthest from the cervix, and most accessible after B. Just now, on the sonogram, D happened to be visible, moving and waving. "D is really active. That's what I hate to see," said Jane, who had woken up in the middle of the night, she said, worried about the "karma of what we are doing."

The women faced this head-on. They intently watched the four fetuses on the ultrasound. At this stage it was possible even without the 3-D picture to see limbs, heads, tooth buds. Evans prepared two syringes, swabbed Emma with Betadine, put the square-holed napkin on her stomach. Then he plunged one of the needles into Emma's belly and began to work his way into position. Looking periodically at the ultrasound, he began talking to Jane using doctor-to-doctor lingo, saying things like "You can see a partial pulmonary silhouette . . . floating . . . just under the diaphragm."

He injected the potassium chloride and B, the first fetus to go, went still. "There's no activity there," he said, scrutinizing the screen. B was lying lengthwise in its little honeycomb chamber, no longer there and yet still there. It was impossible not to find the sight affecting. Here was a life that one minute was going to happen and now, because of its

location, wasn't. One minute B was a fetus with a future stretching out
before it: childhood, college, children, grandchildren, maybe. The next
minute that future had been deleted. The world, or a portion of it, would
unfold differently. One thought of chaos theory. A butterfly flutters its
wings and the effects are felt thousands of miles away, years later. A butterfly doesn't flutter its wings: that has an aftereffect as well. Something
small happens, or doesn't happen, and the result is diffuse and, after a
point, untraceable. Somebody who might have married somebody now
would marry somebody else. A conversation that might have happened
never would, and a different conversation would happen. A child that
might have been born in the year 3000 now would not be. Alternately,
four children that might have been born premature and disabled in the
year 2006 now would not be born premature and disabled. Instead, two
children would be born blooming and healthy, most likely. Who knew,
exactly, what path had just not been taken? The patient, Emma, looked
at the screen, mesmerized. "It truly is amazing, what medicine can do,"
she said.

Evans plunged the second needle into Emma's belly. "See the tip?"
he said, showing the women where the tip of the needle was visible
on the ultrasound screen. Even I could see it: a white spot hovering near
the heart. D was moving. Evans started injecting. He went very slowly.
"If you inject too fast, you blow the kid off your needle," he explained.

After he was finished injecting, D moved for a few seconds, then
went still. "That little heart beat for a while," said Jane, who had
been watching, her eyes wide and bright. Now, as we watched, there was
something called the effusion: a little puff. "When I see that effusion,
I know it's done," Evans said, taking "one last look at D before I come
out," to make sure D was gone.

"Want to see your twins?" he asked the women, who did. On the
ultrasound he showed them the twins, moving vigorously in their sacs.
The women thanked him profusely. "You made an unpleasant experience pleasant," they told him. "Thank God there are people like you,"
said Jane.

"I'm sorry we had to meet under these conditions," said Evans.

Greenbaum meanwhile was working on giving them 3-D photos. On
the screen, C seemed to be sucking his thumb. They decided to rename
C, calling him B, so that the remaining fetuses hereafter would be
known as A and B, as in a standard twin pregnancy. She showed them the
babies from all sides: the little legs, the little heads, the little feet tucked

fetally under, the kind of glorious images parents cannot get enough of. Six months later—I learned, from an e-mail the women sent—Emma went into labor. The boy and the girl were born vaginally, both healthy, both immediately and deeply cherished. The women e-mailed photos of both of them, nestled together on what looked like a hand-knitted blanket, wearing on their heads what looked like hand-knitted caps. The children are exquisite. The women said they could not believe the babies were theirs. Their good fortune seemed incredible to them. "We could not be happier," they wrote on the baby announcement. During the course of the pregnancy, they had been able to see traces of the two reduced fetuses in the ultrasounds; Jane said that she felt it was important to continue to own up to what they had done. After the birth, Jane inspected the placenta for the remains of the reduced two, which were visible in the placental tissue. She said "goodbye to the two that we were unable to carry," Emma wrote me in an e-mail, "which I know was very helpful to her."

CHAPTER TWELVE

TWINS:

THE NEW SINGLETON

Tammy LaMantia is showing me how she learned to get down the stairs. First, she would take one baby—Ellie, say—and tuck her into the crook of her right arm. Then she would tuck the other baby, Edie, into the crook of her left arm. Then Tammy would sit on the topmost stair of her house and scoot down the steps on her rear, working her way down the staircase until she and the twins got to the bottom. It seemed safer than walking down with two infants. And bringing the girls down one at a time was impossible, once they were old enough to object to being the one left behind. Somebody in Tammy's twins group told her about the butt-scooting method, which is known among twins parents as "giving them a ride."

Twins: what Tammy LaMantia ended up with, after the terrible delivery of her dead and dying triplets. Just months after their funeral, Tammy, to her amazement, found herself back in an IVF clinic. Immediately after the delivery, she recalls, "I remember clear as a bell saying, to Steve, 'We will never do this again. It will be just the two of us for the rest of our lives, or we will adopt.' And I am not kidding you, within two weeks of having them, it was: I have to have a baby. When I look back on all of that, it is amazing to me how strongly you can feel about something."

The doctors urged Tammy to wait a year. She waited four months, which was as long as she could stand. A subsequent IVF attempt failed, which she thinks now was a good thing. In 2001, Tammy and Steve went through it all again, the drugs, the retrieval, the how-many-to-transfer decision. Her doctor—saying her endometrium, the lining of her uterus, was too thin for his liking—wanted to transfer two, and they agreed. "We went for the ultrasound," Tammy says. "And there were two sacs." Her doctor reassured her, she recalls, by saying he heard only one heartbeat, and that the other sac—empty, doubtless—would be reabsorbed by her body. Tammy didn't believe him. "I said: I know there's two."

There were two. This time, Tammy LaMantia developed placenta previa, a dangerous, potentially fatal condition where the placenta grows over the opening of the cervix. Placenta previa has been shown to be six times more common in IVF pregnancies than in naturally conceived ones, perhaps because the embryos are placed in the uterus manually, and implant lower than they would naturally. At nineteen weeks, Tammy became a long-term camper in Inova Fairfax Hospital's high-risk maternity ward, where she proceeded to spend more than a hundred days on well-enforced bed rest, lying supine from February to June 2002, missing spring completely, or rather watching spring unfold out the window of her hospital room. Tammy spent her time watching TV and weeping. She was convinced, she says, that the past would repeat itself: the babies would be born premature, the babies would die. "It never felt normal," she says. It *wasn't* normal. She was a woman who had lost three babies, and now she was flat on her back trying to save two. Her weeping was so intense that her doctor tried to get her to talk to a hospital psychiatrist, thinking she might need to be medicated.

Tammy did what she could, short of taking drugs. She learned how to do yoga using her facial muscles, the only part of her body she was allowed to move. She did deep-breathing, which helped. She talked over the phone to a therapist, who told her she would feel better when she passed the week at which the triplets had been delivered. Her therapist was right. At thirty weeks, Steve brought a cake into the hospital, and they celebrated the fact that the twins had made it into the window of likely survival. At thirty-three weeks, four days, Tammy sneezed and her water broke. Her doctors gave her the steroid shots that help premature babies' lungs mature quickly. Tammy was scheduled for a C-section forty-eight hours later, to give the steroids time to work. But then she developed an infection and was in terrible pain. A nurse rushed in and said the procedure had been moved up: the stitch in Tammy's cervix was ripping. They brought her into the delivery room for a C-section, and "Edie was pulled out at 7:30, and Ellie at 7:31."

And the twins were fine. Marvelously fine. And now they are talking, Edie and Ellie, upstairs in the bedroom where their cribs stand side by side. They have been napping, the girls, who are two years old at the time of our interview. Tammy and I listen to them over the baby monitor. Just one of them seems to be awake, making soft noises and singing. Listening, Tammy says that it's Edie, who for some reason almost always wakes up first. The girls are fraternal, meaning that they developed from different embryos. Basically they are just regular sisters who happen to be exactly the same age, give or take one minute.

"Ellie! Ellie! Where are you, Ellie?" Edie is saying. Given that Edie can undoubtedly see Ellie in the crib just a few feet away from her, "Where are you?" seems to have a more cosmic meaning: Where is your consciousness, just now? Are you dreaming? When will you be here for me to talk to? When will you surface? Tammy goes upstairs to find both girls now awake. Ellie is blinking sleepily and Edie is wearing a New York Yankees baseball cap and saying "uppy, uppy," stretching her arms out to her mother. Tammy lifts them out of the cribs and prepares to give them a ride, only this time—now that they can walk—Ellie wants a ride and Edie prefers to walk down the stairs by herself.

It's one of innumerable ways in which the girls have begun to assert their individuality. Downstairs it's time for a snack, so Tammy prepares soy milk for Ellie and cow's milk for Edie, each girl having decided she hates the kind of milk the other one likes. Both girls are gorgeous. They have long glorious hair and beautiful faces, even now, or maybe especially now, when they are soft from sleep and recovering from colds and the ear infections that always seem to follow. Between the two of them, the LaMantia twins had twenty-one ear infections in the first ten months of life. "We could have bought stock in antibiotics," says Tammy. Though their DNA is not identical, both girls inherited a shape to the Eustachian tubes that makes them prone to infection. Ellie ended up getting tubes inserted in her ears; one tube is falling out. It is Ellie who has the ear infection just now. She doesn't want to take her medicine, so Tammy matter-of-factly lies her down and squirts it in her mouth with a syringe, and minutes later Ellie is up, pain forgotten, dancing to the Wiggles.

When I compliment Tammy on her competence and aplomb, she dismisses the praise, saying that as the mother of twins she is plagued with feelings of inadequacy. "When it's just one I feel like I can manage it," she says. But when both girls are sick, or both girls are crying, she worries she's not giving either girl enough attention. "If I can comfort one, I feel okay, but if it's two—you can kind of comfort one. That's been my struggle with twins generally." This was something echoed by so many twins parents I interviewed. One of the great challenges, they say, is learning to live with the fact that you're not going to be able to lavish attention on each child the way modern parents feel they should do. Somewhat ironically, the upsurge in twin pregnancies coincides with a well-intentioned but sometimes misleading new emphasis on the vital importance of a child's first three years, both in terms of a child's emotional development—hence the need for early and frequent physical contact, starting the moment the delivery nurse puts the bare skin of the child on the bare skin of the mother—and intellectually. Today's endless shelves of educational toys and videos are all meant to help a child reach crucial milestones within a window of time that, we are told, will quickly close. The modern parent should be constantly, we are told, stimulating her child, touching her child, working to cement what is considered the most important emotional goal of all: attachment.

And twins parents do try to cement this. Heroically, twins parents try. And yet twins parents have to let go, a little bit, of the idea that to be a responsible parent you have to meet your child's needs every waking moment. Yet often they cannot entirely let go of it, resulting in a really pretty superhuman effort to be all things to two children. As a twins parent, you get used to doing the best you can. You get used to the struggle over you. "Sibling rivalry—I didn't realize that it exists so early on," Tammy reflected. "I think—the only thing I can come up with is that it's because there were two of them, and neither one has ever had my undivided attention. It's like they fight over every little thing. They can be fighting me up and down that they don't have to go potty, but if I get one to go, the other is trying to get her pants down faster and sit on the toilet first."

"You develop these techniques for dealing with two," agrees Linda Gulyn, parent of two sets of twins, who clearly knows whereof she speaks. "Sticking them in car seats and propping bottles. Holding one, and then holding the other. You just get used to this idea that you're not going to be able to care for both of them the way you think you should." But you never stop worrying about it. "I started thinking, oh, is this going to affect their attachment," said Gulyn, a child psychologist. Eventually she told herself, "You know, this is the best I could do. The best I could do. I knew that I just had to give up that guilt and get through." One truth that quickly emerges is that the impact of twins falls as heavily on the parents as it does on the children. "I come from a big family," Gulyn said. "The kids didn't get enough attention. It was a serious problem in my family. You had to compete for your parents' attention. I am very conscious of that, every minute, with my boys. You try to figure out which one needs you at the moment. It's not always right."

The New Normal

And that, for hundreds of thousands of parents, is the new normal. Two babies at the same time are the new one baby. While the triplets rate in this country may decline someday—or not—it seems clear the rate of twins will keep rising. What we are seeing is a whole new challenge to the reproductive system nature spent millions of years refining, a test of how much the fit, yoga'ed, Title IX'ed female body can handle. "The human uterus was not made to carry more than ten pounds," estimates Mark Evans, yet since 1980, the twinning rate has almost doubled, rising from 18 twins per 1,000 births to 31.5 per 1,000 births in 2003. (The average twin weighs about 5½ pounds.) Every year the number of twins born in the United States has set a new record high. With inevitable consequences; some of the cases have been quite dramatic. In April 2006, in Forrest City, Arkansas, a fifty-three-year-old single mother of ART twins suffered a uterus eruption a month before her due date; struggling for life after their birth, she was airlifted to an intensive care unit while her newborn twins were taken into foster care. Then there is Ann Nelson, gestational carrier for Doug Okun and Eric Ethington, hemorrhaging after the birth of their two girls. Her resultant hysterectomy is not uncommon among twins mothers. I interviewed several women who traded their uteruses for their twins. "I would do it again in a minute," said one.

In suburban Washington, D.C., where I live, the influx of twins is such that among the sixteen children who made up my son's first-grade class in Arlington, Virginia, two of the boys were twins, and they weren't twins with each other; their twins were in other classrooms. Delicate questions of etiquette confront all parents: invite both twins to your child's birthday party? Or just the boy your child knows? In New York City, 1995 saw 3,707 twin births in all five boroughs, a number that had risen to 4,633—an increase of 20 percent—by 2005. The surge in twin pregnancies has intensified competition for admissions into Manhattan preschools: "I tell [twins] families that they may increase, hopefully double or triple, their options, by telling schools they are willing to separate their children," said one education consultant, Emily Glickman of Abacus Guide Educational Consulting, in an article in the *New York Times*. The article told the tale of a father, Usman Rabbani, who in trying to get his two sons into preschool slots was obliged to write essays profiling each of his eighteen-month-old twins. Talk about Solomonic. Of one boy he wrote, "He knows that birds like to sit on rooftops when they are not on the ground." Of the other, he wrote, "He is happy to point out his body parts when asked."

A twins birth is riskier than many people realize. Babies born as twins are hospitalized twice as long as singletons, and their medical costs are three times as high over the first five years. Evidence of how fragile a twins pregnancy can be is provided by the Center for Loss in Multiple Birth. According to the bereavement support group's website, "Our founder is Jean Kollantai, whose fraternal twin son Andrew died just before their due date, after a good pregnancy." After the loss of one twin, "Jean began to search for others who had lost a twin and soon . . . met Lisa Fleischer, whose twin son Teddy—born on their due date weighing 8#5—died of the effects of a prolapsed cord when he was 10 days old. Lisa and Jean sadly soon met Becky Crandall, whose twin daughters Julia and Emily were stillborn at 37 weeks along from no known cause."

So no. Having twins isn't safe. But they are—many doctors and patients believe—safe enough. "I have to admit, I'm equivocal about twins," sighs Sam Thatcher, the Tennessee doctor whose clinic has worked to lower its triplets rate. Thatcher allows that "recent information suggests that twins and triplets are much closer in their risks during the pregnancy than twins and singletons. On the other hand, even though they are more difficult, the chances of that twin pregnancy and the mom ending up in good shape are very high. If you're taking someone in their mid-thirties who wants two children, and looking at the costs of coming back and doing IVF [a second time], I'm not so sure twins are bad."

Similarly, at Maryland-based Shady Grove Fertility, which also has worked hard to bring down its triplets rates, "we look at twins as a reasonably acceptable complication," says one of the head doctors, Robert Stillman.

In IVF, it remains true that two embryos are more likely than one to result in a pregnancy. In Europe, movement toward single embryo transfer, especially in women thirty-five and under, is increasingly catching on. SET, as it's called, can be very successful, both in terms of pregnancy rates and healthy deliveries. But the pregnancy rate is not quite as high, yet, as with double embryo transfer. As Cornell embryologist Lucinda Veeck Gosden points out, if you transfer two embryos into women under thirty-four, half of those who get pregnant are going to end up with twins. But if you transfer only one, "the pregnancy rate is down." And so in this country, it's still rare to transfer one. In 2005, a report was released comparing worldwide trends in IVF from the year 2000. Around the world, the mean number of IVF embryos transferred in 2000 was 2.5, while in the United States it was 3.0. The success rate in the United States was also notably higher: 38.5 percent of IVF cycles resulted in a pregnancy, compared to 26.7 percent worldwide. It's a considerable difference, particularly given the fact that in the United States, patients are more likely to be paying for the cycles out of pocket. In this country, people are obliged to buy their children in bulk.

Then there is the puzzling fact that IVF embryos are much more likely to twin than regular embryos. Monozygotic, or identical, twins are created when a single embryo splits. These pregnancies tend to be riskier than fraternal twin pregnancies, because the two fetuses are more likely to be interdependent. For IVF embryos, the rate of monozygotic twinning is *ten times* that of naturally conceived pregnancies. Moreover, when an IVF embryo is taken to the blastocyst stage—allowed to culture for five days—the twinning rate is even higher. Nobody knows why this is, though one expert, the Canadian scientist Judith Hall, thinks it's because the culture medium makes the embryo unhappy, and an unhappy embryo, like Rumpelstiltskin, has a tendency to express its unhappiness by splitting in half.

So IVF pregnancies by their nature are going to produce more twins. So are later pregnancies in general. Add to that the growing number of women using sperm donation along with a dose of Clomid or shots of gonadotropins. A 2006 article in the *Wall Street Journal* noted the increased use of Clomid and fertility drugs among women who are only in their twenties, a trend that greatly increases even these young women's odds of engendering twins. Other studies have suggested that obesity, dairy products, and increased consumption of folic acid are all fueling the escalation of twin pregnancies. "Twins are not normal," insists Alfred Khoury: nevertheless, twins are here to stay.

"We can't stop this train," said Meredith Kramer Hay, who in August 2005 was sitting in her sun room, trying to stay upbeat. Contrary to what many doctors assert, many IVF patients are *not* hoping for twins. Hay already had a two-year-old son from IVF (she attributed the success of her treatment, in part, to meditation and a positive outlook) and in going through a second cycle, she tried to channel the same outcome. But this time, she says, "when we found out I was pregnant, they called me with the number score on the blood test, and it was exactly twice what I'd had on my son, and my husband and I looked at each other, and he said, 'Okay, can you meditate on this, can you make sure we're going to have one?'" This time meditation failed her, meaning that she was now eligible for a niche multiples parenting group known as "three under three." She called her mother, lamenting that she would never again shop at Target; only Costco carts can fit three small children. While purchasing the inevitable minivan, Hay wondered how best to load three small children in and out of the car. Fortunately, her fellow three-under-three parents had answered the question. The way to do this (pay attention, class; you may need to know this someday) is to remove the "captain's chair" behind the driver's seat. The parent can then lean into the space where the door slides open, place one baby behind the passenger seat, and the other two in the rear.

When we talked, Hay was doing what every conscientious mother of multiples finds herself doing: trying to keep her twins in utero as long as possible. Most mothers feel pressure to provide the perfect prenatal environment. But the pressure is intensified for mothers of twins, who often feel it's their own fault if they cannot keep eleven or twelve or fourteen pounds of children gestating in a uterus built for ten. "There is a sense if you cannot carry them as long as you would like to, you have failed," allowed Hay, a slender woman, just five-foot-six and 115 pounds normally. "Every day you can stay pregnant is two days not in the NICU." A devout student of Barbara Luke, she was forcing herself to down as much protein as possible: peanut-butter protein bars, Ensure plus protein, yogurt or cheese with every meal, every protein-rich substance short of, say, worms or caterpillars. During our interview she kept a water bottle on the table alongside her procardia, a heart medication that, it has been found, can help stave off premature labor. Now approaching her thirty-third week, she had gone into labor the week before. Twice a day she was obliged to hook herself up to a home contractions monitor, which reported her results to the hospital over the telephone. "It's very stressful," said Hay, who found it infantilizing whenever her report was bad and the nurses would call and order her to drink water. The day after our interview, Hay would go into labor two months early. The children would be healthy and catch up developmentally, but the first year would be, as she wrote later, "challenging." Meanwhile, there was the need to prepare the house. "I told my husband, we are going to need a bouncy seat on every floor, maybe two; a swing and a bouncy seat on every floor."

And indeed, one of the first questions parents of twins face is, how are we going to get these babies home, and where are we going to put them? The twins rate is also rising at precisely the moment in U.S. history when children have to be maintained in top-notch safety conditions around the clock. In a way, twins would have been easier for mothers to manage back in the 1960s, when it was socially permissible to stick them in the backseat and let them wallow on the vinyl seats untethered while mom sat in the front, smoking a cigarette and driving. Now, mothers scoot down steps on their rear ends, so as to not risk a fall. Now, mothers engage in elaborate prenatal time-and-motion studies. "I spent hours obsessing before I had the kids, how was I going to get them in and out of the house," said one twins mom who lives on Capitol Hill in Washington, D.C. For the urban mother, the problem was this: Say you have both twins in the car, and you find on-street parking near your house. You carry one twin inside, in the baby carrier. What do you do with the other? Leave it in the car? To be kidnapped? Take one twin out; carry it six feet; get the other twin out; carry it six feet; and so on and so forth, the inchworm solution to baby-ferrying?

"I ended up developing very strong forearms," said this mother, who at first would lug both carriers at the same time. But as the babies grew heavier she found it necessary to drive into her basement, which opens directly into the house. That way, she found, she could remove the babies from the car seats, load them into a big twin stroller, and push them into her basement. Listening to sagas like this, it's easy to see why parents were so grateful when in 2003 Baby Trend came out with a twins version of its popular Snap N Go, a lightweight frame onto which two car seats can be clipped, turning it into an instant, easily assembled twins stroller. The advent of the double Snap N Go meant twins parents no longer had to load the children out of the big bulky car seat and into a big bulky stroller that they had to lug along in the trunk whenever they went anywhere.

"We had so many consumers looking for it," said Chip Whalen, the general manager at Baby Trend; he estimated that twins parents make up about 10 percent of his company's business. Needless to say, products catering to twins are becoming widespread. While helpful—most of them—they also serve to subtly normalize twinship. In addition to double strollers there are coordinated clothing lines, as well as lines of extra-tiny infant outfits, with names like "Preemi-Yums." "Preemie dolls" also can be bought by parents who want to memorialize an infant who did not survive, or remember how unimaginably tiny their children were when they emerged. A seamstress I talked to who has a busy Web-based business making preemie dolls pointed out that she needs only one fabric shade for the most premature ones: babies born really early tend to have the same reddish skin shade, no matter what their race or ethnic background. Other manufacturers aren't so scrupulously realistic. At one point one company was marketing a preemie doll that dismayed a mother posting on a March of Dimes website. "I am just sick looking at this," she wrote. "The doll is pink, chubby and healthy looking. My 28 weeker is still so thin he has no fat whatoseover. All this doll does is reinforce the concept that preterm infants are just like full term, only smaller and cuter."

To me, the most disconcerting example of twins commerce was a carrying device I found at www.4coolkids.com. The device's name, Maxi-Mom, suggests some sort of monstrous-sized menstrual pad. In fact it is a carrier that enables a woman (there did not seem to be a MaxiDad) to carry two babies at once, like a milkmaid, or a yoked mule. The way MaxiMom works, one twin gets snuggled on the front and the other gets snuggled on the back and straps go over mom's shoulders. Having myself struggled to fit one recalcitrant baby into one Snugli, with all the holes and compartments, and frequently failed, I can only imagine that by the time you stuff both twins in, it's time to unstuff one and change it or feed it or try to calm it down.

Indeed, so challenging is the dressing and changing of two babies that some twins parents decide they will stay home until the twins are ambulatory, or at least bigger. "I didn't leave the house for three months," said one twins mother, Rachel Haas, who delivered her two babies in winter, a season when even the shortest excursion entails rustling up four booties, and four mittens, and two pacifiers, and two hats and two coats, and endless diapers and blankets and wipes and burp cloths. "By the time I got them changed, dressed them, put their booties on and into the car seat, the other one would need to be changed and fed, and it wasn't worth it. So we just stayed home, and prayed that someday this would pass."

Even basic tasks present a logistical challenge: twins demand a level of competence that would challenge any mother in any era, but seem particularly ill-suited for this generation of mothers, many of whom have spent the past decade, or two, surrounded by cubicles and colleagues, rather than very small humans with very big needs. Remember, the twins rate is growing most rapidly among women older than forty, the precise cohort for whom baby care is a vague memory, left over from that long-ago time when they were teenagers who babysat.

Or maybe teenagers who didn't babysit. "I had never held a baby, and now I had two babies," Nancy Thiel remembered. When her two boys were born, they were put in the NICU as a mild precaution. Expecting them to be under the competent care of hospital staff for several weeks, Thiel planned to use the time to recover and study up on how to care for them. But the NICU staff kept upgrading the boys' condition, and the window of hospital care kept getting narrower. "They said, 'It's not going to be eight weeks, it's going to be seven'; and then, 'It's going to be six,' and after six or seven days, they said, well, 'Andrew is going home tomorrow.' And I said, 'He can't! We don't have diapers! We don't have a car seat!' " Among twins moms, trying to talk the hospital into keeping your babies longer is also not uncommon. Rachel Haas recalls delivering by C-section on a Thursday morning. "The doctor came to see me on the following Saturday, and she said, 'You're doing so well, I think you can go home. I was like 'Oh no, please God, let me stay here for two years.' "

It's another perfect storm, isn't it? The fact that hospitals send a woman home the minute she is stitched up; the fact that most of us have only the foggiest notion of what to do when we arrive home with a newborn; the fact that the feature we've bought tickets to often, now, turns out to be a double billing. "Say you're focused on one child; she had a rough night and she was crying, and then all of a sudden it hits you: I have another one!" says Haas.

Things get easier, physically, as the children get older. There even comes a day when having twins seems convenient. The day I spent with Tammy LaMantia, her girls did a lot of puttering around together, the kind of busy toddler play that was a pleasure to watch. "I love having twins," one mother said, echoing many who gloried in having a built-in playdate at home. Others could see it both ways. "The first year we were always barely staying above water," said one mother. "I think I've missed out on a lot of people's experiences, especially with their first child, being very, very bonded. And even now, I spend a lot of time when I'm with them planning ahead. Who's going to be hungry next? When are we going to eat? What are we going to eat? I don't have a lot of time to relish them. I don't have that, and in some way I'm sad about that, because I feel like that would be fun and maybe a little easier and less stressful. On the other hand, I have two wonderful babies."

And as the children grow, the challenges change. Parents worry about twins who seem intensely competitive, as some twins do. They worry about children who reach milestones—walking, talking—at different times. They worry about the things most parents worry about, such as identity and emotional well-being and, of course, bonding. And everybody struggles with sibling rivalry. Rob and Nancy Thiel spent a morning talking to me about their then five-year-old boys, Andrew and Michael, who were extremely good during our conversation. The boys played in the yard, played inside, and occasionally eavesdropped in that way children have, when they pretend they aren't paying attention but are, in fact, listening acutely. The Thiels talked about learning to deal with the fact that one twin routinely dominates the other. The family joke is that when the boys were in utero, Michael was positioned so that he was kicking his brother in the head, and that Andrew had been getting him back ever since.

"Andrew is bigger, strong, and much more aggressive" said Rob. "You think it won't happen with twins, but there really is an older brother. I'd describe these boys as reluctant brothers: absolute separate humans who happened unfortunately to be in the same womb at the same time. They are radically different from each other; they occasionally like hanging out with each other, but for the most part are preferably separate." At one point Rob asked Michael, who happened to be listening to us: "How many kids do you want?"

"Ten," Michael replied, taking his time to answer. That, his mother said, is characteristic: he is a thoughtful, deliberate child, and likes to reply after a lot of cogitation. The problem, they said, is that Andrew, will "jump in and answer" before Michael has time to say his piece.

"We're looking for an environment where it's okay not to answer," said Rob, who was thinking it would be best to separate the children when they got into school.

And that is the question all twins parents eventually face; put the children in the same classroom, or split them up? Rachel Haas, a teacher, started her girls off in the same preschool class, but after the end of the first year, one girl had evolved into the spokesperson for the pair, and was becoming so dominant, the other so subservient, that Haas and the teachers agreed it would be better to separate them the following year, which currently is the most common pattern. Haas and her husband, Jeff, worried that each child was molding her personality in reaction to the other: "Sarah wants to make sure that she has and does everything that Abby has and does. She wants to have for breakfast what Abby is having for breakfast. That's one of the main reasons I think they need to be split." Which is not to say that you can ever fully grasp the complex dynamics of twin relationships. "Just when you think you've got them down pat, they do something that is completely the opposite of what you would expect," said Haas, who thinks this shape-shifting is "the two of them exerting their individuality." Because the girls were endeavoring so hard to be different from each other, she and her husband found "blanket parenting" doesn't work: they could not use the same strategy with each girl. "In our head we always treated them the same," Haas said. "We loved them equally, kissed them equally, fed them equally, clothed them equally. We were told: they're two different people, and you have to treat them differently. Being fair doesn't work."

Of course, before a parent can decide whether to keep twins in the same class or to separate them, the parent has to get them into school at all. "I got up at five o'clock in the morning," said Haas, whose neighborhood preschool, like most preschools in the D.C. area, is hopelessly oversubscribed. Parents wait in line to register, and some camp out the night before. "There were people, I am not kidding, there were people in the parking lot at five o'clock the Friday before the Saturday that it opened at eight. They slept in their cars." Haas got the last two spots.

How Do Twins Do?

For generations, twins have been useful for scientists because they offer a ready-made laboratory in which to study nature versus nurture; cases of identical twins reared apart offer a way for social scientists to try to get at which traits result from upbringing and household environment, and which traits are purely genetic. And twins are just uncanny and interesting to think about. Whenever identical twins are reunited and it's found that lo and behold, both hold a beer can with the pinky underneath, it's considered proof of the power of genetic inheritance, even though some reunited pairs have it found it difficult to develop much of a relationship.

Though it often turns up striking similarities, twins research also uncovers siblings who share genetic similarities but feel no sense of profound union. Nancy Segal, a fraternal twin, describes in her book, *Entwined Lives: Twins and What They Tell Us About Human Behavior,* the sense that she and her twin were "ships passing in the night": they had different friends, different abilities, rivalries exacerbated by their parents and the school they attended, which, in trying to keep them separate, put only one of them in the classroom for the highest achievers. It took years, she writes, for them to become close. "We were not natural allies."

The other line of study is how twins fare intellectually: the vexing question of whether twins are at a disadvantage. As Isaac Blickstein, the Israel-based physician and expert on multiple pregnancy, observes, there has long been a fallacy that twins have lower intelligence than singletons. "People assume they are stupid," is how one mother put it to me. It is true that in discordant pairs of twins—where one twin is born much larger—the smaller twin does not fare as well, even in later life. It is true that twins have a higher rate of handicap, mortality, and cerebral palsy. Twins on average are born a month premature, and have a 7 percent incidence of long-term sequelae (some lingering negative outcome, often owing to prematurity) as compared to 2 to 3 percent in singletons.

That leaves 93 percent of twins who do not display long-term sequelae. Among these children, according to B. Alin Akerman, a Swedish researcher, studies have shown that the long-term development is not different from that of singletons. Language does develop later, probably because twins hear less language from their parents. A twin whose language develops more quickly than his or her mate may become the spokesperson for the pair. But by far, the main factor in twin development is prematurity. Not surprisingly, one major study found a significant difference in mental development between premature twins and twins that went to term, the premature twins scoring lower.

Studies have also found that one twin—the so-called alpha twin—does tend to be more dominant. Among boy-girl pairs, it is often the girl who assumes the spokesperson role, and, according to Akerman, she tends to "keep it later in life." Fascinatingly, studies have found that twins use the word "I" later than singletons, apparently thinking of themselves for quite some time as a collective unit. "If the parents forget to emphasize regularly that the twins are two separate individuals, it is not inconceivable that they will look upon themselves as one person," Akerman notes, showing just how far the thinking has come since twins were dressed as pairs and paraded in what Louis Keith, a leading twin expert and a twin himself, referred to as "our twinny clothes." In adolescence, Akerman writes, twins must separate psychologically, not only from their parents, but from each other. "The process of liberation often results in violent fights and disputes between twins who are often tied into a kind of love-hate relationship." Yet Akerman adds that twins, even when mad at each other, "usually like to be together."

Now the pendulum is beginning to swing back. New theories are calling into question the ascendant idea that twins should be separated in school. Maybe, the evolving thinking goes, too much emphasis has been placed on encouraging twins to develop independently. "People wrongly equate twin closeness with a lack of individuality," argues Nancy Segal, who is director of the Twins Studies Center at California State University, Fullerton. Segal was quoted in a 2006 *New York Times* article about a movement in which some parents of multiples are starting to demand that their children be schooled together. Not only are these parents showing up in administration offices; in a number of states, twins parents have approached legislatures, asking for laws giving parents of multiples the opportunity to request that their children be kept together. Some educators applaud the movement; others deplore it as still more micromanagement from a generation of parents who are already much too overinvolved in their children's classroom experiences. The *Times* article prompted passionate letters from parents of twins as well as adult twins themselves. "It is vital that [twins] are raised with a sense of self that is separate from the other," insisted one. "Often parents, not experts, know what is best for their children," wrote two adult twins, who supported keeping twins together. "It probably doesn't matter all that much," ventured another reader, both a twin and parent of twins. "The relationship of twins, like water, seeks its own level."

CHAPTER THIRTEEN

SOULS ON ICE:

AMERICA'S FROZEN
HUMAN EMBRYO GLUT

"As time goes by, I've gotten more of an inclination to use them," muses Tammy LaMantia, who is now thirty-four, and with her husband, Steve, trying to decide what to do with the six frozen human embryos left over from their series of IVF cycles. The question imposes itself on their consciences quite regularly, when—along with the cable bill, and the Internet bill, and the phone bill, and all the other bills that float through the mail slot of the modern suburban household—they receive their quarterly frozen embryo storage bill.

Actually, the LaMantias get two frozen embryo bills. One comes from the Virginia clinic where, more than five years ago, Tammy and Steve conceived the triplets who did not live. Another comes from the Maryland clinic where they conceived Edie and Ellie, their twin daughters. All told, Tammy reckons, they are paying more than $2,000 a year to keep six embryos stored in liquid nitrogen. Deciding what to do with them may be the hardest issue they have faced, which is saying a lot, given what they have gone through to achieve their family. The range of choices is dizzying: Should they donate these excess embryos to another couple to gestate and bear? Their own daughters' full biological siblings, raised in a different family? Should they donate the excess embryos to scientific research? Or should they authorize both clinics to remove the glass straws containing the embryos from the liquid nitrogen tanks, and let the embryos, finally, lapse?

Or should Tammy go through IVF again? Knowing how difficult pregnancy is for her? Under ordinary circumstances, Tammy would be content to stop at two children and count her considerable blessings; while it would be wonderful if a pregnancy were to happen naturally, she would not choose to go through fertility treatment again. But the existence of excess embryos complicates that decision. The embryos are there, and something must be done with them. "I think about it a million times over," says Tammy, who with Steve recently visited her fertility doctor, just to talk about the possibilities. "I can't seem to sign the papers. I can't say, let's just throw them away. We worked so hard to get them, and I feel like they're part of us. The alternative would be: we have to use them." Yet if they were to use them, which embryos would they use? How many would they transfer? Just one embryo this time? How would they decide which one? And what would they do with the others? Every solution raises a new dilemma. "If we didn't have them, we would be done" with treatment, Tammy says. "But it's different, knowing that they are there."

Tammy and Steve are by no means alone, either in having frozen embryos they must figure out what to do with, or in the paralysis they feel, surveying the landscape of choices. In fact, they are part of a large and growing group, one that includes, in a way, all of us. Assisted reproduction has given humankind many changing legacies. It has transformed families, their size and structure and makeup. But it has also dramatically impacted our thinking about human life: when life begins, what life is, and morally what can and should be done with it. Embryos are like ultrasound and other technologies that enable us to study and visualize human life in its earliest moments of development; they make it clearer than ever before what life looks like at its beginnings. In medieval times, when gestation was invisible, it was widely held by theologians that the soul entered the body at "quickening"—when the fetus began to move and kick, providing the first real evidence of its presence—and the general assumption was that what a woman did with her pregnancy before then was her business. Now we know so much. We see so much. Science tells us so much. Now, couples scrutinize their earliest ultrasounds; hang photos of embryos in their babies' bedrooms. And because we do this— because we have, now, the possibility to make life, to see life, study it, determine its fate—we are finding that embryos present a new and urgent moral problem. What are human embryos exactly? How should we think about them? What can we legitimately do with them?

It's worth pointing out, yet again, that what we are really talking about when we talk about embryos is a conceptus or a pre-embryo: a newly fertilized egg. But that doesn't seem to make the discussion easier. In the United States, the accumulation of leftover IVF embryos is approaching massive: In 2002, SART, the research arm for U.S. fertility clinics, decided it was time to find out how many unused embryos had accumulated in the nation's 430 clinics. The RAND consulting group, hired to do a head count, concluded that 400,000 excess frozen embryos existed—a staggering number, twice as large as previous estimates. Given that hundreds of thousands of treatment rounds have been performed since the count was done, it seems fair to conjecture that the number of embryos in limbo in the United States alone is closer to half a million. Among them are Tammy and Steve's frozen embryos; frozen embryos belonging to Doug Okun and Eric Ethington; embryos frozen for Janis Elspas, medical director for the Triplets Connection, who has fourteen embryos, more than ten years old, stored in a California facility whose whereabouts are unknown to her.

In a way, we are faced with a host of new moral questions because embryo storage itself turned out to be oddly easy. Early on in the development of the science, the freezing of human embryos proved surprisingly successful, and the embryos themselves remarkably durable. Not all IVF embryos survive the freezing and thawing process, but many do, even when frozen for years. And science is producing more embryos all the time: as fertility drugs have gotten more powerful and lab techniques more sophisticated, more embryos are created during the average IVF cycle—and fewer, hopefully, are transferred into the mother. The upshot is that the nation's frozen embryo glut will continue to build, and build, and built, until . . . what? Who knows? In the process, all Americans—not parents, not just politicians—are being obliged to reconsider whatever they thought they believed about life and death and choice and reproductive freedom. Thanks to assisted reproductive technology, issues many people thought they had long ago resolved have been revived and reconsidered in a whole new emotional context. Our thinking about life has been exploded—or, at the very least, refined.

"I was like: I created these things, I feel a sense of responsibility for them," is how one mother put it over lunch in Bethesda, Maryland. "I created them, I made them, and I want to give them the best home I can." This patient, who is politically liberal and strongly pro-choice, found that once her own treatment was concluded, she could not rest until she located a woman—actually, two women—willing to bring her leftover embryos to term. She had created them using both an egg donor and a sperm donor, anonymous individuals whose résumés were so stellar that she felt it would be a tragedy for these embryos *not* to be born. "These could be superstar embryos," said this patient, who made sure they went to college-educated parents. "I didn't want to put them with high school graduates; you have the product of a doctor and a lawyer, and I wanted them to have some benefit of being around people like them."

Like a woman facing an unwanted pregnancy, this patient had made her own, deeply personal decision about what do with unwanted embryos, a choice based on her own beliefs and priorities. And that's what all patients must do. The presence of embryos for whom (for which?) they feel a certain undefined moral responsibility presents tens of thousands of Americans with a dilemma for which nothing up to now has prepared them, and they do the best they can based on reflection and gut instinct, as part of a very private, very isolated process. A world away from the exigencies, mitigating circumstances, and carefully honed ideologies that have grown up in and around abortion clinics, it is people like the LaMantias and other determined, devoted parents who are being compelled to think, hard, about when life begins, and when it is— or is not—right to terminate it. They are in this position, ironically enough, not because they don't want a family, but precisely because they do. For millions of IVF patients, deciding the fate of frozen embryos is known as the "disposition decision," and it is one of the hardest decisions patients face, so unexpectedly problematic that many decide, in the end, to punt, a choice that is only going to make the embryo glut bigger, the moral problem more looming and unresolved. This is the future assisted reproduction has given us—all of us. What *are* we going to do with this accumulation of frozen human life?

"Are They People? Aren't They People?"

"Until recently, I don't know if any of us were aware of the scope of the disposition dilemma," the doctor and sociological researcher Robert Nachtigall told colleagues at the 2005 ASRM meeting, underlining just how hard it is for individual patients to know what is the right thing to do with potential lives they feel responsible for having created. Several years ago, Nachtigall and colleagues at the University of California, San Francisco, were doing a study of patients who had conceived using egg donation, and during interviews they were struck by unprompted comments by several couples. Hard as it was deciding whether to go ahead with egg donation, these parents said, it was harder still deciding the fate of their leftover embryos. Nachtigall and colleagues decided to do a new study, published in 2005 in *Fertility and Sterility,* looking explicitly at the way patients think about their unused, iced-down embryos.

Strikingly, Nachtigall found that even in one of the most progressive regions of the country, which is to say, among patients living in and around the San Francisco Bay Area, few patients were able to view a three-day-old conceptus with anything like detachment. "Parents variously conceptualized frozen embryos as biological tissue, living entities, 'virtual' children having interests that must be considered and protected, siblings of their living children, genetic or psychological 'insurance policies,' and symbolic reminders of their past infertility," his report noted. Many seemed afflicted by a kind of *Chinatown* syndrome, thinking of their embryos simultaneously as Children! Tissue! Children! Tissue!

An earlier study, done by psychologist Susan Klock and colleagues at the Northwestern University School of Medicine, found that many patients begin IVF with some notion about how they will dispose of surplus embryos. The choices boil down to five: use them; donate them for research; donate them to another infertile person; freeze them indefinitely; or dispose of them, usually by thawing. Klock found that many couples changed their minds once treatment was over. More than half the couples who had planned to dispose of their embryos decided to use them or donate them to research or to other patients. Conversely, seven of eight who had planned to donate them to research decided to use them or dispose of them. Nearly all who planned to donate their embryos to another couple found that, when push came to shove, they could not relinquish their potential genetic offspring. All in all, 71 percent changed their minds about what to do.

Also striking: only about half of patients with embryos stored for more than three years could be located.

The rest were incommunicado.

Nachtigall's study elaborated on these findings. Couples, he found, were confused yet deeply affected by the responsibility of deciding what to do with their embryos. They wanted to do the right thing, if only they could know what the right thing is. All of the fifty-eight couples in his study had children as a result of treatment, so they knew well what three-day-old embryos can and do grow into. (Nachtigall is now studying a much larger sample of couples, where both egg and sperm come from the patients. It should answer the question of whether couples who use donor eggs are in any way distinct in their thinking about embryos.) "Some saw them as biological material, but most recognized the potential for life," Nachtigall told colleagues. "For many couples, it seems there is no good decision; yet they still take it seriously morally."

For all patients, he found, the disposition decision was torturous, the end result unpredictable. "Nothing feels right," he reported patients telling him. "They literally don't know what the right, the good, the moral thing is." Some try to talk themselves into a clinical detachment. "Little lives: that's how I thought about them," said one woman, "but you have to switch gears and think, they're not lives, they're cells. They're science. That's kind of what I had to switch to." Others were not able to make that switch, thinking of their embryos as somehow sentient. "My husband talked about donating them to research, but there is some concern that this would not be a peaceful way to go," said one woman. Another said, "You start saying to yourself, every one of these is potentially a life."

Many were troubled, Nachtigall said, by the notion of donating embryos to research or to another couple, thereby losing control over their fate and well-being; they seemed to feel a parental obligation to protect their embryos. Other studies confirmed his finding; a substantial number of couples are plagued by the fear that if they donate their embryos to science, the embryos might instead find their way to another couple and end up being born. For many patients, this thought is intolerable. "I couldn't give my children to someone else to raise, and I couldn't give these embryos to someone else to bear," said one woman in Nachtigall's study. One woman described her embryos as a psychic insurance policy, providing "intangible solace" against the parental terror that an existing child might die. "What if [my daughter] got leukemia?" said another, who considered her frozen embryos a potential source of treatment for a sick daughter. A patient put the same idea more bluntly: "You have the idea that in a warehouse somewhere there's a replacement part should yours get lost, or there is something wrong with them."

For others, embryos carried a price tag that made them seem like a consumer good; a few parents considered destroying them to be a "waste" of the money spent on treatment.

"You weigh what's best," Nachtigall quoted one parent as saying, but what's best is often not clear. This parent continued: "Are they people? Aren't they people? In part of my mind, they're potential people, but the point is, it seems odd to me to keep them frozen forever. It seems like not facing the issue." A patient who had decided to donate embryos for research said, "We've agreed that it's the right thing for us to do, but the final step is to get the forms notarized, and we haven't done it. I will honestly say that it will be a day of mourning." For those couples who did reach a decision, the resolution came as a great relief, bringing with it, Nachtigall's report notes, "a profound sense of completeness and resolution."

Patients also come up with novel ways to dispose of embryos that fall short of plug-pulling. In a version of the rhythm method of contraception, some patients solve their dilemma by having leftover embryos transferred into the woman's uterus, at a time in her monthly cycle when implantation would be unlikely. Others bury embryos. Still others cannot bring themselves to dispose of them at all. "We'll have a couple more pregnancies and we'll just grow the whole lot," one father told Nachtigall and his team.

The average couple in Nachtigall's study had seven frozen embryos in storage. The average embryo had been in storage for four years. Even after that much time had elapsed, 72 percent had not decided what to do, and a number echoed the words of one patient: "We can't talk about it." The embryos keep alive the question of whether to have more children. "I still have six in the bank," said one woman. "They call to me. I hate to talk about it. But they call to me." Her words are reminiscent of a comment made by Celine Dion, who underwent IVF in 2001 and later said, in describing her plans for a second child: "This frozen embryo that is in New York is my child waiting to be brought to life."

"Pets, Natural Resources, and Pieces of Art"

And patients aren't the only ones who are agonizing; the confusion felt by parents is shared by the nation's family law judges, among many others. As University of Wisconsin law professor and bioethicist R. Alta Charo pointed out at the 2005 ASRM meeting, the frozen embryo is increasingly emerging as a point of dispute in divorce cases; when it does, a judge must inevitably decide what an embryo is. Are embryos children? Aren't they? Tracing the confused path of judicial decision-making, Charo offered one situation where a Tennessee court ruled that frozen embryos were potential children and—in the court's traditional role of acting in the best interests of children—awarded a batch of disputed embryos to the parent who intended to bring them to term. The decision was reversed at the appeals level, by a court that proposed to divide the embryos like any other marital property. But ultimately the Tennessee Supreme Court awarded them to the ex-spouse who did not want to gestate them. Commonly, Charo says, courts take pains to avoid situations where one person will bring the embryos to term against the wishes of the ex-partner, privileging the right not to procreate over the desire to do so.

In general, courts treat frozen human embryos as property, but property with an elevated moral status, property whose well-being society bears some responsibility for, "like pets, or natural resources, or pieces of art," as Charo put it. In Louisiana, the embryo has been designated a "juridical person." "No one knows what this means," Charo said, comparing the status of Louisiana embryos to that formerly assigned to slaves: not fully human under the law, but deserving of some rights. One thing "juridical person" does mean is that in Louisiana fertility clinics may not dispose of embryos. They are directed to act in the best interests of the embryos, whatever that may be: a kind of guardian ad litem of the embryo.

And now, of course, officials at the highest level of government are regularly forced to wrestle with the same questions. In 2001, in one of his first major domestic policy decisions, President George W. Bush was called upon to revisit the federal government's position on human embryo research. Very soon after taking office the president was faced with deciding whether to provide federal funding for new stem-cell lines made from just these embryos: frozen IVF conceptuses that been accumulating for two decades in U.S. clinics. Many scientists believe these embryos could be an invaluable source of material (and lab practice) for embryonic stem-cell research. In the face of pressure from scientists and patients' groups to free up funding, Bush nevertheless chose not to fund any new embryonic stem-cell lines. The president said that to fund research that would destroy embryos with "the potential for life" would be crossing a "fundamental moral line."

Bush thus cast his lot with social conservatives on the farthest reaches of the Right, his political base, many of whom hold that a newly fertilized egg is the same thing as full-fledged human life; that a pre-embryo merits as much protection as a baby in utero. James Dobson, the Christian Right leader who heads Focus on the Family, has equated discarding an embryo with "abortion at an earlier stage of life." Richard Doerflinger, representing the U.S. Conference of Catholic Bishops, testified before Congress that the human embryo should be "protected, in much the same way as other human subjects, from being harmed or killed in federally funded clinics." During a congressional debate in 2005 Tom DeLay, then House majority leader, referred to human embryos as "living, distinct human beings," while a conservative columnist referred to them as "Microscopic-Americans."

Some of this is rhetoric; some of it is doubtless deeply felt. And some of it is a well-thought-out strategy to enlist embryos in the nation's culture wars; to use embryos as another way of enhancing the moral status of the earliest forms of human life and thereby undermining support for abortion rights, much as conservative legislators have enacted "fetal rights laws" purportedly to protect fetuses from acts of homicide. Embryo rights are another weapon in the attempt to defeat *Roe*. As *Slate's* William Saletan has pointed out, pro-life lawmakers periodically threaten all-out war on the liberty enjoyed by infertility patients, threatening to seize embryo stores as part of this new embryo-rights effort. Republican Representative Chris Smith of New Jersey hinted at this when he said "the public policy we craft should ensure that the best interests of newly created human life are protected." Senator Sam Brownback (R-Kansas) has suggested that government should limit the number of embryos created to one or two per IVF cycle. In 2005, the National Association for the Advancement of Preborn Children (NAAPC) sued to stop the $3 billion stem-cell research initiative funded by the state of California; the name is a deliberate echo of the National Association for the Advancement of Colored People (NAACP), an effort to build the idea of "civil rights" for embryos. These efforts are all part of what the family lawyer Susan Crockin astutely describes as "an increasingly visible and multi-pronged strategy, at both the federal and state levels, to elevate the embryo or fetus to constitutionally protected personhood status." As Samuel B. Casey, head of the Christian Legal Society, put it in a 2003 interview with the *Los Angeles Times*, "In as many areas as we can, we want to put on the books that the embryo is a person."

For this reason, pro-choice groups—even as they are worrying over the fate of *Roe v. Wade*—are keeping a nervous eye on the reproductive technology arena, aware that at any moment, a case could emerge that could have consequences as far-reaching as *Roe*. In Chicago, a couple sued an IVF clinic that inadvertently destroyed their embryos, charging the clinic with "wrongful death." The judge ruled that the couple had standing to sue, saying that "a pre-embryo is a human being." The case is on appeal. The outcome matters: if a court does rule that embryos constitute human life deserving legal protection, anti-abortion activists will certainly seize upon it as a precedent. Similarly, if a conservative lawmaker does succeed in asserting control of frozen embryos, that could do a great deal to support the idea that the U.S. government has a legitimate say in the fate of early human life. If *Roe* does topple, it could be reproductive technology that provides the push. "For the moment couples still have dispositional control, but I predict that that is going to be challenged very soon," R. Alta Charo said at the 2005 ASRM meeting, speaking to doctors and clinic staffers. Arguing that pro-life advocates can taste "total victory" after "an ongoing nibble at the edges battle," Charo predicted that somewhere, soon, "some obscure legislature" will propose to seize control of frozen embryos; the measure will be challenged, and the ensuing lawsuit will end up in the U.S. Supreme Court as a test of who gets to decide what to do with early life. Traditionally, she pointed out, abortion rights involves weighing the interests of the woman against those of the fetus, and up to now the interests of the woman have been considered paramount. But now the interests of the embryo, or fetus, or potential child can be distinguished from those of the mother. Now, the constitutional right of the embryo can be considered alone. This, she says, is a watershed development.

For conservatives who want to overturn *Roe v. Wade,* Charo told the fertility specialists, "you guys are the perfect opportunity to separate the question of embryos and best interests, and the woman's right to direct her body. You take a law like Louisiana, saying that personhood begins at conception, and that you cannot discard embryos. Now the Supreme Court has the ability to look at the status of the embryo, not as compared with the woman's right to control what she wants to do with her body. There is no bodily interest. It's entirely possible that the first real challenge to *Roe* will be looking at the embryo in isolation. The question about discard is very, very important. This will be where they start their litigation strategy, to chip away at *Roe*."

But things, not surprisingly, aren't simple, even for the chip-away crowd. Even the federal government is riddled with inconsistency and doubts. In 2002, the U.S. Department of Health and Human Services began distributing grants to raise public awareness about what the Bush administration likes to call "embryo adoption." Also known as embryo donation, this is a process whereby embryos are relinquished by whoever created them and handed over to another couple, or person. In most states this is a property transfer, not adoption, but the Bush administration prefers "adoption" for the obvious, personhood-conferring reasons. To dramatize his opposition to stem-cell research, Bush in May 2005 posed with a group of "Snowflakes" babies, children who started life as leftover IVF embryos and were donated to other couples, thanks to the brokerage of the explicitly Christian, explicitly pro-life "embryo adoption" group, Snowflakes.

Inconveniently for the president, at that very moment the U.S. Food and Drug Administration was defining the human embryo as biological tissue. For years the FDA had been working on overhauling its tissue-donation guidelines, a designation that includes blood as well as sperm, eggs, and, by default, embryos. The new guidelines impose strict advance-testing requirements that threatened to halt the donation of embryos, since infertility patients never think of themselves as "donors" when they start out and don't get the newly required testing. Clinics feared they would have to close down their donation programs. At the last moment, an exemption was made for embryos, and embryo donations were permitted to go forward. The infertility lobby was delighted and a little smug, not just because doctors and patients' groups support embryo donation (which they do), but because "tissue" remains the designation conferred on human embryos by the FDA, a view shared by the infertility lobby, which prefers to have them thought of as multicelled clump of tissue.

When it comes to the moral status of embryos, pro-life conservatives may be the most torn of all. There is another strain of conservative thought, which says that maybe a pre-embryo really *isn't* a human being. Doubtless, conservative thinking, at least the pragmatic thinking of legislators facing re-election, is influenced by widespread public support for embryonic stem-cell research. But some of it may reflect real moral doubt, and with it schisms in how pro-lifers themselves think about life. In May 2005, in a dramatic break with the president, the Republican-led U.S. House of Representatives passed a bill approving funding for stem-cell research using leftover IVF embryos, and in the summer of 2006 the U.S. Senate did the same, with some notable pro-life conservatives, including Republican Majority Leader Bill Frist, supporting the bill.

Frist is allied in this with such Republican stalwarts as former first lady Nancy Reagan, who has become a vocal proponent of using IVF embryos in stem-cell research, arguing that such research someday may cure patients suffering from Alzheimer's, the disease that killed her husband, the late president Ronald Reagan. Ironically, it was the Reagan administration that did so much to enact the ban on embryo research that the former first lady is now objecting to.

The day after the Senate vote, President Bush chose to be resolutely consistent: he issued his first presidential veto, defying his fellow party members in Congress. Bush's press spokesman, Tony Snow, said the president considers it "murder" to destroy a human embryo. Snow later had to withdraw that statement as too extreme, but the gist was true: the president does believe that early human life is life, period. As this book went to press, a similar clash was brewing between Bush and Congress, newly under Democrat control.

Once a person contemplates an embryo—really looks at it, under a microscope or in a photograph—his or her opinion is often changed, and not in any consistent or predictable direction. This can be true for the staunchest conservatives. In the summer of 2004 I interviewed California Representative Dana Rohrabacher, a reliably antiabortion Republican member of the House. Rohrabacher was one of some fifty Republicans who defied the president by voting in favor of federal funding for stem-cell research using IVF embryos. For Rohrabacher it was not abstract: he and his wife, Rhonda, went through IVF and have triplets as a result, as well as frozen embryos. Going through treatment, Rohrabacher told me, his thinking about life and its origins was changed. "For a long time I've been pro-life, and I still consider myself to be pro-life," he reflected, sitting on the front porch of his Huntington Beach bungalow, which, inside, had been taken over by the demands of triplet care. A small house, it now looked like a well-organized day-care center, with family members holding babies while Rhonda sat in a chair with a laptop projector, making a list of supplies they would need for a trip to France. The list included things like "200 burp cloths" and "105 nipples."

"I have done a lot of soul-searching but also a lot of rethinking about reality, and what's going on here, and I have come to the conclusion that I'm—first, I'm still pro-life," Rohrabacher said. "But I always said that life begins at conception . . . I was predicating that on the idea that life begins at conception when conception begins in a woman's body."

Now, Rohrabacher realizes, conception can take place outside the human body. For him that is a meaningful difference. Is the embryo in the womb, or is it in a lab? "I don't think that the potential for human life exists in a human embryo until it's implanted in a human body. So you are not destroying a human life by basically not using a fertilized egg." Rohrabacher has taken heat for his views, but he has also been thanked by constituents who agree, including a man whose granddaughter has diabetes. Rohrabacher says this man implored him to support stem-cell research, saying of IVF embryos that "some of them will be adopted, yes, but no matter how many they give out, Dana, an honest person has to recognize that most of them will be thrown away. And if that's the choice you've got, I don't see how a compassionate person can choose to be consistent to principle to the point of absurdity, that you throw away something that could be useful in scientific research to somebody else." Rohrabacher says he found this argument persuasive. "I said, 'Let me think about it,' and as I talked to other people, I came to the conclusion: These were not potentially human lives until they were implanted in a body. Left alone they will not become a human being. When they are implanted in a female body, they have a chance to become a human being, so I would still be opposed to abortion."

"By the way," he added, "Ronald Reagan—I think he would be right with me on this. Ronald Reagan understood technology and how technology can change the reality of what you're talking about. What was real in terms of stem-cell research fifteen, twenty years ago is not what's real today."

Not long after I interviewed Rohrabacher, I visited the offices of Snowflakes, the pro-life embryo adoption agency, which is headquartered in a nondescript first-floor office suite in Fullerton, California, though it has expanded to the point where it has offices on the East Coast as well. Snowflakes is operated by a pro-life organization called Nightlight Christian Adoptions, which also brokers adoptions of children from foreign orphanages. Though small, Nightlight Christian Adoptions has proved ideologically mighty. As Stephen Hall describes in *Merchants of Immortality*, the group has been instrumental in fighting federal funding for embryo research, and at key times when President Bush has made speeches opposing embryonic stem-cell research, including at his 2006 veto, Bush has surrounded himself with Snowflakes babies, whose parents seem glad to travel to Washington at the drop of a hat and proffer their babies as examples of what embryos can grow up to be, if they don't grow up to be stem-cell lines.

The embryo adoption branch, Snowflakes, founded in 1997, is ably directed by Lori Maze, a likable, down-to-earth attorney who confided that she had been your basic pro-choice career gal until she found herself infertile, at which point—and this is not uncommon—she began to view every aborted baby as one she could have adopted. "I've done a severe 180," says Maze, who ended up adopting a son the old-fashioned way, and who loves her family but would never be a stay-at-home mother: "I need to do this job. I am called to do this job."

Maze oversees a very efficient operation. Her ability is one reason Snowflakes remains the flagship agency for embryo adoption. (There are a number of outfits that broker the donation of embryos, including the National Embryo Donation Center, some clinics, and at least one Yahoo message board, Snowbabies, on which couples can arrive at their own ad hoc arrangements.) Technically, embryos cannot be adopted, but Snowflakes nevertheless does a conventional adoption home study and tries to make sure the embryos will end up in an "appropriate" family.

And this is where, it seems to me, the pro-life message also gets fuzzy. You might think the goal of an embryo-rights organization would be to get all embryos adopted and gestated and born as soon as possible; you might think Snowflakes would be out on the streets, recruiting women with uteruses waiting to take in these microscopic Americans. But Snowflakes is determined to keep dispositional control with IVF patients. Every family that relinquishes embryos to Snowflakes may dictate what kind of home they want their embryo raised in. If the genetic parents wants their embryos adopted by Chinese Buddhists, or by Republicans, or by short rich people living west of the Mississippi, Snowflakes will store the embryos until they find a match. If it takes ten years, it takes ten years. Naturally, certain patterns emerge. "We get a ton of stay-at-home-mom preferences," Maze said. "Or they'll say part-time work is OK, but the child care has to be provided by a family member." Maze grants that "there are some things I can't promise. If they say, I want a family that won't get divorced, I say, I can't promise you that." The drift of preferences is clear. The typical Snowflakes donor is unlikely to favor gays, lesbians, or single mothers. "Most want a mommy and a daddy for their embryos."

In other aspects, the organization takes a very hard-line pro-life stance. Snowflakes does not permit selective reduction, nor does it permit thawed embryos to be discarded. Once thawed, embryos must either be transferred into the "adopting" mother's uterus or refrozen. If your embryologist doesn't have refreezing technology, then all the embryos must be transferred. "If you need to implant all five, recognize that you need to be prepared to be parents of quints," Maze warns. What about selective reduction to save the life of a mother? What if a woman cannot physically bear quints? What if the pregnancy will kill her? Or damage the babies? She gets this question sometimes. "They say, 'What would happen if my life were in danger, if it were the baby or me?' I tell them: 'If you were worried about that, this is not the program for you.'"

I asked Maze if she would save every embryo in the world, if such a thing were possible. Every pre-embryo, even. What about fertilized eggs that are naturally conceived, but don't implant? Would Snowflakes use a strainer to catch every fertilized egg a woman sloughs off, and "rescue" them? How far does a pro-life commitment to three-day-old, eight-celled human life extend? "I don't know," she said. "I've never been posed that question before. There's a natural miscarriage rate. As a Christian woman myself, to a certain extent God does what he does. In terms of this program, really, all we're trying to do is let God be God. If God chooses to—obviously, if the embryos are just thawed and nothing is done, they're going to die, but we have at least given them the opportunity to be implanted. We don't feel like we're playing God. We're letting God be God. He gets to choose how many of those embryos are going to be born on earth. It's his decision. It's his plan."

On the other side of the ideological divide are many other groups. Throughout the stem-cell fray, advocacy groups for the infertile have argued in favor of stem-cell research, and—in the face of threats to seize embryo inventories—have maintained that it is the right of patients to decide what to do with unwanted embryos, just as it's the right of women to decide what to do with unwanted pregnancies. At a press conference in 2005, Sean Tipton, spokesman for ASRM, said that "patients control and make the decision about what happens with those embryos, and that's the way it is now, and it's important that that's the way it stays." ASRM is joined in this by pro-choice groups, progressive science groups, liberal bioethicists, and many churches. The Unitarian Universalist Association of Congregations has endorsed stem-cell research using IVF embryos. So has the General Synod of the United Church of Christ, the Episcopal Church, and some Muslim leaders. Interestingly, the Mormon Church believes that "personhood begins when the spirit unites with the physical body, an event that does not necessarily occur at the time of conception," according to a report that also points out that major branches of Judaism, including Orthodoxy, "while viewing the embryo as worthy of respect, respond to the pressing need to seek healing and support all forms of stem-cell research."

But the confusion at the federal level and the current ban on federal funding make it very difficult for patients willing to donate their embryos to scientific research actually to do so. Currently, no lab that receives federal funding for any purpose can accept frozen human embryos for research. This creates an enormous impasse, and it's one reason why, in the United States, our embryo glut is ceaselessly building. At the ASRM press conference, Pamela Madsen, head of the American Fertility Association, talked about the difficulty the ban poses for patients who want to donate their embryos to research. Madsen has two IVF sons, Tyler and Spencer, who were sixteen and twelve, respectively, at the time. Both boys were present, and spoke charmingly about what it's like to be them. "It's a little surreal to be talking about the way I was conceived and everything, but really I'm very thankful for all the technology and all the money put into, um, me," said her older son, Tyler, while her younger son, Spencer, who as an embryo had been frozen, said, "I just think it's really great that we have the technology that I can be here."

Madsen said she and her husband have four leftover embryos. She would have liked to use them to have another child, but her husband wanted to stop at two. Madsen said they cannot bring themselves to relinquish excess embryos to another family, because any child "would be a full sibling to our children, and that makes it very difficult." They want to donate the embryos to stem-cell research, but even Madsen—a professional in what could be called the embryo field—cannot find a lab that will take them. When people call ASRM's Sean Tipton asking how to donate their embryos for research, he tells them, "It's almost impossible. I can only tell them to call their congressman and tell him to lift the ban. It's frustrating. These people have gone through an incredible emotional and physical hardship, and they would very much like something good to come of it."

"People Do Not Want to Inherit Embryos"

In a strange sort of poetic justice, the nation's embryo glut also presents an urgent problem for the very people who helped create them: fertility doctors. Many doctors are at wits' end trying to figure out what to do with all these embryos they have made: embryos that have fallen, willy-nilly, under their moral, medical, and possibly legal purview. When patients agree to have embryos frozen, they sign forms agreeing to pay for their storage. Patients also state what should be done with the embryos should they themselves divorce, disappear, or stop paying storage fees. Many patients at some point do abandon their embryos. Doctors technically are free then to dispose of them. But many are reluctant to take that step. They are terrified that at some point a patient will come back and sue them for—well, for something.

"Nobody does it [destroys abandoned embryos]," says Alan DeCherney, a reproductive endocrinologist who is now at the National Institutes of Health and who is the editor of *Fertility and Sterility*. "It's a hot topic. People think the risk of holding them is less than the risk of destroying them."

And the risk of holding them is considerable. "I have tons of embryos, and I can't track down the owners," said Vicken Sahakian of Pacific Fertility Center, sitting in his Wilshire Boulevard office. Sahakian practically had his head in his hands, thinking about all those embryos. "It's one of the main problems I have. I have thousands of embryos from patients who have been through this program for, what, ten, twelve-plus years, changing addresses, and never called back, never paid storage fees, you can't track them down." Sahakian does the best he can to whittle down his own embryo inventory; he runs a strong embryo-donation program, encouraging patients to donate embryos to other patients. He has also hired a collection agency to track down patients and force them to make a resolution. His "biggest nightmare," he said, is that he will be unable to sell his practice when he is ready to retire, because no doctor will want to buy a practice that comes with a closetful of unclaimed embryos and the vague, terrible responsibility they entail. "The person buying it does not want to buy the embryos. That's the rule," he says. "People do not want to inherit embryos. So what do you do with them? I have embryos that have been here since 1992."

The overages have grown to such proportions that companies now exist solely to manage embryo inventory. Back in 1990, a sperm-bank employee named Russell Bierbaum had a vision of the future, and what he saw were lots and lots of frozen embryos. So he founded a company called ReproTech, which can be hired to assume and maintain doctors' embryo inventory, as well as to handle transport, a tricky process in and of itself. Business is booming: he has two facilities, in Minnesota and Florida, and is constantly adding new storage tanks. Bierbaum prefers to assume responsibility for embryos soon after their creation. His staff stays in touch with patients, keeping addresses current, periodically calling to say hello and review options. In a few instances, he says, he will take over abandoned embryos and attempt to track patients down. It is therefore people like ReproTech staff members—rather than, say, ministers or psychologists—who often are the ones discussing with patients fundamental questions touching on birth and death and life and reproduction, all the essential questions of humanity. "We end up being the counselors without the credentials," acknowledges Bierbaum, "just answering the questions, being available."

It's hard to know how the embryo glut will go away. The RAND study found that only about 3 percent of unused embryos are earmarked for research. In Britain, policy makers have done a better job of biting the bullet and taking actions that in this country would make both pro-choice groups and pro-life groups livid. In Britain, unused embryos are destroyed after five years. Done. Gone. No argument. This did not occur without controversy: the first time embryos were set to be destroyed, a group of pro-life advocates staged protests. The deadline was extended, but eventually the embryos were allowed to "succumb." Other countries, such as Germany and Italy, forbid the freezing of embryos. In Italy this is the result of moral suasion from the Catholic Church; in Germany it has more to do with post-Holocaust unease with any lab technique involving humans and experimentation. In the United States, at some point the American public may have to recognize that retaining an absolute right to determine what happens to your frozen embryos just isn't workable, particularly when the most common patient decision is not to decide.

Barring government seizure, what are we going to do with all those human embryos? Robert Nachtigall believes that with better patient counseling and logistical coordination between fertility clinics and research labs, many more unused human embryos could be directed toward stem-cell research. "I think it's a mistake to call it a glut," says Nachtigall. "I mean, these embryos are created in a process as hundreds of thousands of couples attempt to overcome infertility, and their presence is perhaps an unanticipated side effect of the use of advanced reproductive technology. But there is nothing inherently negative or wrong about their existence, and as we turn our attention to them, we may find that indeed they could be a tremendous resource for science, the country, and for mankind, for that matter."

So how many unused embryos have to accumulate before it would be considered a glut? A million? Two million? It won't take long to get there. And the problem is that few fertility clinics treat embryo disposition as a serious moral issue. The American fertility industry has not served its patients well by encouraging the idea that human embryos are merely multi-celled clumps of tissue. Embryos are multi-celled clumps of tissue, it's true, but they are also far more complicated and emotionally fraught. One of the powerful findings of Nachtigall's study was how isolated patients felt in making the embryo-disposition decision; how they longed for counseling, advice, an honest out-loud moral conversation between people who had been through, and thought through, the same issues. A conversation about science, life, birth, humanity, and the legacy Patrick Steptoe and Robert Edwards have given us. Whether the reproductive rights community might ever encourage such a conversation seems unlikely in the charged political atmosphere. But it would be useful, to put it mildly.

Meanwhile, the technology itself is so new that nobody knows what the expiration date on embryos might be. In 2005, a San Francisco woman gave birth to a baby who, as an embryo, had been frozen for thirteen years. "I have a family who is pregnant, actually they just gave birth, the embryos were frozen in '93," Lori Maze told me in 2004. "We're talking eleven-year-old embryos. That's an incredibly long period."

CHAPTER FOURTEEN

CHOICE REVISITED:

ETHICS, FEMINISM, AND ART

The Ascendancy of the Egg

More than once while writing this book, I was asked what it feels like to have eggs. The question always came as a surprise. It was not something I had given any thought to. I mean, I had often reflected how marvelous it is—as Natalie Angier points out in her book *Woman: An Intimate Geography*—that women are born with all the oocytes they will have; that each girl child comes into the world with some 2 million ready-made gametes, nestled dormant in her ovaries. When my daughter was small and just starting to ask questions about conception, I would tell her how she had in her tiny belly millions of tinier oocytes. She, too, found this wondrous: already inside her were the seeds of the next generation, the human germ line, yearning ineluctably toward the future.

But these questioners weren't asking what it felt like to be born with one's full complement of reproductive cells—unlike men, who have to wait until puberty for theirs to be manufactured. They were asking what it felt like to carry around one of the most coveted natural resources in modern medicine. It is a question I—and, I think it's safe to say, women's reproductive rights groups—don't know the answer to. Don't know how to think about. And it's only one of innumerable ways in which reproductive science is challenging feminist thinking.

One of the immense fringe benefits that IVF medicine has given us in the past thirty years is a far more detailed understanding of how the reproductive cells go about their business. Scientists can now watch how the conception of life happens, all the steps, from the hardening of the shell of the newly fertilized egg to the march of the chromosomes as they organize themselves into a set of instructions for a human being. In the process, scientists have revised the old notion that the sperm is the source of reproductive energy; they have abandoned thinking—and some did once think this—that the sperm was where the animus of life resided, while the oocyte was just this passive receptacle floating in the Fallopian tube like Sleeping Beauty, waiting to be awakened.

We are living in the age of the oocyte ascendant. The human egg is now understood for what it really is, mistress of the universe: big, mysterious, powerful, transformative, a little domineering, possessing the ability not only to direct the development of the embryo, but also—this is another fringe benefit—to alleviate human suffering. If embryonic stem-cell research is ever to realize its promise of offering cures to people with diseases like diabetes and Parkinson's, it will do so, many scientists believe, thanks to a form of embryonic stem-cell research known as therapeutic cloning. And if therapeutic cloning is ever to work, science will need thousands, perhaps millions, of human eggs, retrieved from the drug-swollen ovaries of thousands, perhaps millions, of living women. "Right now, there's no way you can make embryonic stem cells without an egg," David Keefe pointed out in 2005, on one of the days I spent with him at the fertility clinic at Women and Infants Hospital in Rhode Island. In the long run, Keefe said, science may be able to come up with an artificial substitute. But for now, "there will be a massive demand for eggs."

The two of us had just come from the lab Keefe ran in an academic building at the Marine Biological Laboratory in Woods Hole, Massachusetts, where members of his staff were experimenting on discarded eggs from IVF treatment, trying to understand what makes oocytes age sooner than many other parts of the body. Scientists in his lab, and elsewhere, have spent a lot of time looking at eggs and talking about eggs and reflecting on eggs and—not an easy thing—trying just to *get* eggs. They have developed, as a result, an almost religious reverence for their complexity, their spherical glory, their creative talent. "Eggs really are dramatically beautiful," Keefe mused at one point. "They are beautiful cells."

"I meant to ask you," he asked, "as a woman growing up in a man's world, does it make you feel different, to know you have the most powerful cell in the world? Like—you could, in the not too far distant future, be able to restore parts that were missing in my body?

"Women have been brought up to think the world revolves around men, that the man was bigger and stronger," he continued. "But biologically, women are way stronger than men."

It's true. In many ways men are biologically more fragile than women: male fetuses, for example, are more likely to miscarry than female fetuses. Similarly, by the time conception happens, the sperm is this exhausted, wrung-out creature that barely has the energy to move, this incapacitated, shut-down thing whose development the egg promptly takes over. "What happens with sperm, it has DNA just like the egg does, but it's all packaged and shut down," Keefe explained. "The sperm has to swim like crazy. So you shut down everything you need; lose all your excess weight; that's what a swimmer does, and that's what a sperm does. A sperm is really just a guy in a Speedo, all shaved down. It loses its ability to transcribe genes; it can't really make proteins, it can't express its genes. It's sort of quiescent. The chromosomes are all rolled up and packaged. The egg has an extreme ability to open the sperm up; to get the male to open up. It gets him to open up and start expressing." Actually, he amended, nobody knows whether the egg is getting the sperm to open up, or changing it. Conception is like marriage that way. "A lot of these billions of dollars in California will go to [figuring that out]," Keefe said, talking about the $3 billion that California has authorized to encourage stem-cell research. Once scientists understood the way the egg gets to work on the sperm, Keefe said, "they thought: if an egg can do that to a sperm, maybe it can do it to a somatic cell."

What this means is that once scientists understood that the egg unfolds the sperm's chromosomes and persuades them to express their messages, they wondered what would happen if you took the nucleus from a regular body cell and let the egg get to work on that. One great hope of stem-cell science is that the oocyte, with its boundless directorial energy, someday can help rebuild nerve cells, rebuild brain cells, make injured and aging bodies whole again. The hope is that scientists can make new systems and body parts that will not be rejected by the immune system, because the part has been tailor-made from the nucleus of a cell that came from the patient's own body, a cell rejuvenated, redirected, by an egg. The hope is that the ability of the egg to produce new human life can be harnessed to reverse human aging.

The oocyte: it can make the young grow, and it can, perhaps, make the grown young.

"You can create replacement parts for organs that previously were accepted—their failure was accepted as inevitable," Keefe said. "People assume that as you age, your brain will give up its memory capacity. How about if you can take those dying neurons and replace them with stem cells that you had generated from an egg that had been denucleated? That's very close to being able to be done. How about if you had a heart attack, and you take the infarcted part, the dead part, and create a viable replacement?"

Earlier that day, one of Keefe's embryologists, Jim Trimarchi, had been rhapsodizing along the same lines. He pointed out that at their beginnings, human cells are totipotent. Any cell can turn into anything. But as life goes on, cells—like humans—become specialized. Paths of life are chosen. Other paths are closed off. Options narrow. As a cell matures, it stops listening to all the messages the DNA is communicating, and hears only the messages telling it to be, say, a skin cell. What the egg has the power to do, Trimarchi points out, "is erase, from the skin cell, all knowledge that it is a skin cell. Over years and years all these other cellular messages have been silenced so that the only message this particular cell listens to are ones saying: be a skin cell." What the egg can do is make the cell hear, once again, the full symphony of human possibility. "You take the nucleus out of the skin cell and put it in the egg, it erases all that silencing, it erases all the silencing of the DNA, so that now it can read all that."

That's what happens in therapeutic cloning. Like a good mother, the egg whispers to a skin cell: Honey, you can be anything. You don't *have* to be a skin cell. You can be a brain. You can be a spinal cord. You can be a heart cell. You can even be a baby. Those few scientists who are interested in human reproductive cloning may someday try to use this technique to make a baby that would be a near genetic replica of the person from whom the skin cell came, a prospect many scientists abhor and would gladly see outlawed. What most scientists want to do—what most Americans want science to do—is use cloning for therapeutic purposes. While IVF embryos could be very helpful in basic stem-cell research, it is therapeutic cloning—in which an entirely new embryo is created from an egg and a body cell, and, from that embryo, a cell is plucked which can then be coaxed into a "replacement part"—that could possibly provide truly tailor-made new body parts and systems. And this great hope of stem-cell science can be traced back to Patrick Steptoe and Robert Edwards. Back to Louise Brown. Assisted reproduction helped open the way for this potentially revolutionary form of medical treatment.

More specifically, it is the cytoplasm—the viscous stuff surrounding the egg's own nucleus—that has the ability to carry out this crucial deprogramming and reprogramming. It is the cytoplasm that Jim Trimarchi likes to think about. "The cytoplasm is not like tea or milk— it's not watery, it's far more structured than that," Trimarchi said. "But at the same time, it's not like gelatin, it's far more liquid than that. And it's where everything—everything—lies. All the magic. Cytoplasm is the fountain of youth. It's that simple. You take a cell that is a totally differentiated cell, for all intents and purposes a skin cell, this thing churning out proteins that are only appropriate to being a skin cell; it's connected to other cells that are skin cells; they have special skin-cell-to-skin-cell contacts. You take it out, and put it in an egg, and it makes a *whole sheep*, with liver and brains and a heart. It's the egg cytoplasm. It's magic."

It was a heartfelt homage to the oocyte, and it wasn't the only one I heard. "Have you meditated on the phenomenon that you are carrying around cells that have the complete ability to create a *human being from scratch*?" I was asked by Carl Gunderson, deputy director of United Cerebral Palsy. We had been discussing how marvelous it is that most children do turn out perfectly constructed, and his thoughts had turned in the direction of how that happens. I never did come up with a good answer. I wasn't sure whether to feel empowered or alarmed. Anyone living in a country like, say, Nigeria knows that being rich in natural resources, particularly resources that must be extracted from deep, hard-to-reach places, does not always benefit that country's inhabitants. Being rich in natural resources attracts outsiders, wildcatters, dictators, trouble. Civil war. Corruption. Mayhem. Plunder.

"The egg is such a powerhouse," David Keefe said. "It's going to be one of the biggest powerhouses in medicine. And the thing is, who controls it? Women!"

Making Abortion Look Easy

But do we? Do women know what to think about reproductive science and the new relationships it creates, the challenges it poses? Do we own this science? Not remotely. To me, one of the most striking things about assisted reproduction is the way it creates dilemmas that are so much harder for women's groups to grapple with than abortion, which seems, in contrast, almost easy to think about. For or against, people can make their minds up about abortion. Few people can make their minds up about reproductive science. The ethics of having babies is so much more complicated than the ethics of not having them. There are prominent feminists—such as Judy Norsigian of the redoubtable women's organization Our Bodies, Ourselves, which has done so much to promote women's health over more than four decades—who oppose egg donation for stem-cell research. Norsigian and others worry that women, especially poorer women, may be attracted by financial compensation to donate eggs without understanding the risks. To donate eggs, women must take powerful hormonal drugs, which may or may not (nobody quite knows) increase their susceptibility to ovarian cancer. Then they are subjected to retrievals that carry a 1-in-100 risk of infection as well as the even higher risk of ovarian hyperstimulation syndrome, in which the ovaries swell and fill with fluid. It's a real worry. "We have a donor in the hospital right now for hyperstimulation; she's semi-sick," I was told by Robert Stillman, a doctor at Shady Grove Fertility, talking about an egg donor in their fertility program. "One of them is going to die one day," Robert Nachtigall says.

Recently the world was handed an example of just how badly science wants the oocyte, and just how far some scientists are willing to go to get it. In early 2005, the lionized South Korean stem-cell researcher Woo Suk Hwang published a paper in *Science* announcing a breakthrough in therapeutic cloning. He and his team had, he said, created eleven individualized stem-cell lines (that is, stem cells with the ability to proliferate indefinitely), using just seventeen oocytes per line. It was a stunning increase in efficiency from 2004, when Hwang reported that his team needed 248 oocytes just to make a single stem-cell line. Both times, Hwang asserted that the eggs had been voluntarily donated. It was heartening news: perhaps, people allowed themselves to hope, science wouldn't require a massive drift net to pull in millions and millions of eggs. How, everyone wondered, had Hwang done it? Which denucleation technique did the Koreans use? The slice? The squish? Then scandal hit: investigative journalists in Korea learned that Hwang had not created any of the new stem-cell lines that he described in his 2005 paper. Zero. Zip. He had faked the data. Moreover, he had used, according to a former colleague, at least 1,100 eggs. More than one thousand human oocytes, excavated from who knows how many women— for nothing. And Hwang had paid some donors. He also reportedly pressured a junior associate to "donate" her own eggs.

It's not surprising that in the United States and elsewhere, scientists—even as they covet oocytes—are very nervous about the circumstances under which they are obtained. As a result, the National Academy of Sciences recommends that women donating oocytes for stem-cell research should not be paid. Speaking to biologists at the Marine Biological Laboratory in 2005, the University of Pittsburgh stem-cell researcher Gerald Schatten articulated the rationale for this view. "I am terrified of the financial reimbursement of egg donors," Schatten told them. "Don't think for a minute that this can be swept under the rug or minimized. This is different from donating blood. This is different from donating sperm. Ovarian hyperstimulation is a life-threatening risk. If California moves superfast in stimulating thousands of women, when that first woman dies for the sake of cells in a plastic dish, this is going to be a nightmare. And I am seriously, seriously worried."

Schatten spoke passionately, even emotionally, about the dangers egg retrieval posed to women, particularly poor women and women of color. At the time, Schatten was a rock star in stem-cell circles; he was identi-fied as a senior author on Hwang's 2005 cloning paper, which had not yet been exposed, and every young scientist in the room was hanging on his words, hoping he would reveal how the Koreans got so good. Soon after-ward Schatten sought to detach himself from Hwang's paper, saying he had learned some parts of the experiment were misrepresented. After the full scandal broke, the University of Pittsburgh rebuked Schatten for serving as a senior author on the Hwang paper, without knowing more about the data and how they were obtained. Or, presumably, the oocytes and how *they* were obtained.

But why shouldn't women be paid for donating eggs? Why is egg extraction from poor women acceptable as long as they are not paid? Let's not forget the *other* great hope of stem-cell research: that it will yield, in addition to cures, pots and pots of money. One reason states like California and now Illinois, Maryland, and many others want to facilitate stem-cell research is because of the tax revenues and other financial benefits that may ensue, as private companies settle their operations in research-friendly venues. Given how much money everybody else wants to make, it seems reasonable that egg donors not only be paid, but be offered a percentage of the profits from stem-cell-related patents.

Kathy Hudson, a biologist who directs the Genetics and Public Policy Center, a think tank operated by Johns Hopkins University, does think egg donors should be paid. Hudson reviewed the risks as part of a task force operated by the International Society for Stem Cell Research, which is trying to draw up rules scientists all over the world can follow. While the society itself remains equivocal about payment—one envisions these male scientists who really, really want those oocytes, and feel really, really guilty about it—Hudson published a policy paper giving her own sensible viewpoint. She thinks oocyte donation can be done safely, with informed consent and careful oversight. She also thinks it's inconsistent, even patronizing, not to pay donors. I agree. If egg donation for science is going to be done at all, it seems perfectly acceptable to pay the donors.

"Research participants undergoing an MRI for research or those donating blood, urine, sperm, or bone marrow are compensated financially in recognition of their personal contribution. In my opinion, to not do so for women undergoing hormonal stimulation and invasive oocyte retrieval research is just plain unfair," Hudson argued in her paper, putting herself at odds with the likes of Norsigian. It's just one of many issues on which women who have women's health in mind part ways when confronted by new technologies.

Choosing Teams

"Unless Planned Parenthood can grapple with the bioethical issues of reproductive life in the twenty-first century, it's going to be left behind," Paul Root Wolpe, a bioethicist at the University of Pennsylvania, warned summer-attired pro-choicers at a retreat Planned Parenthood held in July 2003. The meeting had been called to talk about any number of issues posed by the newer reproductive technologies. This is, after all, Planned Parenthood, and no parenthood is more planned than parenthood pursued with the help of a doctor, an embryology staff, therapeutic counselors, and the global pharmaceutical industry. For three days, during talks and meals and hikes at the Snowbird resort in Utah, top administrators as well as local staff of the country's best-known women's health organization tried to decide what to think about commercial surrogacy, egg donation, a child's need for a mother, a child's need for a father, embryo research, donor offspring and their right to their genetic identity. Over and over, they asked themselves whether every choice made possible by science is a choice pro-choicers should welcome into the broad philosophical tent of choice.

The meeting had a special urgency. Rightly or wrongly, there was an ominous sense that conservatives were massing, and, worse, that conservatives *knew what to think about everything*. Early in his tenure, President George W. Bush appointed the conservative bioethicist Leon Kass to head a commission that would look at stem-cell research and related issues. Kass, an early foe of IVF, used the opportunity to reopen even the oldest debates about assisted reproduction. In the summer of 2003, the President's Council on Bioethics issued a draft overview of IVF technology that contained recommendations with ramifications for reproductive freedom as a whole. The draft repeatedly referred to the human embryo as "nascent human life," "nascent life in vitro," and even a "child-to-be," language that sounded very much like a backdoor antiabortion sally. Like measures aimed at embryo protection, the wording in this draft looked, walked, and quacked like another pro-life push to enhance the moral status of the embryo. Kass and Company also took pains to lay out the ways in which IVF technology sometimes pits the interests of a potential parent against those of a potential child—for example, by giving parents the ability to choose certain embryos for transfer, while passing over others. To pro-choicers, these all looked like alarming precedents.

The draft also suggested policy options that would erode reproductive privacy. It proposed that every embryo be tracked and records kept of its fate; that a federal registry be created of IVF children; that arrangements like paid surrogacy should perhaps be prohibited. But after the draft was made public, ambivalence set in. Members of the president's council argued with other members; patients' groups vehemently challenged the council. The language was watered down. The council's final report lost much of its language suggesting that an embryo was practically a child; abandoned proposals for embryo inventories; and proposed only mild regulatory measures, such as closer monitoring of clinic outcomes. Contrary to what some pro-choicers had feared, even like-minded conservatives had trouble reaching consensus, in part because of the paradoxical nature of what they were doing. Regulating an industry under a Republican president with a record of letting industries flourish unfettered is not an easy thing to pull off. IVF involves reproduction, it's true, but it also involves business. Within the conservative fold, "there is a very strong internal tension between listening to the quasi-religious conservative voice and the libertarian market voice," as the bioethicist Arthur Caplan has put it. "There is no way they are going to regulate that industry."

But there was also a broader social context to the Planned Parenthood retreat. In the last decade, at least some women's groups have recognized that public thinking is shifting in a way they need to pay attention to; that support for abortion may be softening, or at least changing; that the abortion debate itself is stagnating as the two sides endlessly revisit old arguments; that younger women want to know how the movement can help them face new concerns and challenges. Also in 2003, *Newsweek* published a cover story on the "fetal rights" movement. The article pointed out that the latest reproductive technologies—providing, as they do, the ability to see human embryos, and cultivating, as they do, an atmosphere in which pregnant women happily scrapbook those eight-week ultrasounds—have created a real image problem for reproductive rights groups. Now that we do know what the developing fetus and the laboratory embryo look like, so vividly, it is no longer possible to argue, as pro-choice leaders once did, that abortion is not taking some form of human life. Because of technological advances and new social realities, women's groups are realizing that they need to overhaul their approach. As Kirsten Moore, president of the Reproductive Health Technologies Project, put it, the *Newsweek* piece "kind of prompted us to realize, oh my God, our movement's messages suck."

In response, Moore's group has spearheaded a movement to "reframe" abortion rights language in a way that recognizes that many women do want to have children. "Pro-choice, pro-child," she argues, should be the new, family-friendly reproductive rights slogan. She believes the abortion-rights movement should even move away from talking about "choice"—since choices sound selfish—and argue instead that abortion is a "decision" women have to make, "decision" being a concept with greater moral gravity. Other liberal organizations have followed suit. In 2006 the Center for American Progress issued a report that attempts to expand the definition of reproductive rights to include gays and lesbians, as well as both people who do and people who don't want children. Part of this overhaul involves thinking about reproductive science and how it fits into the pro-choice, pro-child philosophy.

It is significant that liberal attendees at the Planned Parenthood conference had every bit as much trouble coming up with positions as the President's bioethics council. And many of the sticking points were the same. Should fertility medicine be more closely regulated? Should the health of egg donors be better protected? What about surrogacy? Surrogacy for gay men? Feminists with an eye toward social justice were just as skeptical about paid surrogacy as Kass and his conservative colleagues, while others equivocated. Arthur Caplan stated his view that children born through egg and sperm donation do have the right to know the identity of the donors. *Children's rights!* Taking precedence over the rights and desires of their parents! And the thing is, nobody disagreed. There they were, the brain trust of the pro-choice movement, collectively accepting that rights of children sometimes trump the rights of the adults who begat them. At one point, one attendee told another that she sounded like Leon Kass. People laughed. It was true.

Local staffers got up and begged for position papers: some sort of publication, with bulleted points, telling affiliates how to think and what to say about assisted reproduction and its mounting repercussions. Around the country, state legislatures regularly propose laws regarding not only embryos but surrogacy and egg donation. Legislators in several states have tried to put through measures to ban IVF and sperm donation for unmarried women, thus far unsuccessfully; one Virginia measure was referred to, by right-wingers, as the "get a man" bill. That's unreasonable, clearly, the group agreed, but what about when conservatives put forth measures to improve informed consent for egg donors? Doesn't that sound—well, reasonable? Where should Planned Parenthood be on these issues? "Every state is already proposing and passing legislation, and so far, when the legislators turn to us . . . and ask what is our position on surrogacy, and any number of other issues, our response is: nothing," said Patti Caldwell, the CEO of Planned Parenthood of Southern Arizona. "We have no comment. We have no position. It is deeply difficult for each of us in the field. There's 120 of us who speak to our press on a regular basis. . . . The time is now. Our legislatures are meeting now. We need to decide fairly quickly if there are things we can take a position on, and if there are things we can't, that we are clear on why we can't."

Yet there was no pro-choice consensus to be had. "There are times when you have to choose teams, like with abortion," said a clinician from Poughkeepsie, New York. "But there are times we have to acknowledge that we don't quite know which team we're on. With most reproductive technologies, I think that most of us are genuinely unsure."

For women's reproductive rights groups, new technologies raise an infinitely expanding array of questions. "Choice" can be used now to justify almost any procedure; the movement has lost ownership of the term. Reproductive liberty can be, and is, invoked by patient-advocacy groups who want to discourage any law that would limit the number of embryos a doctor can transfer. Reproductive liberty can be, and is, invoked by fertility doctors who want to justify performing every new lucrative procedure, no matter how untested. It can be, and is, used to justify selecting for sex, one of the most difficult aspects of the new technology for women's groups to reach consensus on.

Around the country, use of sex selection is growing. Not all fertility clinics will select for gender before transferring embryos, but many will. A recent study found that 40 percent of patients, given the opportunity, would select the gender of their IVF embryos. At the 2005 ASRM meeting, Sandra Carson, a well-respected doctor at Baylor College of Medicine, made the case for sex selection, arguing that it has an ancient pedigree. And it's true; there have always been would-be parents who want a boy, or want a girl, and up to now they have sought out all sorts of unreliable techniques, including vinegar douches. Baylor now has an NIH grant to survey the physical and psychological health of children who have been sex-selected through IVF. That is truly a watershed moment: it represents the transition of sex-selection technology from fringe treatment to something approaching a mainstream one.

Some women's groups believe all sex selection should be banned. As Mark Evans pointed out, selecting for sex has traditionally been sought by parents from cultures where boys are valued more than girls. It reeks of culling females, and so is very hard for women's rights activists to stomach. On the other hand, in the United States and other first-world countries, most patients are not trying to get rid of girls, but to obtain one baby of whatever sex they don't have already. Moreover, anecdotal evidence suggests that in this country, sex-selection clients more often than not are women who want that longed-for girl. "Most of the time it's the woman who's the driving force," says Keith Blauer, who at the time of our interview was medical director of the Genetics and IVF Institute in Fairfax, Virginia, a clinic that has patented a sperm-sorting sex-selection technique. "More than 75 percent of the sorts we do are for girls. Women want girls. You hear about men, they want a son, but not as badly as the women want daughters."

So what's a pro-choice activist to think about this new item on the menu? Does wanting girls make sex selection okay? As Kirsten Moore points out, selection for any sex is based on stereotyping—some inexact, probably inaccurate notion of how a relationship with a child of this or that sex is likely to play out. I think she's right. In the sex-selection chat rooms I looked at, there were lots of women looking forward to dressing little girls in pink outfits and putting pretty bows in their hair. Any gender stereotyping is something feminists should guard against, Moore argues. "We think it's a slightly different situation in the U.S. than in China," she said—meaning that here, we don't abort female fetuses—but even so, "people's decision to have a boy or girl reflects social biases about what girls should be and what boys should be, and that's hooey."

That said, Moore does not think sex selection should be banned. When I interviewed her in the summer of 2006, she had just discussed sex selection with her board of directors, arguing that "the way to change the stereotypes is not banning the technology but really trying to force that conversation." Her fear is that any effort to direct any reproductive decision made by any individual is to call into question all decisions made by all individuals, including, of course, the decision to abort.

I am inclined to disagree. In Britain, the government bans sex selection for nonmedical reasons, yet abortion rights still exist. (In Britain, sex selection is permitted only for patients who are carriers of diseases that affect children of one sex or the other.) It seems to me that in the United States, abortion rights should not be the precedent determining every reproductive issue. It should be possible to (1) accept a woman's moral right to choose whether or not to continue an unintended or unwanted pregnancy, and (2) reject an infertility patient's right to infinitely select desired traits in offspring. Banning sex selection for what's called "social reasons" seems a reasonable place to draw a line. This is particularly true when a couple want to use pre-implantation genetic diagnosis, the sex-selection method that is most accurate, but also highly invasive. It's wrong to tamper with embryos in ways that may be harmful to them—we don't know—for a reason that just isn't good enough.

And many of the same issues are at stake, for women's groups, in thinking about larger issues of family-making associated with PGD, which was invented as a way of identifying embryos carrying genetic diseases by tugging out one cell and examining its chromosomes. Among the afflictions that can now be identified in a human embryo are hemophilia; fragile X syndrome; neuromuscular dystrophy; cystic fibrosis; Tays-Sachs; Down syndrome; and hundreds of other genetic disorders, some of them fatal, some of them fatal after months, even years, of suffering. Genetic testing gets more comprehensive every day; as Mark Evans noted, it inevitably happens that a medical treatment invented to solve life-or-death situations ultimately becomes more commonplace, its uses liberalized. It is now possible, for example, to use PGD to identify embryos carrying a disease that will not manifest itself until adulthood, sometimes late adulthood: a predisposition to breast cancer, for example, or Alzheimer's. PGD has even been used to cull embryos that carry a risk of colon cancer, a disease that doesn't occur until the forties or even fifties, and that is highly treatable, if diagnosed early. Some couples also use PGD to create "savior siblings": unaffected children who can be a tissue-match donor for an older sibling suffering from a fatal disease.

Sorting embryos this way remains extremely controversial, and it should be. For one thing, PGD, like so many diagnostic tests, is not always reliable. Many very early embryos demonstrate "mosaicism," meaning that the chromosomal makeup of every cell is not identical because the embryo is developing so rapidly, and testing only one cell, which is what happens in PGD, doesn't always provide an accurate assessment of the embryo as a whole. There is also the real possibility that PGD itself does damage the embryo, a question that will remain unanswered as long as embryo research remains unfunded. And more broadly—doesn't life that will be cut short nevertheless have value? Does a propensity to breast cancer mean a person should never be born? Do certain kinds of life now have more value than others? While many disability rights groups welcome a technique that can eliminate the suffering associated with, say, sickle cell disease, it's clear that PGD is, and will increasingly be, used to eliminate embryos with disabilities that aren't fatal. Or even, maybe, painful. Pro-choice groups, which typically like to align themselves with social justice causes, have to decide how to think about a technique that could someday eliminate, say, embryos that carry a gene for blindness.

Again, Kirsten Moore, one of the women's rights leaders paying closest attention to these issues, is inclined to let couples using PGD make their own decisions about which embryos to transfer and which to discard. The important thing, she believes, is to talk about all these issues; to keep the conversation loud, and keep it going. But she would stop short of regulating the uses to which PGD can be put. "If you're talking about parents raising kids, and deciding they want to raise this kind of kid, and they don't think they can take care of that kind of kid, who am I to tell them they're wrong?" she asks. "We've never as a society said, this is a good enough reason, this isn't a good enough reason."

This argument feels more persuasive. Many bioethicists—liberal and conservative alike—rightly worry about PGD and the uses to which it may be put, both now and in the future. Academics and thinkers whose job it is to ponder this field often worry that PGD, and other commercialized aspects of the science such as selection of sperm and egg donors, go too far in permitting parents to over-determine the child being born, encouraging them to lose sight of one of parenthood's fundamental imperatives: to be deeply grateful for the child one gets. Many bioethicists worry that parents will be led, by science, to forget that unconditional love is a signal requirement of being a parent. They worry that couples using PGD may select out embryos for diseases that will manifest themselves late, or not at all; and that someday the same technique may enable people to create a genetically enhanced baby. It's important to understand that science cannot do this now. PGD can weed out afflicted embryos—more afflictions every day—but it cannot insert traits parents desire. It is not possible, yet, to select embryos based on IQ or hair color or eye color. Nor is it possible to add these attributes to an existing embryo, like pouring vanilla flavoring into custard. Babies designed to this degree are not here, and some scientists think they never will be.

I do think that ethicists are right, to a point, to worry about parents engineering offspring. But the reporting I've done suggests that when it comes to PGD, pickiness is not what's motivating most patients. Another parental imperative—forestalling suffering in one's child—is. Often, couples seeking PGD are people who want to use this very, very early form of genetic testing as a way to avoid later prenatal tests, and, in so doing, avoid having to decide later whether or not to abort. These are people who know that their offspring are at risk for genetic disease, because one or both parents are carriers. And the most likely reason these parents know they are carriers is that someone in their family had the disease, and possibly died of it. Often, these are people who have watched a loved one—a parent, a sibling, a child—die of some awful degenerative disease. But for many of these patients, to actually abort a fetus carrying this disease would feel like aborting the person they love, or loved. That's why people go to the considerable trouble of doing PGD on lab-made embryos: because they could never abort a fetus. One doctor told me about a patient who wanted to use PGD to eliminate embryos that carried hemophilia, a genetic disease that travels on the X chromosome and usually afflicts boys. Her brother, a hemophiliac, had contracted HIV from a blood transfusion and died of AIDS. She did not want a hemophiliac child, but to abort an actual fetus would feel to her like aborting her brother.

"Most of them don't want to consider amniocentesis, because they don't want to terminate," I was told by Kelly Pagidas, medical director of the PGD program at Women and Infants Hospital in Providence. "Most of them are here either because they have a sibling who was affected, and they would feel they were killing their sibling [by aborting an affected fetus]. Or they've had a child who was affected, and they would feel like they were destroying that child." PGD enables them to avoid having to go through that. At the same time, it is wonderfully pro-choice, and wonderfully pro-life. "I've never come across a patient who wants to design their baby," Pagidas said. "And I've seen hundreds and hundreds of patients."

"The Put-Them-In People"

There is another schism illuminated by the new technologies, between two kinds of groups that would seem to be natural allies: old-line pro-choice groups, and newer infertility-patients-rights groups, of which there are a number now in this country and around the world. Both groups have the same mission and the same anxieties. Both are dedicated to helping women (and men) control their childbearing; both oppose federal involvement in any reproductive issue. "We don't want the government touching, monitoring, getting involved with our eggs, our sperm, our embryos," said Pamela Madsen, the head of the American Fertility Association, in an interview. "It's ours, not theirs. Get their bloody mitts off of them."

That is, pretty much, the pro-choice position. And yet infertility groups have received little help from traditional pro-choice groups, who of course are busy fighting state measures that aim to chip away at abortion rights with things like mandatory waiting periods and parental notification laws, and meanwhile trying to make sure the U.S. Supreme Court doesn't overturn *Roe v. Wade* while developing a contingency plan in case it does. It's a lot to think about. Yet Joseph Isaacs, head of Resolve, the oldest U.S. infertility patients' advocacy group, says his group has not received so much as moral support from reproductive rights groups. Infertile women aren't on their radar screen. All-too-fertile ones are.

"Their mandate is not to support or protect reproductive rights when it comes to creation—only when it comes to abortion and birth control," says Pam Madsen, more bluntly. Yet Madsen herself is at pains to differentiate her mission from that of abortion rights groups. "We're about creating life, not ending it," she says virtuously. "We're the put-them-in people, not the take-them-out people."

There is also an inconvenient deviation of message. Infertility groups want women to know they don't have all the time in the world, while some pro-choice groups are still lukewarm about getting out that message. The tension IVF exposes is between pro-choice groups that have encouraged the idea that women can indefinitely control their childbearing, and newer infertility groups made up of women who, sadly, know better. Biology sometimes is destiny, and women's groups are going to have to figure out how to accommodate that essentialist truth. "We need to know that we have a limited amount of time," Madsen emphasizes.

Living Under a Deadline

Many young women today do know that. They may not know exactly when fertility drops, but they know it doesn't last forever. For the first time, there is a generation of young women coming of age who realize that they are living under a childbearing deadline. "I see a lot of women including my aunts who waited and weren't able to have babies," Cari Elam, twenty-three, told the *Wall Street Journal* in 2006, explaining why she had already spent $5,000 on two years of unsuccessful treatment. "Women in their twenties and thirties are probably more worried that their eggs are going to age than whether they're going to be able to obtain an abortion," says Kirsten Moore. While this may be an overstatement, it's true that women know that they've got a lot to get done in a short period of time. Get a job, get a partner, have babies, not necessarily in that order. And quietly, without much fanfare, this combination of new technology and new fertility anxiety has produced a new generation of activists, young women hoping that science may someday solve the problem of jobs and careers and children and timing.

One of these is Lindsay Nohr Beck, who was diagnosed with cancer in the late 1990s, when she was just twenty-two, living in San Francisco and doing business development for a health-care dot-com. Just a year out of the University of Colorado at Boulder, Beck learned that she had a rare throat cancer. Even now, Beck still has a girlish, fresh-out-of-college way of speaking. When I interviewed her, she said things like "When I was sick? And I saw this need? That experience is very similar to writing a business plan? What I had learned translated very well to solve this problem?" She is a slender woman, pale and fragile-looking, but she demonstrated a remarkable tenacity in finding a way to ensure that if she lived, she would be able to have children.

The first time Beck was diagnosed, she underwent aggressive radiation treatment, which beat the cancer back. When she was twenty-four, however, her cancer returned, and this time it had spread to her lymph nodes. She would need surgery, radiation, and chemotherapy, the last of which put her at risk of going into early menopause. Not that her doctor told her this. "He went through a laundry list of side effects, down to there's a 1 percent chance I might go deaf, but he didn't tell me there's a huge chance I might be infertile," Beck recalls. "I was like, 'What about infertility? You didn't mention that.' And he was like, 'Oh yeah, there's a high likelihood you'll be infertile from the procedure.'" As Beck recalls, "The thought of being sterile was almost as devastating as my cancer itself."

Beck wasn't married and didn't have a partner. She could go through IVF and freeze embryos, but to do that she would need to use donor sperm. Clinics encouraged her to look through some donor catalogs, but she didn't want to conceive a child from donated sperm. If she did ever have a baby—and she very much wanted to—she wanted to conceive with a husband, or adopt.

So Beck investigated the possibility of freezing her own eggs, before chemotherapy, for use in an IVF procedure later. Egg freezing was, and is, a technique in its infancy. Unlike both sperm and embryos, human eggs have proved stubbornly hard to freeze and thaw successfully, in part because of the size and complexity of the oocyte, the way the chromosomes float untethered throughout the cell. Very few clinics had an experimental egg-freezing program going. But Beck was so motivated that—after surgery to remove a third of her tongue, and lymph nodes in her throat—she managed to telephone clinics, lying in her hospital bed, sometimes calling the same clinic one, two, three times, hoping to reach somebody who would give her the answer she wanted. She finally got through to a doctor who did have an egg-freezing program, but warned her it was only for cancer patients. "For the first time," she notes wryly, "I was glad to be a cancer patient." Her own doctor discouraged her. She didn't care. It was her body. Her choice. "The key is, I got to make my own decisions."

After freezing her eggs and finishing cancer treatment, Beck decided to found a nonprofit organization that would help other cancer patients raise their chances of having children later. There are at least 10 million cancer survivors in America now, many of them young, many of them likely to be affected reproductively by their treatment. By the time she got started, the dot-com bubble had burst; the start-up Beck had worked for was bankrupt, there was plenty of free labor available. "Everyone was unemployed, so the people I worked with before worked as advisers," says Beck. Fertile Hope, the charity she founded in 2001, has become a formidable patients' rights organization, whose aim is to spread the word about egg freezing as well as other experimental techniques. A Cornell doctor, Kutluk Oktay, for example, has tried surgically removing one ovary before a cancer patient undergoes chemotherapy. After treatment, pieces of the ovary (ovaries can actually function as portions) are grafted just beneath the skin, in a place unaffected by treatment. Drugs can be given, and oocytes retrieved, rather easily, with a special needle. In one case, Oktay implanted portions of the ovary into the forearm of a breast cancer survivor. There have been only a few pregnancies from ovarian grafting (none yet from the forearm graft) but their origins were debatable: the women may have resumed ovulating from a remaining, intact ovary. Grafting, were it to work reliably, could help cancer patients who do not have time to undergo IVF between diagnosis and treatment. It's quite a concept: After taking fertility drugs, the woman with the forearm graft noted that a patch of her forearm was hard and spiky. That's because her arm was ovulating.

Beck's other goal is to get cancer doctors to regard infertility as a side effect that matters. "Stats in the market show that 90 percent of cancer patients are at risk of infertility, but less than 10 percent of their oncologists inform them," she says. Moreover, she adds, "doctors have admitted to cherry-picking patients based on sexual orientation and socioeconomic status. They'll say, I didn't think my patients could afford fertility treatment, so I didn't tell them."

Beck has also persuaded drug companies to underwrite egg freezing at a low cost to patients who otherwise could not afford it. In this, she joined the ranks of a very, very few women who have taken social justice as far as paying for fertility treatment for poorer women, who are more likely to need such treatment than anybody, and less able to afford alternatives such as adoption. Two others are Nancy Hemenway, a former schoolteacher who runs a charitable IVF program as part of her fertility group, the International Council on Infertility Information Dissemination; and Madeleine Gordon, a Cincinnati socialite who runs a charity, Gordon Gift of Life, to pay for IVF for blue-collar and lower-income Ohioans. These women have pursued their projects without any help or endorsement from larger women's groups, who still tend to regard low-income women as most in need of pregnancy termination, not pregnancy enablement.

A Woman Helping Women

Another new reproductive activist is a woman who wants to do good while doing well. She aims to capitalize on young women's by-now-well-stoked fertility anxieties as well as their enhanced earning power; to offer women a chance to buy their way out of an unfair social problem. Her name is Christy Jones, and she wants to offer all women—at least, affluent ones—the chance to freeze their eggs for use later, when their work schedules and/or social lives permit childbearing. "Now [egg freezing] has a range of success rates: the low end is 15 percent, the high end is 40 percent," Jones said in early 2005 to a group of working women who had gotten up early to attend a breakfast meeting about egg freezing, a technique that has improved somewhat since Lindsay Nohr Beck started looking into it. To Jones, it seems entirely possible that research into the oocyte can do more than save lives; it can enable women, all women, to delay childbearing—and still bear their own genetic children.

Jones is a reproductive entrepreneur who wants to bring egg freezing to the Title IX generation, to use new technology to level the playing field in every way. "There's more than one in three women in their thirties today who are single and childless. That compares to one in ten in the 1970s," Jones told women seated around a table in an office suite above a bank in downtown Boston. She had turned the dreaded Sylvia Ann Hewlett statistics into a sales pitch. It was one of scores of conference calls, cocktail parties, "fertility freedom receptions" and alumnae events Jones attends, spreading the gospel of egg freezing to groups like 85 Broads, a group of Goldman Sachs alumnae.

Jones is another Silicon Valley product; it may be no accident she and Beck both worked in the land where technology reigns. At nineteen, when she was an undergraduate at Stanford University, Jones co-founded a software company called Trilogy. She traded her shares to found a spinoff, PCOrder, which she then took public and sold back to Trilogy for $100 million. As noted in a 2004 article in *Elle,* Jones by then "was 31, had been on the cover of Forbes Magazine three times, and was ready to step back."

Stepping back for her meant pursuing an MBA at Harvard. Before moving to Cambridge, Jones was hanging out at home with her mother, talking about the career-and-family conundrum. She had always assumed that by her thirties she would be starting to have children, but she had worked so hard and single-mindedly that she didn't start dating until she was twenty-nine. Her mother said maybe she and her sisters should think about freezing their eggs. Jones decided to do more than that; she decided to start a company that would recruit women interested in egg freezing and connect them to a network of cooperating clinics. The plan she and some Harvard classmates developed, Extend Fertility, won the Harvard Business School's Annual Business Plan Contest in 2004, beating out a plan for a company that would provide interactive children's entertainment on DVDs and one for a venture to provide software for cell phones.

And now here she was, her vision realized. "Can we ask embarrassing questions?" said one woman at the Boston breakfast meeting. The woman, who worked for an environmental agency, wanted to know how long a woman can safely wait to use eggs that have been frozen. Her mother and grandmother had both gone into early menopause, at forty-two. If she froze her eggs, and stopped menstruating in her early forties, would a pregnancy still take? The response was yes, thanks to the miracle of hormone replacement.

The women were intrigued, but skeptical. "If somebody had come at me when I was thirty or thirty-five, and said, 'Freeze your eggs, it's $10,000,' I would have gone, 'First of all, I don't have $10,000. And second of all, everything's going to be fine, I'm going to meet a guy, we're going to live in the suburbs,'" said one woman who was in her early forties—too old for egg freezing. She herself was planning on becoming a single mother through adoption. "How receptive are people?"

Jones—who was wearing a dark, expensive suit; dark, expensive shoes; a pearl necklace, and matching pearl earrings—said that Extend Fertility had been contacted by 450 women, just 50 of whom had decided to go through the $10,000 egg-freezing process. Most queries had come from women thirty-nine or forty, for whom it was already too late. "We aren't profitable yet," she admitted. "We—I think we're beginning to see it begin to be more accepted by the right audience, women in their mid-thirties. I think it will take some more time for this to become really accepted."

There were about a dozen women at the meeting. Four or five of those present worked for Extend Fertility. What the others seemed most interested in was the fact that Jones, by then thirty-five, had frozen her eggs—and gotten married just a month later. Judging from the discussion, these women didn't *want* to delay childbearing until they were forty-five. They wanted to have children now. Or two years ago. And given the choice, they would prefer to have children the old-fashioned way. "My girlfriend two years ago went to something like this, and within two weeks, met a guy, got engaged, and had a baby," said one woman present, hopefully. Another woman asked Jones how she met her husband. She replied that they both worked in the software business, that they dated for more than five years, and that marriage came after she'd frozen her own eggs. The women perked up: maybe, like washing your car to bring on rain, egg freezing could raise the likelihood that a life partner would show up in time after all!

Since then, Jones has persisted. When I checked with her in late 2006, she said that 2,000 women had inquired, of whom 200 had frozen their eggs. Six babies had been born and Extend Fertility was "closer" to turning a profit. Jones has also developed a website, www.laterbaby.com, that aims to arm young women with "the information necessary to navigate the often overwhelming choices surrounding family planning and career." The website offers a "fertility quiz" women can take, and online chats among a "community of women who care about their future fertility." Jones is, in a way, the antithesis of Kim Gandy: She wants to not only promote awareness of reproductive aging, but to spell out the details. "On average, a woman's fertility begins to decline in her 30s" is one of the test's trickier true-or-false questions; the answer is "false," because the decline actually begins in a woman's late twenties. Currently, according to her website, only 7 percent of quiz-takers get this one right.

Not everybody thinks egg freezing is ready for prime time, however. Many doctors, while they will freeze eggs for cancer patients, will not offer the option to fertile women; they argue that the science is still not advanced enough to warrant charging so much for a procedure that may or may not work. "Ethically, it's wrong," said Vicken Sahakian of Pacific Fertility. "I would have made a lot of money, but ethically it is wrong. You freeze a hundred eggs, two survive." The fear that egg freezing still doesn't work reliably enough to offer it was a reservation I had anticipated; I heard another, unexpected one from Michael Feinman, a doctor with Huntington Reproductive Center in California, a clinic that is part of the Extend Fertility network, and who believes that Jones is going to make it "respectable."

Making egg freezing respectable, Feinman argues, raises its own moral problem: Is it responsible for doctors to help women delay having children? For medicine to normalize waiting? "Should we offer it even if it works?" Feinman said. "Obviously, I can't defend offering something that doesn't work. We hope that what we're offering is going to. But I think there's an ethical debate about offering it even if it works." Isn't there enough pressure already on women to wait? What if law firms someday require that female partners freeze their eggs and wait even longer? Whose interests does egg freezing serve? The woman's? Or that of an ambitious, still pretty unforgiving work culture that doesn't really ever see childbearing for female employees as convenient?

And is there an optimal time to have children? The message of Extend Fertility is that there will come a day, maybe in your mid-forties, when the magical childbearing window will open up. My own experience has been that there is no wide open window. There is never an easy time to have children. Life gets more complicated as you go along. It's true; many parents in their twenties don't make enough to pay for child care; I didn't. Having a child in your twenties, and working, or going to graduate school, is very hard to manage. But that doesn't mean things get much easier later. One day I was meeting with my friend Melanie, whose treatment saga was narrated in chapter 2. It had been more than six months since we last met. Melanie looked great; she had a short haircut, and was wearing a sleeveless shirt that showed off her tan. By now Melanie had given up on the idea of IVF with her own eggs. If she was going to conceive, she had accepted, it would have to be with an egg donor. And that meant she still had time to wait. "Time's on my side," she said. "We're going to get some young chick's eggs, so I don't have to be racing." Because now wasn't a good time for her, either. Since we last met, Melanie had gotten a well-deserved promotion to the number one spot. She was head of a major nonprofit corporation, doing exactly what she wanted to do all her life. Traveling. Making hiring decisions. Pursuing a career she believed in. "It is," she said, "really fun." So fun that she wasn't sure she wanted to take the time off for maternity leave. "I'm at the peak of my career," she said. "Do I really want to be reading *Pat the Bunny*?"

Melanie had told her husband she would still be willing to adopt. But if he wanted to have his own biological offspring, then he—not she—would have to find the egg donor. "I said, 'If you want to do this, do it. I'll tell you what my calendar is, and you fit it in.'" Eventually they did find another donor, but in an unrelenting series of disappointments, the egg donor cycle didn't work. Melanie's husband agreed to explore adoption. "Finally!" she wrote me in an e-mail. "Of course, we may be fifty and too tired to deal, once we complete the paperwork."

The idea that there is a convenient time to begin a family, and that the time will materialize later, seems a risky concept to advance. The world many of us live in rarely delivers an ideal moment.

So what's a woman to do? "What *are* we doing to do with educated women? When *are* they going to have a family?" the fertility doctor Robert Nachtigall asked, rhetorically, in an interview. Good solutions are not obvious, though some younger women, noting the plight of their predecessors, are harnessing technology in creative ways. One of the more novel solutions I encountered was devised by a San Francisco doctor, Lisa Sterman, a lesbian who had always wanted children, but did not want to raise them alone. When I spoke to her, in spring of 2005, she was in her late thirties and unpartnered, after a relationship had ended. "I found myself at thirty-seven going, 'Oh shit, what am I going to do?' I want to have kids, and I don't want to have them out of the context of a stable relationship." Rather than freezing her eggs, she decided to use a sperm donor and freeze the resulting embryos. In fact, she used two sperm donors, one dark-haired, one fair. She went through egg retrieval, had half her eggs inseminated with sperm from one man, and half with sperm from the other. This way, when she met a woman who wanted to settle down, they could select embryos based on her partner's coloring. More than a year later, when I checked back, Sterman did have a partner. Since neither of them had a preference, she was planning to have four embryos transferred, two from each man. Twins, she felt, would be fine. "We're going to cook them up in a couple of months," she said. After that, it would be her partner's turn. Of her years, now, of planning: "I was just trying to be proactive."

REPRODUCTIVE SCIENCE AND THE FUTURE OF OUR FAMILIES

1-800-EGGS

Not long ago a friend of mine was on vacation at a beach in New Jersey, and she looked up to see a plane flying overhead. Behind the plane stretched a banner whose message read:

WOMEN EARN $8,000
AGE 20–30
888-968-EGGS

When my friend told me about this, I experienced a moment of clarity, when it was possible to look a reproductive phenomenon straight in the eye and say: Well, this does not seem like a good thing. There's something about seeing the essential ingredient for a child solicited on an advertising vehicle associated with all-you-can-eat seafood buffets and wet T-shirt contests that feels, dare one say it, wrong. Even given that there are women (and men) who want and need those eggs, and even given that many of these men and women will be loving parents, trolling for oocytes from the ranks of New Jersey shoregoers does not seem like a good way to go about creating modern families.

In many other instances, I have to say, I found sorting "right" from "wrong" to be not nearly so easy. It was not my goal to do that; not my goal to declare that all this is bad, or all this is good, or all this should be permissible under the banner of reproductive freedom, or all this should not be permissible under any circumstances, or even that there should be a moratorium while we all give it more thought. It was my goal to help readers understand why certain changes in the family are taking place and what their likely consequences might be. Why there is so much demand for donor eggs, now. Why there are so many more triplet sets than there once were. What life is like for those triplet parents. How embryo research and embryo politics are influencing our thoughts on human life and its origins. What is the real, rather than the imagined, impact of medicine and science on families and culture.

Still, it was impossible not to develop a few opinions, and one opinion I developed is that David Keefe is right in saying that what is at work in assisted reproduction is often not science but business; that commercialization does make some clinics far too willing to offer patients an array of services whose implications and consequences are often not spelled out or even understood, really, by anybody. How nice it would have been, I often thought, if the U.S. gamete trading industry had developed along the lines of The Sperm Bank of California: nonprofit, committed to research, flexible-minded but with some sense of boundaries. And at the same time that one can legitimately, I think, feel uneasy about medical providers who contract with pilots to carry beach banners, one also can sometimes feel uneasy about the demand side of this equation. There were moments when it seemed to me that two modern ways of thinking about parenthood—an absolute commitment to reproductive liberty under all circumstances, and a trend toward compulsive micromanagement of every aspect of our children's lives—sometimes persuade would-be parents that they are entitled to pursue any avenue of reproductive technology, and to determine every aspect of the outcome.

And the ironic thing is—for me, this was a key revelation—determining the outcome is exactly what cannot be done. Assisted reproduction is, often, the least controlled form of reproduction imaginable. That's what we should be taking away. Situations never turn out the way patients expect them to. There is always a surprise, waiting, somewhere. There is always that emergency C-section delivery. The problematic presence of the surplus frozen embryos. Meeting a person at a party who used the same sperm donor you did. Designer babies: What a laugh! So many of these situations are the epitome of undesigned.

It also seems safe to say that in the United States, some excesses of the field could and should be reined in. In this country, science and medicine tend to be less closely regulated—or rather, they are permitted a much greater degree of self-regulation—than in countries with more socialized forms of medicine. Even so, the federal government could do much more to encourage and direct the science; it could and should fund embryo research, as well as infertility research. And it should limit the number of embryos transferred. This is a matter not of reproductive liberty, but of public health. States, too, could do much more to regulate gamete trading; they could require licenses for gamete brokers, establish acceptable practices, and enforce them. And people should talk about these issues: In Britain, one of the positive offshoots of government regulation is that whenever the HFEA issues a decision, it is vehemently protested by somebody, and the upshot is that people know about the decision, and debate it, and develop an opinion, and are more likely to reach a social consensus, and to be better informed. When I was in Britain in 2004, the front pages of the mainstream papers carried news of an HFEA decision permitting "saviour siblings," while the tabloids were full of the saga of an Indian woman who served as the gestational carrier for her own daughter's twin children, but gave birth to the babies in India, making it unclear which country the twins were a citizen of. OUR LITTLE MIRACLES, the *Evening Standard* screamed over a photo of the parents with their children, whose nationalities were undetermined. The issues weren't easy, but at least they were being aired.

And of course, that plane flying over New Jersey raises another question. If this is where we are, now, where will we be tomorrow? Medically? Morally? How will reproductive science continue to drive the way in which families are made and maintained and structured?

I wonder about this all the time. I wonder about it in the context of my own children, who face a world of choice-making so much more complex than the one I faced twenty-five years ago. My daughter, who is eleven, is already thinking about how her life will be organized: "When I'm an adult I . . ." is how she begins many sentences. When she was little, she used to say that when she was an adult she wanted to be a pediatrician, so that when her own children got sick, she could just bring them to the office with her. Now, she thinks that she wants to run a pottery-making business. Then again, maybe she wants to be a police detective. Whatever she does, she wants to have four babies—maybe two sets of twins. She understands that she doesn't want to have children too early: twenty would be young, but thirty, she feels, would be getting late. She wants to make sure I'm around to help with all those kids (she also plans to have a lot of dogs, and they're going to need watching, too) while she's off potting and detecting, and since my generation got started late, hers is going to have to get started earlier. Twenty-eight, she figures, is about right. For that first set of twins.

I love these conversations. I love to think about her future, and I love the fact that she envisions me, her mom, as an integral part of it. At the same time, I am struck by how many dilemmas her generation will face in making families, so many of them involving, or growing out of, new technologies and the decisions they present, the options they dangle. For college women now, it's a real question whether to call that 1-800 egg number from your cell phone, to help defray your tuition payment. Now young women have to think about how they would feel if their genetic offspring were raised by another. How might their own children, arriving later, feel about these pre-existing half siblings? They have to think, from the get-go, really, about when they are going to have children and whom they will likely have them with.

While writing this book I often thought of a day when I was an undergraduate, in about 1980. My friends and I—about eight of us—found ourselves sitting around talking about our futures, wondering how we could possibly realize our professional aspirations, and our desires— most of us did desire this—for family lives. We knew it was going to be hard. We just didn't know how hard it would be. We didn't know yet about the deadline. Some of us would go on to have careers and children, while others, even those who married not long after graduation, would have trouble conceiving. Some would marry and divorce and remarry and then get started. Some would have, yes, twins. Some would not have children. Thinking about that conversation, which I remember vividly, it strikes me that all these technologies were just materializing on the horizon. Louise Brown had been born in England, but Elizabeth Carr was still to come. Everything—all of this science, all of these changes—has occurred in my adult lifetime. And I'm not really, I like to think, so old. What will the next ten years bring? The next twenty? For my daughter? For my son? My son is eight, and makes gagging noises every time he hears mention of ladies and their oocytes. (I am afraid my children have had to endure a number of these conversations.) Does he think he's going to be exempt from these issues?

Will he be exempt? What sort of families will our boys grow up to make? What sort of fathers, if they do become fathers, will our sons turn out to be? If you take the argument of the maverick mom ideologues to its logical conclusion, it follows that thanks to reproductive technology, a new group of boys is being incubated by lesbians and single mothers, who are procreating with anonymous sperm even as you read this. Twenty or even just ten years from now, these brave new boys will have grown into maturity. Released from the tyranny of pernicious old male role models, they—one might expect—will be different from men at any other time in history. They will be sensitive, empathetic, and, pre- sumably, reproductively willing. They will be domestic and workplace egalitarians. Perfect husbands. Perfect fathers. They'll be glad to have children young, and they'll help with the child-rearing every step of the way. Maybe today's single mothers are brewing the obsolescence of sin- gle mothers. Maybe today's moms are creating a reproductive solution for tomorrow's. Tailor-making the new, willing man! How nice of them!

Will it be easier for my children's generation to figure out the way to make everything work? Will balancing careers and households get eas- ier? Will workplaces get more child-friendly? Will all problems eventu- ally be solved by science? For the rich? For the poor?

Why Few People Would Clone

There's no way to know; it's always dangerous projecting future repro- ductive scenarios. No matter how plausible, they always end up looking ridiculous twenty years later. But it does seem safe to predict that sci- ence will continue to fuel alternative family-making; it will continue to enable the creation of new families that are nuclear, in some cases, but not exactly traditional. When people talk about the future of reproduc- tive science, one thing that always comes up is human cloning: the possi- bility that a person might someday breed a near genetic replica of himself or herself, a little genetic lab-made mini-me. The assumption seems to be that there are a lot of people who would duplicate them- selves if science could help them do that. Doubtless, some would. But the cloning controversy ignores two things. One is that mammals remain incredibly difficult to clone. Scientists who work in animal cloning have found that it takes thousands of embryos to produce one live offspring, and those that survive tend to suffer from growth abnormalities such as "large offspring syndrome." It's clear to cloning scientists, maybe more than anybody, that it's still healthier to have two genetic parents.

What the cloning controversy also ignores, I think, is the fact that what most people want to do is procreate *together*. People want to have a baby, if they can, with the person they love. Everything we've learned from reproductive science confirms that this is what most people are after. And so to me what's more interesting than cloning is whether science might someday let us manufacture gametes: take an ordinary skin cell and turn it into a reproductive cell. Scientists are now trying to do this. They are trying to take an ordinary cell, get it to divide its forty-six chromosomes to twenty-three, and then combine it with a sperm or egg cell. If science can ever manufacture the human germ line, so many revolutionary social scenarios would follow. Finally, women really could have babies at any age. When the time seemed right they could scrape off a skin cell, get a gamete constructed, combine it with a partner's sperm, and have that baby.

Moreover, gay and lesbian partners could do the one thing they cannot do now: produce together a child that combines their genes. One lesbian partner could provide the egg; the other could provide the cell that would be transformed into a sperm, and both would be full genetic parents of their children. Two gay men could do something similar, though they'd still need to borrow somebody's womb, unless science learns to make wombs as well. If this became possible—when this becomes possible—there would be an avid market. We could get away from gender entirely. Whether you are man or woman wouldn't matter. Any two people would be able to conceive together. Mother, father: such distinctions would become meaningless. I mentioned this to Howard Jones, the doctor who made IVF work in this country, and he agreed it's not out of the question, fifty or a hundred years from now. He pointed out that men, because they possess both an X and a Y sex chromosome, would be able to produce both an X-bearing and a Y-bearing gamete, were lab-made gametes possible. Two men could make either sons or daughters. Lesbians, because they have only Xs, could produce only daughters. So science still couldn't make the world fair, after all, for women.

More immediately, it seems clear that science must continue to try to discern whether the children of reproductive science are in any way damaged by the technologies that made them. This has been an open question now for twenty-five years. It remains unanswered in part because there are few registries of IVF children and few organized efforts to track their well-being over time. Interestingly, when it comes to the health of IVF children there is a division between the instinct of scientists and that of medical doctors. Many people think scientists are wild-eyed proponents of risky procedures, but in many cases nothing could be farther from the truth. During the session I spent at the Marine Biological Laboratory in Woods Hole, many young cell biologists were just being introduced to reproductive techniques like ICSI and PGD. They watched a lot of them—particularly procedures that involved the injection or the breaching of any cellular membrane—with horror. "Like stabbing a body with a bayonet" is how one young scientist described ICSI. "I would never put my wife through that," another said. Maybe they just hadn't been acculturated yet. But scientists who do reproductive research often have an outsized reverence for the natural processes they spend their lives studying. It is the physicians in the clinics, often, who tend to be more bullish about babies made this way, and who resist the notion that children are harmed by the IVF procedure.

"Very bloody" is how Mary Croughan, a University of California, San Francisco, researcher who is overseeing a large study of fertility treatment and its impact on children, recalls emerging from a meeting with fertility doctors ten years ago, after she broached the question of the health of children. Similarly, Richard Schultz, a biologist at the University of Pennsylvania, several years ago came up with an experiment to test the effect of culture media on the embryo. He exposed some mouse embryos to culture, then examined certain behaviors of the adult mice that resulted. He found that culture-grown mice behaved oddly in a couple of ways. They had less fear of open spaces, which is not good when you're a mouse, and they seemed to have less spatial memory. When he presented his findings to a group of IVF doctors in 2001, the reception was so hostile, Schultz recalls, that he was glad afterward to get on the train and get out of town. "They were not received well," he says.

But lately many in the medical profession have come to accept the idea that IVF children are different, and not just children born as multiples. It is a known fact that ICSI is allowing male infertility and other genetic anomalies to be transmitted. Moreover, it has consistently been found that even singleton IVF babies are more likely to be born prematurely, and to weigh less, than those conceived naturally, and that IVF mothers are more likely to suffer from pregnancy complications such as placenta previa. In 2002, a report published in the *New England Journal of Medicine,* using information from Australian public health registries, found that children of assisted reproduction are twice as likely to suffer from cardiovascular, urogenital, chromosomal, or musculoskeletal birth defects. This study was controversial; there was the predictable scientific assault on the methodology. Other studies have found no elevated risk of birth defects. But the evidence for some negative outcomes is mounting. That IVF children are more likely to be born smaller and premature has been confirmed over and over, including in a 2005 study published in *Obstetrics and Gynecology*. The cause could be the underlying infertility of the father, or the mother, and related genetic problems that are being passed along, or it could be some aspect of treatment. Some scientists suspect that fertility drugs cause too many oocytes to mature too quickly, and that sped-up growth is not good for the egg. Others point to the tendency of IVF embryos to split, or twin, the longer they're left in culture, as evidence of the impact culture media must have. And several years ago, two scientists who study a rare disorder called Beckwith-Wiedemann syndrome found that among the cohort of children they were studying, there was a larger proportion of IVF babies than in the general population.

The work of these scientists, Michael DeBaun and Andrew Feinberg, suggests that IVF may affect "imprinting," an early stage of embryonic development. "Imprinting" occurs when the embryo is dividing into two entities: the mass of cells that will become the fetus, and the mass of cells that will become the placenta. What's going on during imprinting is a genetic battle of the sexes: it is thought that the genes of the embryo's father are driving placental growth, with the aim of building the biggest, healthiest baby possible, and ensuring the perpetuation of the father's gene line, while the genes of the embryo's mother are trying to hold the baby to a manageable size, so that the mother is not killed in the birthing process. While all this is occurring, certain genes—the mother's in some cases, and in others the father's—must be shut off so that the growth of baby and placenta are perfectly regulated. When imprinting goes awry, growth disorders result. Children with Beckwith-Wiedemann syndrome suffer from overgrowth in certain parts of their bodies; they are more prone to tumors, and they have a propensity to cancer, another form of uncontrolled cellular growth. Scientists now wonder whether IVF interferes with imprinting. They point to "large offspring syndrome" among cloned animals as evidence of the effect culture media may have on growth regulation. What Feinberg finds troubling is that the companies that make culture media aren't required by the FDA to reveal their media's contents, so it is impossible to test the impact of this or that ingredient. "It is particularly worrisome that many IVF clinics purchase media from companies that do not divulge the formulation of the media they market," he and colleagues wrote in a 2004 paper published in the *American Journal of Human Genetics,* adding, "Long term follow up is critical." Specifically, Feinberg recommends that childrens' birth certificates should indicate whether fertility treatment was involved, making it entirely feasible to track health outcomes. This is a great idea.

While doing research for this book, I encountered many parents whose IVF children were absolutely perfect. Most were, of course. Most were gorgeous. One of the great pleasures of doing the research was meeting so many children. It's important to remember that all of these risks, even if they do exist, are very small. But I also spoke with two mothers who conceived children with severe malformations of the gastrointestinal organs. One of these children, a triplet, died in utero; the other, also a triplet, died after nine months, all of them spent in the NICU. I interviewed another mother who gave birth to twins: the girl was fine, but the boy was afflicted with hypospadia, a malformation of the male genitalia. The boy had a particularly severe form known as penile-scrotal transposition. The scrotum was attached to the side of the penis rather than at the back; the urethral opening was at the base of the penis rather than the tip; and the penis itself was pointed so that, his mother said, if he could have peed out of the tip, "he would have peed right into his nose." This woman had worried all along about the fact that IVF treatment routinely involves taking progesterone supplements to support the pregnancy. This mother asked her doctors if it was safe, and was reassured. But when her son was born, his genitalia were so disfigured that doctors at first thought he might be a hermaphrodite; they warned her she and her husband might actually have to choose his sex. In the end, he was clearly male, but he needed multiple operations to repair the damage.

To make that repair, this mother found her way to John Gearhart, a pediatric urologist at Johns Hopkins University. Like the Beckwith-Wiedemann researchers, Gearhart has found that his patient population—boys with penile malformations—contains a higher proportion of IVF babies than the general population. It's true that hypospadia seems to be increasing in boys overall. But in 1999, Gearhart published a study in the *Journal of Urology* that found that IVF boys have a fivefold risk of hypospadia: 1.5 percent, as compared to 0.3 percent in naturally conceived children. Gearhart thinks it's very likely that progesterone supplements are to blame. He explains that when progesterone is administered to the pregnant mother at a certain early stage of fetal development, the construction of the male urethra is disrupted. The instructions being sent by the boys' DNA are garbled by the extra dose of this female hormone. "This is not an extreme price to pay for having kids," Gearhart emphasizes, but he does think further study should be done and that patients should be warned of this risk. All of this evidence is something the U.S. government is taking seriously. In the fall of 2005, the National Institute of Child Health and Human Development held a major two-day conference on fertility treatment and its outcomes. One federal researcher, Germaine Buck Louis, stood and rather passionately told IVF doctors that emerging data suggest fertility medicine should be regulated like any other industry that may produce toxic pollutants.

More benignly, what this research confirms is that environment and nutrition may reprogram the human genome. Genes, scientist now suspect, are not immutable after all. The genetic blueprint we inherit is revised and changed by its surroundings. Nurture affects, and changes, nature. One scientist, David Barker, has studied the adult development of babies gestated by women who were malnourished while pregnant, and found that these children are more prone to obesity as adults. Because they had been starved as fetuses, the children's bodies got used to making do with less. The idea that uterine environment has a powerful effect on fetal genetic development is now known as Barker's hypothesis. It suggests that genes aren't static and unchanging. The genome can be reprogrammed by the environment—including, possibly, culture media—in which it is unfolding. The branch of science that studies the impact of environment on the unfolding genome is called "epigenetics," and it's changing the way we understand genes. Studies of IVF children will help fuel this scientific exploration.

And epigenetics, in turn, could affect the way we think about genetic legacies and even, maybe, genetic relationships. Remember those egg-donor mothers whom Clare Murray was so taken by: the women who managed to convince themselves they were the genetic mothers of their children, simply because they carried these egg-donation babies in their womb? Remember how pregnancy enabled these women to forget they used a donor egg in the first place? Well, in a way, these women were right. In a way, they are their children's genetic mothers. By providing the uterine environment for their embryos, they really were affecting how the genes of the fetus played out. They were molding the genome. They were becoming, in a way, their children's *other* genetic mother. If it's true that genes are not immutable, that instructions can be revised, then egg donor mothers do direct the genetic development of their children. There is, you could almost say, a new genetic blending going on, in which one woman incubates and alters the genes of another. Seen this way, mothers using egg donation might feel from the start more entitlement, more of a sense that they are the legitimate mothers of their children.

As for all this genetic trading, all this lively reproductive commerce, all this sperm and egg donating and buying and selling and exchanging of this or that piece of the reproductive process: tawdry as it sometimes is, it seems to me that the willingness of people to trade their genes back and forth casts an interesting light on certain trendy theories of human behavior. Thirty years ago, the scientist Richard Dawkins proposed in his influential book *The Selfish Gene* that what humans are, really, is "survival machines" for our own genome. We are a mechanism our genes have constructed to ensure their own immortality. The better the gene builds us, the more likely that gene is to survive into the next generation. And it is our job—the theory goes—to perpetuate our own gene line. From this fascinating and persuasive theory rose, among other things, the fashionable notion that most of our sexual behavior, male and female, is set up to accomplish this signal task, and that men and women accomplish the task differently. Men, the theory goes, are natural philanderers; it is the nature of men to spread their seed and ensure the survival of their gene line by flinging it around. Women, on the other hand, are more invested in the upbringing of their genetic offspring; it is the nature of women to see that their genetic children grow up and prosper. It seems to be male commentators, for the most part, who really, really like this theory, which seems to justify a male imperative to fool around.

What, then, about egg donation? If it is a female mission to nourish and raise only children produced by one's own gene line, how does one think about Laura Ramirez and other women who are raising, and loving, the genetic offspring of another? How to think about Kendra Vanderipe and other egg donors, women fully willing to spread their own seed far and wide? And not always for money? I talked to many egg donors who were happy to give their genes to another woman. Women, too, are willing to fling their genes around a little indiscriminately. How to think about this fact? And how to think about those men, those sperm-donor dads, who have for generations been willing to raise, and love, another man's genetic offspring? How to think about the man who says, of his sperm-donor children, "How can you not just love them?"

Love? Remember love?

Most of all, what reproductive science has done, I think, is remind us that having children and loving children is an unstoppable urge; that humans, or many humans, have an overpowering need to have—to be— a family. For me, one of the most moving reporting experiences was a day I spent with Madeleine Gordon, the woman who has set up a small charity, funded largely out of her own pocket, to pay for IVF for poor and working-class people in Ohio. So grateful were her recipients that they descended upon her Cincinnati condo to be interviewed and express their thanks, driving in from outlying areas, often after work. The condo complex's hospitality room swarmed with toddlers; Gordon went back and forth, getting them juice and snacks. One prospective father, whose wife was pregnant, was carrying an ultrasound photo inside his police hat. "I've never won anything," said one pregnant mother who still couldn't believe that she had been selected as a candidate for the highly competitive charitable program; couldn't believe that she was actually going to have a child, after fearing she never would do so. "We had babysat for so many nieces and nephews," said another father, recalling how, once he and his wife finally had a child of their own, thanks to Gordon's program, it took a while to get their minds around the fact that this time the child would not be taken away at the end of the evening. This time, they weren't just babysitters. "We couldn't believe we had our own."

And it's worth pointing out that in this rough, confusing terrain, there are individuals who display the most extraordinary grace under the most extraordinary new forms of pressure. This occurred to me when talking to so many parents who were trying to make the right moral decision in any number of unforeseen situations. It occurred to me when talking, for example, to Tammy LaMantia, an amazing woman, an amazing mother, an amazing wife, who with her husband is wrestling with issues so much more complex than many couples confront when forming families. When she and I were chatting, recently, I asked Tammy if her daughters, now four, had weighed in on the sibling issue. Turns out they had: One wanted a sister. One wanted a brother. No problem, they figured! "Only twins," Tammy said, laughing, "would think of having twins."

Meanwhile, she and her husband were weighing what was the right thing to do: by each other, by the girls, by the frozen embryos. The right decision wasn't yet clear. But what was clear is that reproductive science is a testament to what we will do, we women, we men, to have the children we love and long for.

AFTERWORD TO

THE ANCHOR BOOKS EDITION

In the year since *Everything Conceivable* was first published, the trends discussed in it have become even more pronounced. Far from abating, for example, the number of multiple births continues to rise. In 2004—the most recent year for which detailed government figures are now available—America saw the highest number of twins born, ever, in our nation's history. More than 132,000 twins were born in 2004, nearly double the number born in 1980. The twins rate (the proportion of twins born relative to the overall number of births) also rose two percent, meaning that twins are becoming not only more numerous, but more likely. There was a slight decrease in the triplets rate, but very slight: about 7,300 higher-order multiples were born, four times the number born in 1980 and about 700 percent more than the number in 1970. Because the vast majority of multiple births are twins, the country's overall proportion of multiple births also reached an all-time high in 2004; the largest increase was to mothers over 45. The government attributes this rise to fertility medicine, and the prognosis for these children has not gotten better: In 2004, the premature birth rate rose two percent, to 12.5 percent of all births. The rate of babies born with low birth weight also rose. In 2007, at least three sets of sextuplets were born to American parents. All the babies in two sets lived, but in one set, five of the six babies died. As usual, there was a flurry of media attention, and then the families were left to find their way forward as best they could.

In other areas, as well, trends continued to rise steadily. The number of IVF cycles rose in 2005, according to figures just released by the U.S. Centers for Disease Control and Prevention: that year saw 135,000 IVF cycles performed, and, thanks to steadily improving success rates, the number of infants born reached 52,000, more than double the number of IVF children born just nine years earlier, in 1996. The number of children born from lower-tech treatments, such as drugs alone, remains untracked. The number of births to unmarried mothers also reaches an all-time high with each successive year: In 2004, an extraordinary 36 percent of all births were to single mothers.

The increase in IVF usage has also given us the now-predictable parade of extremes: In late 2006, Maria del Carmen Bousada Lara, an unmarried Spanish retiree who lied to a U.S. clinic about her age, gave birth to twins just a few days shy of her sixty-seventh birthday, becoming, as far as anybody can tell, the new "world's oldest mother." Her achievement eclipsed that of Frieda Birnbaum, the 60-year-old American who had to settle for being the oldest woman to give birth to twins in the United States, and who, not content with having the children and being thankful, felt compelled to trumpet her delivery as being "basically about women and empowerment." Possibly the only ART parents less appealing, this past year, were the lesbians living in Australia who filed a lawsuit against a doctor who transferred two IVF embryos rather than one; the women argued that they only wanted one child, and the second, healthy twin constituted a "wrongful birth."

Not to be outdone in the lawsuit arena, in 2007 an embryo named Mary Scott Doe pressed a two-year-old suit against California's stem cell research institute. Or rather, a pro-life group pressed a suit claiming civil rights for Mary and all frozen IVF embryos. The suit was thrown out of court. It was one in a series of lawsuits filed in an ongoing but thus far unsuccessful effort to derail California's $3 billion initiative to fund embryonic stem cell research. Judging from these efforts, the contentious debate over embryonic stem cell research—and over the moral status of the human embryo—also is not going to go away soon. In 2007, the U.S. Congress passed a bill authorizing federal funding for stem cell research on frozen IVF embryos; President George Bush again vetoed it. Hillary Clinton has said she will approve funding if elected president. But even presidential approval will not stop the argument: rather, such a move would likely be followed by a round of lawsuits like the one filed on behalf of little embryo Mary, holding up research and keeping uncertainty alive. This is despite a new study showing that when it's made easier for couples to donate their IVF embryos to science, they are much more likely to do so, and tend to be quite relieved to have the opportunity to make a useful contribution to possible cures.

The plot thickened further in November 2007, when two teams of scientists announced that they had achieved startling success "reprogramming" skin cells, which is to say, returning them to an embryonic state without actually creating an embryo. Instead, they used viruses to introduce genes into the skin cell that effectively did what an egg can do: made the cell "forget" it was specialized, returning it to a pluripotent state in which it could be coaxed into to developing into another kind of cell entirely—such as a heart or spinal cord. As with embryonic stem cell research that involves therapeutic cloning, the goal would be to create a specialized form of therapy that would be an exact tissue match to a patient suffering from Parkinson's or spinal cord injury. Opponents of embryonic stem cell research were quick to hail this as a terrific breakthrough, which it well may be, and to argue that embryonic stem cell research is therefore unnecessary. In return, scientists and advocates argue that it's important to move forward on both fronts, since many questions about reprogramming remain, including whether introducing genes could introduce a cancer into the new cell line. What's fascinating is that scientists are trying to find a way to replicate the transformative power of the human egg. "The egg has evolved this process over hundreds of millions of years of evolution," says one scientist who believes the new research is promising, but that it remains far from being able to rival the egg.

The broader category of embryo research (as opposed to embryonic stem cell research) remains all but ignored in this country. And yet it's so important: In 2007, researchers in Japan published a study in which they performed time-lapse recordings of IVF embryos. Their goal was to answer the vexing question of why IVF embryos are so more likely to split into identical twins than naturally conceived ones are. And they gleaned a lot from just this one study, finding that IVF embryos are more likely to develop two "inner cell masses" (the part of the embryo that develops into a baby) and that culture medium may be the cause. They performed this research on leftover IVF embryos donated by European patients.

This is crucial research—the better scientists can understand the workings of the human embryo, the better able they are to spot a good one, and the more feasible it becomes to transfer just one, and cut down on multiple births. Embryo research is a *child health issue*, or should be. But such studies remain impossible in this country, owing to the ban on federal funding for embryo research. It's so absurd: IVF science has been in play for more than forty years, yet American scientists remain unable to study the embryos they are so rapidly making. They can use 'em, freeze 'em, discard 'em—they just can't study them. If supporters of embryonic stem cell research had the courage of their convictions—which they don't seem to—they would lobby for all embryo research, not just the profitable stuff that promises high-profile cures. In for a dime, in for a dollar.

Nor, of course, have any of the moral questions involving fertility commerce been resolved. The arena of sperm, egg, and embryo donation continues to be wildly profitable and unregulated, even as the numbers of children born by donor gamete climbs. (Also in the past year or so, a Texas firm announced it would start offering made-to-order human embryos, using both donor egg and donor sperm.) According to the CDC, in 2005 some 16,000 IVF cycles were performed using donated eggs or embryos. This, combined with instances of donated sperm—though nobody is really keeping track, 30,000 donor inseminations every year seems a reasonable baseline—means that each year, in America, there are more children conceived with donated gametes than there are infants adopted, either domestically or internationally.

Far more people are affected than you might think. Not long after this book was published, I gave a talk to an audience of churchgoers who were mostly in their sixties. As I spoke, I reflected that the material must seem remote to these listeners, but I was wrong: afterward, I received an email from a listener whose grandchildren are donor-egg twins. The examples are everywhere, yet the issues of parenthood and conflicting rights between parents and children remain unresolved and, often, undiscussed. From time to time, the industry talks about establishing a voluntary donor registry. But opposition to any registry that would require donor's identities to be known was crystallized by Cappy Rothman, the outspoken medical director of California Cryobank. In an interview with *LA Weekly*, Rothman claimed that there is no problem with anonymous donors. Procreation by half siblings, he argued, is both statistically improbable and no big deal: after all, he ventured, most people used to live in small villages and marry cousins or other relations, so "we're all descendants of incest."

The families interviewed for this book continue to grow and prosper. With some regularity I sit down at my computer and am rewarded with a new set of photos. The twins whose two siblings were reduced are breathtaking, although looking at them, I cannot help but think of the siblings who didn't happen. Laura Ramirez, the working mom of triplets, manages to send regular slideshows: over the summer the family loaded up the car and drove to visit Kendra, who for her part has donated eggs again, this time to a couple who has limited contact with her, but does not want contact between half siblings. This is hard for Laura; it means that despite her efforts at openness, her sons may have half siblings they don't know, proving—again—how hard it is to control the situations the technology sets in motion. Hally Mahler moved her family to Africa, temporarily solving, for the time being, the problem of the half siblings she does know, but doesn't quite know what to do about. Eve Andrews has continued getting to know her donor-dad; after their first meeting, she says, "I can remember looking in the mirror and seeing myself a little differently." Ryan Kramer, whose desire for a sibling led to the creation of the Donor Sibling Registry, finally met a half sister.

And Doug and Eric left Noe Valley and moved to a more suburban-feeling neighborhood, inadvertently joining an exodus of gay families out of traditionally gay neighborhoods, which seem less necessary all the time in the new, more inclusive climate. Of course, San Francisco is still San Francisco: there are gay families in their new neighborhood as well, in numbers that surprised them. If anything sums up their life over the past year, Doug reports, it's how "ordinary our concerns are: childcare, schooling, balancing work and home, coping with aging parents and family members with health issues, etc."

Oh . . . and deciding what to do with the frozen embryos. "We just don't know," he reports. "We're avoiding the question and continuing to store them. But we're not having more kids, so at some point we need to make a decision." He attached a photo of the girls, who are, he says, the "reliable joy machines" in their lives.

And in many realms, science moves more slowly than most people think. When I began reporting this book, in late 2003, egg-freezing was considered an experimental procedure, not yet ready for prime time, but widely hyped as an imminent solution to female reproductive aging. In late 2007, the American Society for Reproductive Medicine issued an advisory saying that egg-freezing is still considered . . . an experimental procedure. Some problems, then, remain more intractable than others. If change is slow, however, it is inexorable. Our world has been altered and so have our families, in ways we had expected, and ways we never did; that, for sure, is not likely to abate.

Liza Mundy
January 2008

ACKNOWLEDGMENTS

This book was a great pleasure to write, in ways I had expected and ways I had not. I would like to thank, first and foremost, the individuals, couples, and families who had the generosity of spirit, the time, and the trust to share their stories, always with the aim of helping others. I hope readers will appreciate their courage. It was a privilege to meet all of them.

I would also like to express my gratitude to those who saw merit in this book idea early on. Chief among these is the Henry J. Kaiser Family Foundation, whose media fellowship in health made possible much of the reporting and travel. I especially thank the fellowship's director, Penny Duckham, for her generosity and lively interest. I also was fortunate to be a fellow in the science journalism program at the Marine Biological Laboratory in Woods Hole, Massachusetts. I am indebted to its administrator, Pamela Clapp Hinkle, to its directors, and to the scientist David Albertini, for permitting me to sit in on the 2005 Frontiers in Reproduction session. It was invaluable and—this is the great secret of lab science—really fun.

Also, among the early believers, I would like to thank Todd Shuster, my agent, who devoted immense energy to pushing this book forward. And of course Jordan Pavlin, an editor who combines every quality a writer hopes for: rigorous intelligence, acute sensibility for the words on the page, and unparalleled warmth and enthusiasm. I am deeply grateful for her belief and guidance.

I would also like to thank the friends and colleagues who listened, advised, and read. Sydney Trent suggested the story idea for the *Washington Post Magazine* that would lead to this book. Susannah Gardiner edited my work there, provided smart guidance, and read this manuscript. Tom Shroder always approved and facilitated. Steve Coll suggested I write a book in the first place, and gave priceless advice on methodology. Jack Shafer shared thoughts that sent me in productive directions. Ann Hulbert helped me develop ideas into articles for *Slate,* always with ready interest, and applied her broad knowledge to a reading of the manuscript. Clara Jeffery published a version of the frozen embryo chapter in *Mother Jones*. Arthur Allen shared the wisdom of experience. And Margaret Talbot was there before the first moment, and after the last; she made suggestions, read the manuscript, and reacted with her usual combination of enthusiasm and incisive commentary. Truly, this book would not exist, and I would have faltered long ago, if not for her wise counsel and unflagging friendship, for all of which I gratefully thank her.

I would also like to thank the scientists, doctors, and other professionals who made time to give interviews. In addition to those named in the book, I thank Diana Broomfield, Michael DiMattina, Michael Thomas, Guy Ringler, Arnold Strauss, Kari Danziger, Gina Davis, Rene Almeling, Joan Rabinor, Ellen Singer, Jennifer Sedlmeyer, Alison Carlson, Meryl Rosenberg, Cori Borjan, Elizabeth Stephen, Laura Schieve, Joyce Martin, Anjani Chandra, Stephanie Ventura, Murray Goldstein, and Eve Harris. For special thanks, I would like to single out Harriet Dolinsky, Madeleine Gordon, Nancy Hemenway, Paul Turek, Marcelle Cedars, Mark Evans, Fady Sharara, Alfred Khoury, Gail Taylor, and Sean Tipton for their help. I owe a debt of gratitude to Arthur Leader, for patiently reviewing with me the history of the field and reading portions of the manuscript, and David Albertini, who also read many sections for accuracy. And I particularly want to thank David Keefe—a dedicated scientist, committed clinician, and deep, clear thinker—for his steady assistance. I am indebted to him and to his wife, Candy Hasey Keefe, for their warm hospitality during a cold winter sojourn.

At Knopf, I would also like to thank Leslie Levine, who knows the answer to every question, for ushering this book—and me—through with efficiency and kindness; Erinn Hartman, for her zest for getting the book out to stores and readers; Gabriele Wilson for her wonderful cover design; Virginia Tan for the elegant layout; Jenna Bagnini for her formidably hard work keeping production moving; and Avery Flück for doing the same with printing. I also thank Jonathan Stein, who did not flinch before the task of fact-checking a large manuscript. Any errors are mine only.

And I would like to thank my family. It takes a village to write a book, and I could not have written this one without the support of many family members. I would like to thank my mother, Jean Mundy, and my father, Marshall Mundy, for their lifelong encouragement. In addition, I thank Lorretta Bradley, Teresa and Barry Bowles, and Monika Mundy, for providing glad and loving child care, which permitted me to take reporting trips with an easy mind. I also thank the parents who gave rides and sympathized. And in a category of his own is my husband, Mark Bradley, who despite the demands of his own difficult job managed to find the energy to support mine; whose own writing and thinking are models of depth and clarity; and who was always willing to lend his sturdy moral compass to discussions of family-making scenarios. He supported this book, and me, from start to finish.

And finally, my own children. Possibly no one takes books as seriously as elementary school students, who are told so often that books matter, and who, wonderfully, believe this to be true. And possibly no one considers the topic of trying to have children as worthy as children themselves. Anna, my daughter, and Robin, my son, were supportive beyond their years. They took a real interest in the subject, never objected to my work schedule, and served as homegrown bioethicists, sharing their views on many issues. Their presence was a constant reminder of why all this matters. I am more indebted to them, for their warmth and love and their existence, than I can ever express.

NOTES

Prologue: An Unexpected Development

xi   One day in early May: Beth Parab, interview, June 15, 2005.

xiv   Researching that article: Liza Mundy, "The Impossible Dream: Are Solutions to Infertility Only for the Rich?" *Washington Post Magazine*, April 20, 2003, p. 8.

xviii   Some of the most prestigious practitioners: A discussion of government regulation of in vitro fertilization technology and related techniques—including the U.S. government's ban on ooplasm transfer, in which the cytoplasm of one woman's oocyte is injected around the nucleus of another's—can be found in a brief issued by the Genetics and Public Policy Center, whose website is www.dnapolicy.org. The federal government regulates very few aspects of IVF; ooplasm transfer is the rare exception. Discussion of experiments that were curtailed can be found in Rick Weiss, "Pioneering Fertility Technique Resulted in Abnormal Fetuses," *Washington Post*, May 18, 2001, p. A3, and Katherine Eban, "Key Work on In Vitro Fertilization Crippled by Federal Oversight," *New York Sun*, June 5, 2002, p. 1.

1. The New Reproductive Landscape

4   The conference was dominated and underwritten: $3 billion is the commonly accepted figure for the U.S. fertility industry. The figure is cited, among other places, in Spar, *The Baby Business*, p. 3.

9   Through the displays: Discussion of Edwards, Steptoe, and their achievement—and specifically, the lack of a Nobel Prize—can be found, among many other places, in Anjana Ahuja, "God Is Not in Charge, We Are," *The Times (London)*, July 24, 2003, p. 6. Richard Paulson of the University of Southern California also pointed out the injustice of the Nobel oversight in an interview, August 17, 2004, and Robert Nachtigall compared their achievement to other twentieth century breakthroughs in an interview, October 28, 2005.

9   Observers differ: Descriptions of the collaboration between Steptoe and Edwards, and the experiments that led to the birth of Louise Brown in 1978, can be found in a number of books and articles, including Challoner, *The Baby Makers*, pp. 13–46; Henig, *Pandora's Baby*, pp. 32–45, 170–172; and Marsh and Ronner, *The Empty Cradle*, pp. 233–238.

9   "I think it's interesting": Robert Nachtigall, interview, October 28, 2005.

10   when commissions: Hall, *Merchants of Immortality*, pp. 100–101.

10   Men are responsible: Zev Rosenwaks, interview, August 14, 2005.

10   as well as women caught up in: Patricia Hunt, "Control of Mammalian Meiosis," lecture delivered for the Frontiers in Reproduction course at the Marine Biological Laboratory in Woods Hole, Mass., on June 2, 2005.

11   In 2006, the International Committee: These data are from the *2002 World Report on Assisted Reproductive Technology*, issued by the International Committee for

Monitoring Assisted Reproductive Technology, an abstract of which was presented at the conference of the European Society of HumanReproduction and Embryology on June 21, 2006. A fact sheet is available at www.eshre.com.

12   And their numbers are rapidly: The numbers that follow are from 2004 *Assisted Reproductive Technology Rates: National Summary and Fertility Clinic Reports*, the annual survey of reporting IVF clinics published by the U.S. Department of Health and Human Services, Centers for Disease Control and Prevention, December 2005, p. 19.

12   About 12 percent: A. Chandra, G. M. Martinez, W. D. Mosher, J. C. Abma, and J. Jones, "Fertility, Family Planning, and Reproductive Health of U.S. Women: Data from the 2002 National Survey of Family Growth," National Center for Health Statistics, *Vital Health Statistics* 23, no. 25, pp. 21–24.

13   To track the impact: The statistics on birth rates, births to single mothers, and multiple births in 2003, cited here and elsewhere in this chapter, can be found in J. A. Martin, B. E. Hamilton, P. D. Sutton, et al., "Births: Final Data for 2003," National Center for Health Statistics, *National Vital Statistics Reports* 54, no. 2, September 8, 2005, pp. 6–21.

15   A paper presented at the 2005 conference: G. L. Ryan, R. A. Maassen, A. Dokras, C. H. Syrop, and B. J. Van Voorhis, "Infertile Women Are Twice as Likely to Desire Multiples as Fertile Controls," oral presentation at 2005 annual convention of the American Society for Reproductive Medicine, October 17, 2005, abstract published in *Fertility and Sterility* 84 (September 2005), suppl. no. 1, p. S22.

16   In part it's because of what James Gleick: Diverse evidence of our appetite for getting everything faster can be found throughout Gleick, *Faster.*

16   Now, one out of every eight: Richard E. Behrman and Adrienne Stith Butler, eds., Committee on Understanding Premature Birth and Assuring Healthy Outcomes, Board on Health Sciences Policy, Institute of Medicine of the National Academies, "Preterm Birth: Causes, Consequence, and Prevention" (Washington, D.C.: National Academies Press, 2006).

17   A number of authors: Various genetic-engineering scenarios are laid out and well debated in Gregory Stock, *Redesigning Humans: Choosing Our Genes, Changing Our Future* (Boston: Houghton Mifflin, 2003); Glenn McGee, *The Perfect Baby: Parenthood in the New World of Cloning and Genetics* (Lanham, Md.: Rowman & Littlefield, 2000); and Francis Fukuyama, *Our Posthuman Future: Consequences of the Biotechnology Revolution* (New York: Farrar, Straus and Giroux, 2002), as well as a publication by the President's Bioethics Council, *Beyond Therapy: Biotechnology and the Pursuit of Happiness* (New York: Regan Books, 2003).

19   When Howard Jones appeared: Henig, *Pandora's Baby*, p. 225.

19   "IVF children are different": Arthur Leader, interview, April 28, 2005.

20   One British fertility doctor: Lockwood is quoted in "British Women Struck by 'Bio-Panic,'" *Bioedge* 222, October 10, 2006.

21   Recently, a boy plugged his genetic: Sam Lister, "How a Donor Sperm Boy Traced His Father Using the Internet," *The Times (London)*, November 3, 2005, p. 5.

21   Scientists who study "paternal discrepancy": Mark Bellis, Karen Hughes, Sara Hughes, and John R. Ashton, "Measuring Paternal Discrepancy and Its Public Health Consequences," *Journal of Epidemiology and Community Health* 59 (2005): 749–754.

21   "Let's face it: donor gametes": Robert Nachtigall, interview, October 28, 2005.

2. Women and the Dilemmas of Modern Motherhood

27   "We limited our cases at first": Howard Jones, interview, February 8, 2005.

27   And there were many such women: In the United States, good infertility statistics
are hard to come by. Infertility is a difficult condition to measure: a woman only
knows she is infertile if she is trying to have a child and failing. If she is contracept-
ing, or not having sex, she doesn't know whether she is fertile or infertile. If she is
trying to have a child and failing, the infertility could be her partner's. And yet, in
the United States, fertility statistics are collected only on women. The data on
women are published in a periodic survey of women and childbearing, the
National Survey of Family Growth (NSFG), conducted by the National Center for
Health Statistics. These statistics are also a little inconsistent: up until 1982, fertility information was collected only on married women, and even now, married
women who are actively trying to conceive and have failed for twelve months are
the *only* people included in the official government category of "infertility."
Clearly, a huge swath of people are therefore being left out. After 1982, the NSFG
added a catchall category called "impaired fecundity," which includes any woman
of any age and marital status who is trying to have a child, and having problems.
With all these caveats, the NSFG statistics for 2002 show that strict twelve-month
infertility among married women declined from 8.5 percent in 1982 to 7.4 percent
in 2002, but the percentage of married women with "impaired fecundity" has increased, from about 12 percent in 1982 to 15 percent in 2002. Of all women
between fifteen and forty-four, 7.3 million have impaired fecundity, or close to
12 percent, an increase of 2 percentage points since 1982. A. Chandra, G. M. Martinez, W. D. Mosher, J. C. Abma, and J. Jones, "Fertility, Family Planning, and
Reproductive Health of U.S. Women: Data from the 2002 National Survey of
Family Growth," National Center for Health Statistics, *Vital Health Statistics* 23,
no. 25, pp. 21–24. There is no government data on male infertility, but the most
common view among doctors is that infertility affects men and women about
equally. As for trends in causes: according to Elizabeth Hervey Stephen and Anjani
Chandra, "Declining Estimates of Infertility in the United States: 1982–2002,"
Fertility and Sterility 86, no. 3 (September 2006): 516–523, between 1982 and
2002, the percentage of women treated for pelvic inflammatory disease, which can
cause tubal problems, dropped significantly. Between 1975 and 1997, incidence of
gonorrhea decreased by almost 74 percent. Chlamydia rates increased, but this
may be due to more testing and diagnosis. The report points out that there is a
"favorable relationship between higher education and infertility," meaning that the
more educated you are, the less likely you are to be infertile. In David Guzick and
Shanna Swan, "The Decline of Infertility: Apparent or Real?" *Fertility and Sterility* 86, no. 3 (September 2006): 524–526, the authors argue that the new NSFG
figures "understate" the prevalence of twelve-month infertility among married
women, citing four different studies of couples who stopped contraception to
become pregnant. Those studies suggested a 15–20 percent infertility rate among
these couples. The authors point to the current NSFG figure of 15 percent for
"impaired fecundity" among all married women as evidence that infertility in the
generally understood sense—not able to conceive, or unable to carry a pregnancy
to term—is on the rise. "I think our estimates from the NSFG have been grossly
inaccurate," the University of California, San Francisco, epidemiologist Mary
Croughan agreed, speaking at a National Institute for Child Health and Human

Development (NICHD) conference on infertility outcomes, September 12 and 13, 2005, in Washington, D.C.

27 In fact, childbirth is a frequent cause: Robert Nachtigall, written summary of presentation on "International Comparisons of Access to Infertility Services," delivered at the Health Disparities in Infertility conference convened by the National Institute of Child Health and Human Development, Washington, D.C., March 10–11, 2005.

28 there is a form of PCOS: Fady Sharara, in a June 24, 2004, interview, talked about the fact that there are a number of different causes of PCOS, and that one cause may be obesity, while another may be underweight.

28 In addition, as Margaret Marsh and Wanda Ronner: Marsh and Ronner, *The Empty Cradle,* pp. 60, 89–92.

28 In the twentieth century: David Keefe, interview, January 21, 2005.

28 Let it not be forgotten either: Bigelow is quoted in Marsh and Ronner, p. 80. Clarke is quoted in "Infertility and Inheritance," *Economist,* U.S. ed., June 10, 2000.

29 For many years: Surgical enlargements are described in Marsh and Ronner, pp. 60–64; John Rock is discussed throughout the book, especially pp. 173–196.

29 "the end of the beginning": Marsh and Ronner, p. 237.

30 The next day, the Joneses received a call: The story of the founding of the Jones clinic is told in Marsh and Ronner, *Empty Cradle,* pp. 240–241, and Henig, *Pandora's Baby,* pp. 175–176, 202–223, 226–227.

30 Still, the pro-lifers exacted: Lucinda Veeck Gosden, interview, December 8, 2004.

31 In the United States, IVF clinics were created: Dorothy Greenfeld, interview, October 18, 2005.

32 "It was pretty crude by today's standards": Arthur Leader, interview, April 28, 2005. The details of Steptoe and Edwards's early experiments also can be found in Challoner, pp. 13–46.

32 "We were always worried": Veeck Gosden, interview, December 8, 2004.

32 One of the most profound retardants: A good description of the torturous deliberations over embryo research funding in the United States can be found throughout Henig, *Pandora's Baby,* and in Hall, *Merchants of Immortality,* pp. 105–122, which trace the convening of commissions, formation of commissions, ignoring of commissions, and dissolution of commissions that occurred during the 1970s and '80s; the resolve of the administration of President Bill Clinton to put funding through for embryonic research; and the passage of the Dickey-Wicker amendment, ending that prospect.

32 "That did an enormous amount": Anne McLaren, interview, July 21, 2004.

33 As early as the mid-1970s, a scientist: The story of Pierre Soupart and his ill-fated funding application can be found in Henig, *Pandora's Baby,* pp. 130–31, 136–137, 223–224; and Hall, *Merchants of Immortality,* p. 99.

33 "Anything funded by the federal government": Phyllis Leppert, chief, Reproductive Sciences Branch, National Institute of Child Health and Human Development, public comments at NICHD conference on "Infertility Treatment and Adverse Pregnancy Outcomes," Washington, D.C., September 12–13, 2005.

34 "We wanted a registry": Alan DeCherney, interview, October 18, 2005.

35 half the time: Arthur Leader, interview, April 28. Details of the early IVF procedures were also provided by Richard Paulson, interview, August 17, 2004; Marcelle Cedars, interview, October 16, 2004; and Lucinda Veeck Gosden, interview, December 8, 2004.

37 "The money ruined a lot of things": Robert Nachtigall, interview, October 28, 2005.

37   In 1986, there were 41: Elizabeth Hervey Stephen and Anjani Chandra, "Use of Infertility Services in the United States: 1995," *Family Planning Perspectives* 32, no. 3 (May/June 2000): 132–137; the 2002 figures are from A. Chandra, G. M. Martinez, W. D. Mosher, J. C. Abma, and J. Jones, "Fertility, Family Planning, and Reproductive Health of U.S. Women: Data from the 2002 National Survey of Family Growth," National Center for Health Statistics, *Vital Health Statistics* 23, no. 25, pp. 21–24.

38   In 1960, the average: "Births: Final Data for 2003," *National Vital Statistics Reports* 54, no. 2 (September 8, 2005). Additional data has been provided independently by statisticians at the National Center for Health Statistics.

39   There is no cutting-off point: Howard Jones, interview, February 8, 2005.

39   According to one study: Henri Leridon, "Can Assisted Reproduction Technology Compensate for the Natural Decline in Fertility with Age? A Model Assessment," *Human Reproduction* 19, no. 7 (June 2004): 1548–1553.

40   She has the most eggs: Rogerio A. Lobo, "Potential Options for Preservation of Fertility in Women," *New England Journal of Medicine* 353, no. 1 (July 7): 2005.

40   Patricia Hunt, a reproductive: Patricia Hunt, lecture on "Control of Mammalian Meiosis," Woods Hole, Mass., June 2, 2005.

41   "Non-Hispanic blacks and other race women": Marianne Bitler of RAND and Lucie Schmidt, of Williams College, in a written summary of a talk delivered at the Health Disparities in Infertility conference convened by NICHD, Washington, D.C., March 10–11, 2005.

42   "You know the old real estate": Marcelle Cedars, interview, October 16, 2004.

42   Studies show that among ART patients: Sigal Klipstein, Meredith Regan, David A. Ryley, Marlene B. Goldman, Michael M. Alper, and Richard Reindollar, "One Last Chance for Pregnancy: A Review of 2,705 In Vitro Fertilization Cycles in Women Age 40 Years and Above," *Fertility and Sterility* 84, no. 2 (August 2005): 435–445.

42   Among women between forty-six: S. D. Spandorfer, K. Bendickson, K. Dragesic, G. Schattman, O. K. Davis, and Z. Rosenwaks, "IVF Outcome in Women 45 Years and Older Utilizing Autologous Oocytes: Success Is Limited to Women at 45 Years of Age with a Good Response," oral presentation at the ASRM meeting, October 19, 2005, abstract published in *Fertility and Sterility* 84 (September 2005), suppl. no. 1, p. S139.

43   "Infertility patients thought": Sean Tipton, interview, June 17, 2004.

43   "We knew it would be a hot potato,": Marcelle Cedars, interview, October 16, 2004.

43   But *Newsweek* picked up on the topic: Claudia Kalb, "Should You Have Your Baby Now?" *Newsweek*, August 13, 2001, p. 40.

43   But the current NOW president: Kim Gandy, CNBC interview, August 7, 2001.

43   "what is essentially a scare campaign": Kim Gandy, NBC *Today* show interview, August 6, 2001.

44   "We were shocked": Robert Stillman, interview, September 1, 2005.

44   "Women are, once again": Kim Gandy, "Campaign Goes Too Far," *USA Today*, September 6, 2002, p. 14A.

44   "Boy, are they pissed": Robert Nachtigall, interview, October 28, 2005.

44   "All we were trying to do": Marcelle Cedars, interview, October 16, 2004.

45   According to Marsh and Ronner: Marsh and Ronner, *Empty Cradle,* describes informal and formal adoption practices, pp. 17–19, 125–128.

45   recalls Paige McCoy Smith: Paige McCoy Smith, interview, June 13, 2005.

46   A government study: Christine A. Bachrach, Kathy Shepherd Stolly, and Kathryn London, "Relinquishment of Premarital Births: Evidence from National Survey Data," *Family Planning Perspectives* 24, no. 1 (January–February 1992): 27–32.

48　As early as 1984: Sanford Rosenberg, interview, February 21, 2005.

48　"Who knew that": Richard Paulson, interview, August 13, 2004.

49　In 1990, Paulson: M. V. Sauer, R. J. Paulson, and R. A. Lobo, "A Preliminary Report on Oocyte Donation Extending Reproductive Potential to Women over 40," *New England Journal of Medicine,* 323, no. 17 (October 25, 1990): 1157–1160.

50　In papers with titles such as: Ruth C. Fretts, Julie Schmittdiel, Frances McLean, Robert H. Usher, and Marlene Goldman, "Increased Maternal Age and the Risk of Fetal Death," *New England Journal of Medicine* 333, no. 15 (October 12, 1995): 953–957. Linda J. Heffner, "Advanced Maternal Age—How Old Is Too Old?" *New England Journal of Medicine* 351, no. 19 (November 4, 2004): 1927–1929.

50　In 1993 they published: M. V. Sauer, R. J. Paulson, and R. A. Lobo, "Pregnancy After Age 50: Application of Oocyte Donation to Women After Natural Menopause," *Lancet* 341, no. 8841 (February 1993): 321–323.

51　"The combined age should not": Vicken Sahakian, interview, August 17, 2004.

51　As early as 1993: M. V. Sauer and R. J. Paulson, "Quadruplet Pregnancy in a 51-Year-Old Menopausal Woman Following Oocyte Donation," *Human Reproduction* 12 (December 8, 1993): 2243–2244.

52　Then, one day, a woman: Paulson tells the story of Arceli Keh in Richard J. Paulson, *Rewinding Your Biological Clock: Motherhood Late in Life* (New York: W. H. Freeman, 1998), pp. 37–38; and Richard J. Paulson, Melvin H. Thornton, Mary M. Francis, and Herminia S. Salvador, "Successful Pregnancy in a 63-Year-Old Woman," *Fertility and Sterility,* 67, no. 5 (May 1997): 949–951.

53　"Retired lecturer": Nigel Hawkes, "Retired Lecturer, 67, Set to Be Oldest Mother, and It's Twins," *The Times (London),* December 31, 2004, p. 42.

53　Before long, to make headlines: "California Woman in Her 50s Gives Birth to Second Child," *Orange County Register,* February 5, 2005.

53　Then there was: Michael D. Shear and Rob Stein, "Woman, 55, Gives Birth to Grandchildren," *Washington Post,* December 29, 2004, p. B1.

55　"all the other people": Laura Brounstein, "I Get Up Every Morning and Smile," *Ladies' Home Journal,* May 2004, pp. 116–118, 124.

55　In 2004, an article: Suz Redfearn, "Did Elizabeth Edwards Use Donor Eggs? All Signs Point to Yes," *Slate,* October 29, 2004. http://www.slate.com/id/2108863/.

56　"You need a huge section": Fady Sharara, interview, March 19, 2004.

57　"We've seen over the past": Robert Nachtigall, public comments at the annual ASRM Meeting, session entitled "Counseling Couples About Collaborative Reproduction: The Ethical, Cultural and Psychological Dimensions Following ART," delivered at the ASRM meeting in Montreal, Canada, October 15–16, 2005.

3. Every Man a Father, Every Man Infertile

61　Paul Turek is mad: Paul Turek, interview, March 2, 2005.

62　"Come in": Rob Ginis, interview, March 4, 2005.

65　"Since then my life hasn't been the same": Stephen Seager, interview, May 10, 2005.

67　"Every time your heart beats": Ina Dobrinski, lecture, "Germ Cell Transplantation," at the Frontiers in Reproduction course at the Marine Biological Laboratory, Woods Hole, Massachusetts, June 3, 2005.

68　So is sitting while exposed to heat: Yefim Sheynkin, Michael Jung, Peter Yoo, David Schulsinger, and Eugene Komaroff, "Increase in Scrotal Temperature in Laptop Computer Users," *Human Reproduction* 20, no. 2 (August 2005): 452–455.

68　The ability of sperm to rebound: R. J. Levine, R. M. Mathew, C. B. Chenault, M. H. Brown, M. E. Hurtt, K. S. Bentley, K. L. Mohr, and P. K. Working, "Differences in the Quality of Semen in Outdoor Workers During Summer and Winter," *New England Journal of Medicine* 323, no. 1 (July 5, 1990): 12–16.

69 "Human males are just infertile, period": Sherman Silber, interview, September 23, 2005.

69 In 1992, a Danish scientist: E. Carlsen, A. Giwercman, N. Keiding, and N. Skakke-baek, "Evidence for Decreasing Quality of Semen During Past 50 Years," *British Medical Journal* 305 (1992): 609–613.

69 Then in 2000: Shanna H. Swan, Eric P. Elkin, and Laura Fenster, "The Question of Declining Sperm Density Revisited: An Analysis of 101 Studies Published in 1934–1996," *Environmental Health Perspectives* 108, no. 10 (October 2000): 961–965.

69 This finding coincides: "Challenged Conceptions: Environmental Chemicals and Fertility," conference paper published by Women's Health @ Stanford and the Collaborative on Health and the Environment, October 2005, p. 13.

70 In a letter: Shanna H. Swan, Irva Hertz-Picciotto, "Reasons for Infecundity," *Family Planning Perspectives* 31, no. 3 (May–June 1999): 156–157.

70 This last point was: Tina Kold Jensen, Elisabeth Carlsen, Neils Jorgensen, Jorgen G. Berthelsen, Niels Keiding, Kaare Christensen, Jorgen Holm Petersen, Lisbeth B. Knudsen, and Niels E. Skakkebaek, "Poor Semen Quality May Contribute to Recent Decline in Fertility Rates," *Human Reproduction* 17, no. 6 (2002): 1437–1440.

71 "The decreasing trends": Niels E. Skakkebaek, Niels Jorgensen, Katharina M. Main, Ewa Rajpert-De Meyts, Henrik Leffers, Anna-Maria Andersson, Anders Juul, Elisabeth Carlsen, Gerda Krog Mortensen, Tina Kold Jensen, and Jorma Toppari, "Is Human Fecundity Declining?" *International Journal of Andrology* 29 (2006): 2–11.

71 For men, as for women: Robert D. Nachtigall, Gay Becker, and Mark Wozny, "The Effects of Gender-Specific Diagnosis on Men's and Women's Response to Infertility," *Fertility and Sterility* 57, no. 1 (January 1992): 113–121.

72 Then in 1884: Henig, *Pandora's Baby,* pp. 26–28.

73 "Let's say": Robert Nachtigall, public comments during a course in "Counselling Couples About Collaborative Reproduction," at ASRM convention, October 16, 2005.

74 "It was a tremendous problem": Lucinda Veeck Gosden, interview, December 8, 2004.

75 "I had never seen": Gianpiero Palermo, interview, July 26, 2005. The account of Palermo's ICSI success is also told in Challoner, *The Baby Makers,* pp. 112–114.

76 But in 1992: G. Palermo, H. Joris, P. Devroey, and A. C. Van Steirteghem, "Pregnancies after Intracytoplasmic Injection of Single Spermatozoon into an Oocyte," *Lancet* 340 (1992): 17–18.

77 "When I started doing IVF": Peter Brinsden, interview, July 21, 2004.

78 "Now we can take": Lucinda Veeck Gosden, interview, December 8, 2004.

78 "All of a sudden": Paul Turek, interview, March 2, 2005.

79 Now he can: David C. Page, Sherman Silber, and Laura G. Brown, "Men with Infertility Caused by AZFc Deletion Can Produce Sons by Intracytoplasmic Sperm Injection, but Are Likely to Transmit the Deletion and Infertility," *Human Reproduction* 14, no. 7 (1999): 1722–1726.

79 So what else is passing into the gene line?: Discussions of what ICSI may be introducing into the gene line can be found in many medical articles, including Joanna Gonsalves, Fei Sun, Peter N. Schlegel, Paul J. Turek, Carin V. Hopps, Calvin Greene, Renee H. Martin, and Renee A. Reijo Pera, "Defective Recombination in

Infertile Men," *Human Molecular Genetics* 13, no. 22 (2004): 2875–2883; Sherman J. Silber and Sjoerd Repping, "Transmission of Male Infertility to Future Generations: Lessons from the Y Chromosome," *Human Reproduction Update* 8, no. 3 (2002): 217–229; Jan A. M. Kremer, Joep H. A. M. Tuerlings, George Borm, Lies H. Hoefsloot, Eric J. H. Meuleman, Didi D. M. Braat, Han G. Brunner, and Hans M. W. M. Merkus, "Does Intracytoplasmic Sperm Injection Lead to a Rise in the Frequency of Microdeletions in the *AZFc* Region of the Y Chromosome in Future Generations?" *Human Reproduction* 13, no. 10 (1998): 2808–2811; and Paul J. Turek and Renee A. Reijo Pera, "Current and Future Genetic Screening for Male Infertility," *Urologic Clinics of North America* 29 (2002): 767–792, which points out that "it is assumed that many natural selection barriers are bypassed with ICSI."

79   Sherman Silber and colleagues: Malcolm J. Faddy, Sherman J. Silber, and Roger G. Gosden, "Intra-cytoplasmic Sperm Injection and Infertility," *Nature Genetics* 29 (October 2001): 131.

80   "I have enormous respect": Paul Turek, interview, March 2, 2005.

83   "Male factor is hugely more": Sanford Rosenberg, interview, February 21, 2005.

85   There is so much about natural conception: A wondrous description of the natural conception process—the arduous passage of the sperm through the hostile environment of the female reproductive tract, the repeated battering of the egg by sperm, the final victory of a single sperm, the way the egg lures that sperm inside—can be found in Boyce Rensberger, *Life Itself: Exploring the Realm of the Living Cell,* (Oxford: Oxford University Press), 1997. Rensberger's vivid description includes an explanation of how the egg essentially grants a single sperm admission: the outer coating of the egg contains thousands of "recognition factors" to assess the sperm; after admitting one, Rensberger writes on p. 148, "Contractile filaments from the egg attach to the sperm and pull it in, tail and all, leaving its empty membrane, like a ghost, outside."

4. It Takes a Village to Make a Child: ART and the Evolving Human Family

88   "None of us has been": Kendra Vanderipe, interview, April 28, 2005.

91   Writing the prayer: Rev. Beth Parab, interview, June 15, 2005.

94   In the United States: Statistics on egg donation and surrogacy are from the CDC's 2003 *ART Success Rates;* 30,000 is a 1987 federal government estimate of annual sperm-donation children, though nobody really knows how many children are born annually from this procedure; the FDA estimates for donor inseminations can be found in Michael Leahy, "Family Vacation," *Washington Post Sunday Magazine,* June 19, 2005, p. 12.

94   "courage, conviction, and strategy": R. Cook, S. Golombok, A. Bish, and C. Murray, "Disclosure of Donor Insemination: Parental Attitudes," *American Journal of Orthopsychiatry* 65, no. 4 (October 1995): 549–559.

95   "The whole concept of family": Lori Maze, interview, August 16, 2004.

96   "The average surrogate": Gail Taylor, interview, April 30, 2005.

96   One woman: Kristine Cicak, interview, September 23, 2004.

97   "It has really changed": Dorothy Greenfeld, interview, October 18, 2005.

99   "Children conceived": R. Cook, I. Vatev, Z. Michova, and S. Golombok, "The European Study of Assisted Reproduction Families: A Comparison of Family Functioning and Child Development Between Eastern and Western Europe," *J. Psychosom. Obstet. Gynecol.* 18 (1997): 203–212. These findings are also contained in Susan Golombok, Rachel Cook, Alison Bish, and Clare Murray, "Families

Created by the New Reproductive Technologies: Quality of Parenting and Social and Emotional Development," *Child Development* 66 (1995): 285–298, which compared IVF and donor-insemination children to families with a naturally conceived child and adoptive families; and in S. Golombok, A. Brewaeys, M. T. Giavazzi, D. Guerra, F. MacCallum, and J. Rust, "The European Study of Assisted Reproduction Families: The Transition to Adolescence," *Human Reproduction* 17, no. 3 (2002): 830–840, a longitudinal study using the same sorts of comparison groups. In these and other studies conducted by Golombok and British and European colleagues, ART parents were found to have fewer marital difficulties, and lower levels of anxiety and depression, and ART mothers expressed more warmth to their children and more emotional involvement, than parents of children who were naturally conceived. Golombok stressed that the control group families were also doing well; it was just that the ART families "were functioning extremely well." What's notable in all of these studies is that donor-insemination families also scored well on all measures, despite the fact that not one family had told the children the truth of their origins. From this Golombok began to develop the view that "genetic ties are less important for family functioning than a strong desire for parenthood"; that families conceived through sperm and (later) egg donation are not analogous to adoptive families; that secrets, though not desirable, do not seem to have a detrimental effect on family functioning; and that in the case of gamete donation, being wanted all along—and never having been relinquished by a birth mother—might enhance a child's well-being. This view, which emphasizes the importance of warmth, attachment, and wantedness, dismays proponents of mandatory disclosure, who do believe that gamete donation is analogous to adoption, that genetic ties are crucial, and that family secrets are inevitably destructive.

99 "These are the most motivated": Robert Nachtigall, interview, October 28, 2005.

99 "When you've had to work": Fay Johnson, interview, August 15, 2004.

100 "We still have a very": Jean Benward, public comments in a lecture, "Adoption and Gamete Donation: Similarities and Differences," at the 2005 ASRM conference, October 15–16, 2005.

101 "I've always looked at this": Bill Cordray, interview, February 20, 2005.

102 "There is no loss": Kim Bergman, interview, May 22, 2006.

102 "Adoption starts in a moment": Robert Nachtigall, interview, October 28, 2005.

102 "challenged deeply rooted": Susan Golombok, lecture given at "Real Families, Real Facts," a conference on gay and lesbian parenting organized by Family Pride, an organization dedicated to same-sex family-building, in Philadelphia on May 22, 2006.

102 Should mothers be: A wonderful exploration of a century's worth of argumentation among child development experts, on all of these issues, can be found in Ann Hulbert, *Raising America: Experts, Parents, and a Century of Advice About Children* (New York: Alfred A. Knopf, 2003).

102 "There is a growing body": Susan Golombok, Clare Murray, Peter Brinsden, and Hossam Abdalla, "Social Versus Biological Parenting: Family Functioning and the Socioemotional Development of Children Conceived by Egg or Sperm Donation," *Journal of Child Psychology and Psychiatry* 40, no. 4 (1999): 519–527.

5. "Sperm Bank Helps Lesbians Get Pregnant!": How Women Changed the Sperm-Banking Industry—and the Makeup of the Family

109 In 1981, Raboy was: The details of the founding of the Sperm Bank of Northern California are taken from an interview with Barbara Raboy, March 15, 2005, and

from an interview with Alice Ruby, current director of The Sperm Bank of California, March 2, 2005.

110   David Plotz in his book: David Plotz, *The Genius Factory: The Curious History of the Nobel Prize Sperm Bank* (New York: Random House, 2005).

115   at least 60 percent: California Cryobank provided this figure in Leahy, "Family Vacation," *Washington Post Magazine*.

115   "Lesbians are very accepting": Charles Sims, interview, August 13, 2004.

115   "We really wanted": Gretchen Lee, interview, March 3, 2005.

117   "embody": David Keefe, interview, January 19, 2005.

118   In a 2006 lecture: Susan Golombok, lecture on "Research on Children of Same-Sex Parents: A Historical Perspective Across 30 years," Family Pride conference, Philadelphia, Pa., May 22, 2006.

118   "this evil must stop": Conservative MP Sir Rhodes Boyson, quoted by Golombok, May 22, 2006.

119   To test whether this applied: Judith Stacey, in a careful and exhaustive review of the literature on same-sex parenting that she published with her research partner, Timothy Biblarz, "(How) Does the Sexual Orientation of Parents Matter?" *American Sociological Review* 66 (April 2001) 159–183, discusses the longitudinal study by Susan Golombok and Fiona Tasker, the results of which were published, in the mid-1990s, in a number of places. Two sources of finding from the longitudinal study itself are Susan Golombok and Fiona Tasker, "Do Parents Influence the Sexual Orientation of Their Children? Findings from a Longitudinal Study of Lesbian Families," *Developmental Psychology* 32 (1996): 3–11, and Fiona L. Tasker and Susan Golombok, *Growing Up in a Lesbian Family* (New York: Guilford), 1997. In their 2001 paper Stacey and Biblarz also summarize much of the rest of the literature on same-sex parenting; my discussion of it is based in part on their characterization, as well as e-mail communications from Judith Stacey in which she characterized the two waves of research: one on lesbian mothers after divorce, the next on donor-insemination lesbian households. Some of the major same-sex parenting studies are as follows: C. J. Patterson, "Children of Lesbian and Gay Parents," *Child Development* 63, no. 5 (October 1992): 1025–1042, concluded that children of same-sex parents enjoy the same sense of well-being as children of heterosexual parents. In 1997, in A. Brewaeys, I. Ponjaert, E. V. Van Hall, and S. Golombok, "Donor Insemination: Child Development and Family Functioning in Lesbian Mother Families," *Human Reproduction* 12, no. 6 (June 1997): 1349–1359, researchers compared thirty children of lesbian DI mothers, thirty-eight of single DI mothers, and thirty of heterosexual families, and concluded that the quality of interaction between children and their lesbian nonbiological mothers was higher than that between children and their heterosexual fathers, and that these lesbian social mothers were considered as much a "parent" as a father. In H. M. Bos, F. van Balen, and D. C. van den Boom, "Planned Lesbian Families: Their Desire and Motivation to Have Children," *Human Reproduction* 18, no. 10 (October 2003): 2216–2224, researchers in the Netherlands explored the motivation of lesbian mothers and found that happiness was more important for them than for heterosexual parents, that they had spent more time thinking about their motives for having children, and that their desire to have a child was stronger. In J. L. Wainright, S. T. Russell, and C. J. Patterson, "Psychosocial Adjustment, School Outcomes, and Romantic Relationships of Adolescents with Same-Sex Parents," *Child Development* 75, no. 6 (November–December 2004): 1886–1898, researchers at the University of Virginia compared forty-four 12-to-18-year-olds with same-sex parents, and the same number from heterosexual parents, and

found that there was no difference in assessments of romantic relationships and sexual behavior.

121 In July 2006: Anemona Hartcollis, "New York Judges Reject Any Right to Gay Marriage," *New York Times,* July 7, 2006, p. A1.

121 Yet some cultural: The idea that "maverick moms" are free to raise a new type of boy is made, among other places, in a book by the gender scholar Peggy Drexler, *Raising Boys Without Men: How Maverick Moms Are Creating the Next Generation of Exceptional Men* (Emmaus, Pennsylvania: Rodale), 2005.

122 Like Golombok, Stacey spoke: Judith Stacey, a professor of sociology at New York University, delivered a lecture, "How Does the Gender of Parents Matter?" at a Family Pride conference on gay and lesbian parenting convened in Philadelphia on May 22, 2006.

123 The Supreme Court of California: August 2005 was a good month for the legal affirmation of lesbian parenting commitments, at least in the state of California. In addition to the case Halm mentioned, the Supreme Court also ruled that a lesbian mother cannot avoid paying child support for her former partner's biological children, conceived when the two women lived together. Again, the crucial point was that both women intended to be parents, therefore both women were parents. Both cases are outlined in Henry Weinstein and Lee Romney, "Court Affirms Gay Couples' Parental Status," *Los Angeles Times,* August 23, 2005, p. A1.

124 Perhaps needless to say: Elizabeth Marquardt, principal investigator, Commission on Parenthood's Future, "The Revolution in Parenthood: The Emerging Global Clash Between Adult Rights and Children's Needs," published by the New York–based Institute for American Values, 2006.

125 Because in truth: Gail Taylor's remarks are from two interviews, August 16, 2004, and June 21, 2005.

6. Two Men and Two Babies: Gay Fatherhood Through Surrogacy

127 "We know more": Interviews with Doug Okun and Eric Ethington took place on March 3, 2005, and April 21, 2005.

130 "reproductive communism": Rothman discussed this concept in remarks at a Planned Parenthood conference, "Beyond Abortion: Critical Bioethical Issues in Reproductive Health for the 21st Century," July 23–25, in Snowbird, Utah.

130 An increasing percentage: Ginia Bellafante, "Surrogate Mothers' New Niche: Bearing Babies for Gay Couples," *New York Times,* May 27, 2005, A1.

130 "A lot of the girls": Laura Fretwell, interview, February 25, 2005.

131 Johnson, a gracious woman: Fay Johnson related the details of her own surrogacy story in an interview, August 16, 2004.

131 In *The Nation:* Katha Pollitt, "The Strange Case of Baby M," *The Nation,* May 23, 1987. Available on the web at http://www.thenation.com/doc/19870523/19870523pollitt.

132 "It's a lot easier": Gail Taylor, interview, August 16, 2004.

133 "If you're looking at beauty": Vicken Sahakian, interview, August 17, 2004.

133 "There's two separate": Michael Feinman, interview, August 17, 2004.

134 Ann Nelson is a graduate: Ann Nelson shared her narrative of surrogacy in an interview on April 20, 2005.

139 There were so many options: The anxiety induced by too many consumer choices is explored in Barry Schwartz, *The Paradox of Choice: Why More Is Less* (New York: HarperCollins, 2004).

7. Single Mothers by Choice, and the Magazine
Article That Made Them

154 Checking back: *Newsweek*'s revision of its earlier reporting can be found in Daniel McGinn, "Marriage by the Numbers," *Newsweek,* June 5, 2006, p. 40.

155 The beginning of: Sylvia Ann Hewlett, *Creating a Life: Professional Women and the Quest for Children* (New York: Hyperion, 2002).

155 As hard as some feminists: Joan Walsh, "The Baby Panic," April 23, 2002, http://archive.salon.com/mwt/feature/2002/04/23/hewlett_book/print.html.

155 Cope had an assortment: Debra Cope related her narrative in an interview, July 28, 2005, and subsequent e-mails.

157 Between 1999 and 2003: These figures from the National Center for Health Statistics are cited in Jennifer Egan, "Looking for Mr. Good Sperm," *New York Times Magazine,* March 19, 2006, p. 46.

158 In 1980: Jane Mattes, interview, May 2, 2005.

160 "There was lots of speculation": Clare Murray, interview, July 26, 2004.

160 This too may be: Mark Henderson, " 'No Father Needed' Under Shake-up of UK Fertility Rules," *Times Online,* July 12, 2006, http://www.timesonline.co.uk/article/ 0,,200-2267114,00.html.

161 "I think it is so hard": Marcelle Cedars, interview, October 16, 2004.

161 In the United States, it's getting risky: The Benitez case is discussed in Elizabeth Weil, "Breeder Reaction: Does Everybody Have the Right to a Baby? And Who Should Pay When Nature Alone Doesn't Work?" *Mother Jones,* July/August 2006, p. 33.

161 The lower income associated with divorce: C. Murray and S. Golombok, "Going It Alone: Solo Mothers and Their Infants Conceived by Donor Insemination," *American Journal of Orthopsychiatry* 75, no. 2 (April 2005): 242–253.

162 Judith Stacey: Stacey, lecture at the Family Pride conference, May 22, 2006.

163 "I haven't ever met": Denise Feinsod made this reflection, and described her narrative of single parenthood, in an interview, March 5, 2005.

164 "It was fun": Lisa Schmidman, interview, August 12, 2005.

166 In deciding that she: Lori Gottlieb describes the thought process that led her to become a single mother by donor insemination in "The XY Files," *Atlantic Monthly,* September 2005, p. 141.

167 "I found myself dating": Julie Sabala, interview, August 11, 2005.

167 There is also no consensus: The narrative of Raechal McGhee and her successful quest to find the sperm donor of her children is told in Leahy, "Family Vacation." Raechel and Mike Rubino later appeared on CBS's *The Early Show* to discuss the unfolding relationship, and the emergence of at least three more Rubino-sired offspring, on August 23, 2005.

169 "It was like": Elizabeth Reynolds is quoted in Dennis Fiely, "Registry Connects Sisters Who Shared Same Donor Dad," *Columbus (Ohio) Dispatch,* April 8, 2006, A1.

169 Call it "really, really, really big love": The saga of eleven women who used Donor 401 is described in Lois Romano, "Multiple Single Moms, One Nameless Donor," *Washington Post,* February 27, 2006; after Donor 401 moms appeared on *Today,* eight more mothers of Donor 401 children came forward, as described in Romano, "Popular Donor's Family Tree May Keep On Growing," *Washington Post,* May 1, 2006, A2. Meanwhile, the accounts of women who had used Donor 150 of California Cryobank were told by Amy Harmon in "Hello, I'm Your Sister. Our Father Is Donor 150," *New York Times,* November 20, 2005, p. 1, and the tale of one donor who came forward to meet nine offspring, Matthew Niedner, a.k.a. Donor

48QAH, is told in Susan Schindehette, "My Life as a Sperm Donor," *People,* June 5, 2006, p. 133.

173   "If you're single": Hally Mahler, interview, November 3, 2005.

8. ART and the Rights of the Child

177   Eve Andrews: Eve Andrews related the story of finding out the truth about her conception in an interview on March 13, 2005, and in subsequent e-mails.

181   "You need to be able to grapple": Olivia Montuschi and Eric Blyth shared their views on donor anonymity in an interview on July 26, 2005.

182   The judge who heard her case: Mark Henderson, "Donor Children Win Right to Find Biological Parents," *The Times (London),* January 22, 2004, p. 4, and Steven Boggan, "Children Born of Sperm Donors Win Right to Take on Government," *The Independent,* July 27, 2002, p. 9.

183   For years, the shame: R. Cook, S. Golombok, A. Bish, and C. Murray, "Disclosure of Donor Insemination: Parental Attitudes," *American Journal of Orthopsychiatry* 65, no. 4 (October 1995): 549–559.

184   "He was strong and athletic": Bill Cordray, interview, February 26, 2005.

185   "I don't like the word 'donor' ": Barry Stevens, lecture at 2005 ASRM annual meeting, October 18, 2006.

186   Australia, Broderick told the audience: Pia Broderick, lecture, "Lessons from Down Under: Can Disclosure Really Be Legislated?" at 2005 ASRM conference, October 15–16, 2005.

188   "We feel that a bit of choice": Juliet Tizzard, interview, July 22, 2004.

188   Ian Craft: Craft, who made an estimated $16 million over one seven-year period, is prominent in a list of highly paid British IVF doctors compiled by Sophie Goodchild and Jonathan Owen, "The Baby Millionaires," *The Independent,* January 8, 2006, which can be found at http://news.independent.co.uk/uk/health_medical/article337250.ece.

189   "The person is thinking of this": Gail Taylor, interview, August 16, 2005.

189   "If there weren't anonymity": Sanford Rosenberg, interview, February 21, 2005.

190   As the psychotherapist: Jean Benward, lecture, "Adoption and Gamete Donation, Similarities and Differences," at 2005 ASRM conference, October 16, 2006.

191   How do donors react?: Jane Dreaper, "IVF Donor Sperm Shortage Revealed," BBC News, September 13, 2006, http://news.bbc.co.uk/go/pr/fr/-/2/hi/health/5341982.stm. News of shortages was reported by Kirsty Horsey, "Cash and Sperm Shortages Threaten UK Fertility Services," *BioNews* 361, Week 30/5/2006 to 4/6/2006, a weekly e-mail summary of bioethics news published by the Progress Educational Trust. The closure of a Scottish clinic was noted in Eva Langlands, "Clinic Closes as IVF Crisis Hits Home," *Times Online,* June 4, 2006, www.timesonline.co.uk./article/0,,2090-2210281,00.html.

192   "We're all just one big happy": Amy Housler, interview, February 17, 2005.

193   As one 2005 study: A. Brewaeys, J. K. de Bruyn, and F. M. Helmerhorst, "Anonymous or Identity-Registered Sperm Donors? A Study of Dutch Recipients' Choices," *Human Reproduction* 20, no. 3 (March 2005): 820–824.

193   Even in Sweden: Actually, in Sweden, one study found that only 11 percent of parents had told their child, though a further 41 percent intended to tell. That study is cited in E. Lycett, K. Daniels, R. Curson, and S. Golombok, "School-aged Children of Donor Insemination: A Study of Parents' Disclosure Patterns," *Human Reproduction* 20, no. 3 (March 2005): 810–819.

193 A 2004 study: Clare Murray and Susan Golombok found that only 29 percent of oocyte-donation families intended to tell their children the truth about their conception, and that not one had done so, in "To Tell or Not to Tell: The Decision-Making Process of Egg-Donation Parents," *Human Fertility* 6 (2003): 83–89. S. C. Klock and D. A. Greenfeld also showed that only about half of oocyte-donation parents planned to tell their children, in "Parents' Knowledge About the Donors and Their Disclosure in Oocyte Donation," *Human Reproduction* 19, no. 7 (2004): 1575–1579. Klock and Greenfeld found that many nondisclosing families had told other people, and that "many regret having done so."

193 Another egg-donation mom: Nancy Hass, "Whose Life Is It, Anyway?," *Elle*, September 2005.

194 In his book *Stigma:* Goffman, *Stigma,* p. 74.

194 as the medical anthropologist Gay Becker: Becker delivered these remarks in a lecture, "Resemblance Talk: A Challenge for Parents Whose Children Were Conceived with Donor Gametes," at the 2005 ASRM conference, October 16, 2005, and she elaborates in Gay Becker, Anneliese Butler, and Robert D. Nachtigall, "Resemblance Talk: a Challenge for Parents Whose Children Were Conceived with Donor Gametes in the U.S.," *Social Science and Medicine* 61 (2005): 1300–1309.

196 "The majority of youths": J. E. Scheib, M. Riordan, and S. Rubin, "Adolescents with Open-Identity Sperm Donors: Reports from 12–17 Year Olds," *Human Reproduction* 20, no. 1 (2005): 239–252.

197 When I met with Alice Ruby: Alice Ruby, interview, March 2, 2005.

198 "There is often a seeming lack of interest": Benward, lecture at 2005 ASRM meeting, October 16, 2006.

201 As for others: A. Gurmankin, A. Caplan, and A. Braverman, "Screening Practices and Beliefs of Assisted Reproductive Technology Programs," *Fertility and Sterility* 83, no. 1 (January 2005): 61–67.

202 "There are many ways of being a good parent": Joann Galst, lecture, "To Treat or Not to Treat: Predicting Parental Fitness," at 2005 ASRM meeting, October 18, 2005.

203 "Are we going to make everybody": David Keefe, interview, January 2005.

9. Be Fruitful and Multiply: The Big Family, by Overnight Delivery

210 "Who has four kids these days?": Linda Gulyn, interview, April 15, 2005.

212 "The vital evolutionary importance": Gillian Lockwood, "The Diagnostic Value of Inhibin in Infertility Evaluation," *Seminars in Reproductive Medicine* 22, no. 3 (2004): 195–208.

212 "If Nature thought it was appropriate": Alfred Khoury's comments, here and elsewhere, were communicated in interviews on November 30, 2004, and February 1, 2005.

212 And yet this is what: Figures on the trends in multiple births are from J. A. Martin, B. E. Hamilton, P. D. Sudon, et al., "Births: Final Data for 2003," *National Vital Statistics Reports* 54, no. 2 (September 8, 2005) as well as statistics compiled by statisticians with the National Center for Health Statistics.

213 These increases are the result: In B. Luke, M. B. Brown, C. Nugent, V. H. Gonzalez-Quintero, F. R. Witter, and R. B. Newman, "Risk Factors for Adverse Outcomes in Spontaneous Versus Assisted Conception Twin Pregnancies," *Fertility and Sterility* 81, no. 2 (February 2004): 315–319, researchers found that delayed childbear-

ing is responsible for one-fourth of the increase in multiple pregnancies since the 1980s, and that the remaining three-quarters of the rise is due to assisted reproductive technologies, including both drugs and IVF, which, depending on the therapy, carry anywhere from a 25 to 40 percent risk of multiples. In Meredith Reynolds, Laura Schieve, Joyce Martin, Gary Jeng, and Maurizio Macaluso, "Trends in Multiple Births Conceived Using Assisted Reproductive Technology, United States, 1997–2000," *Pediatrics* 111 (2003): 1159–1162, U.S. government researchers found that between 1997 and 2000, the proportion of naturally conceived twins decreased from 73.2 percent to 67.3 percent, whereas the proportion due to fertility treatments—IVF or fertility drugs—increased from 26.8 percent to 32.7 percent.

213 Fertility drugs have been around now: Marsh and Ronner, *Empty Cradle,* p. 208.

213 "Estrogen levels": Jamie Grifo, David Hoffman, and Philip McNamee, "We are due for a correction . . . and we are working to achieve one," commentary in *Fertility and Sterility* 75, no. 1 (January 2002): 14.

214 As a result: The decline in multiples in many European countries was noted in a press release, "Three Million Babies Born Using Assisted Reproduction Technologies," issued by the International Committee for Monitoring Assisted Reproductive Technologies during the annual conference of the European Society of Human Reproduction and Embryology, on June 21, 2006.

214 "They are one of our most popular": Kim Richmond, interview, February 2005.

215 multiple births are a staple: Nancy Dillon, "Their Joy Is Fourfold," *New York Daily News,* December 28, 2004; Rich McKay, "Orlando Hospital Triples Its Pleasure Three Times Over," *Bradenton (Florida) Herald,* May 6, 2005; Judith Newman, "Triplet Epidemic: Why New Jersey Leads the Nation in Multiple Births," *National Geographic,* October 2005, p. 120. The answer to why New Jersey has so many triplets is contained in the story: it has more fertility clinics per capita than any other U.S. state.

215 "We apologize for any pain": Elizabeth Leland, "Expecting 5," *Charlotte (North Carolina) Observer,* September 8, 2002, p. 1G.

215 According to a 2006 report: Richard E. Behrman and Adrienne Stith Butler, eds., Committee on Understanding Premature Birth and Assuring Healthy Outcomes, Board on Health Sciences Policy, Institute of Medicine of the National Academies, "Preterm Birth: Causes, Consequence, and Prevention" (Washington, D.C.: National Academies Press, 2006).

215 The March of Dimes: The contribution that multifetal pregnancies are making to the rise in infant mortality and premature births is starkly outlined in a major opinion paper, "Perinatal Risks Associated with Assisted Reproductive Technology," ACOG Committee Opinion No. 324, American College of Obstetricians and Gynecologists, *Obstetrics and Gynecology* 106 (2005): 1143–1146; and in several major March of Dimes press releases, including "Benefits and Risks of High Tech Fertility Treatments Featured at March of Dimes Luncheon," December 9, 2004, which points out that "16 percent of all preterm deliveries in the U.S. were due to multifetal gestations, which are usually a result of ART"; and "U.S. Infant Mortality Rate Fails to Improve: Premature Birth and Low Birthweight Major Contributors to Infant Death," May 8, 2006.

216 "It has all the hallmarks": Louis Keith, interview, June 16, 2005.

217 As a result of prematurity: Many of these statistics can be found in a PowerPoint presentation made by Alfred Khoury, "Multiple Gestations: If One Is Good . . . Why Aren't Two or Three Better?," which he shared with me. These statistics are

also taken from a number of presentations made by perinatologists at the NICHD conference on the outcomes of IVF pregnancies, convened on September 12 and 13, 2005, in Washington, D.C., which included several talks on multiple births. And they come from studies: In 2006, researchers showed that risk of dying during pregnancy was 3.6 higher among with women with twin and triplet pregnancies. Andrea MacKay, Cynthia Berg, Jeffrey King, Catherine Duran, and Jeani Chang, "Pregnancy-Related Mortality Among Women with Multifetal Pregnancies," *Obstetrics and Gynecology* 107 (2006): 563–568. In 2004, Canadian researchers showed that women pregnant with two or more fetuses were thirteen times more likely to suffer heart failure, four times as likely to have a heart attack, and two and a half times as likely to develop deep vein thrombosis, as cited in a December 4, 2004, press release from *Medical News Today.* A United Cerebral Palsy fact sheet, "Multiple Births and Developmental Brain Damage," points to the increases in multiple births as the likely factor "for the increased occurrence of cerebral palsy in the United States in recent years." And in 2005, a major study on the fate of very premature babies, Neil Marlow, Dieter Wolke, Melanie Bracewell, and Muthanna Samara, "Neurologic and Developmental Disability at Six Years of Age After Extremely Preterm Birth," *New England Journal of Medicine* 352, no. 1 (January 6, 2005): 9–19, pointed out that babies born before 26 weeks of gestation display "a level of impairment that is greater than is recognized with the use of standardized norms," particularly cognitive and neurological impairment.

217 According to a 2006 study: William Callaghan, Marian MacDorman, Sonja Rasmussen, Cheng Qin, and Eve Lackritz, "The Contribution of Preterm Birth to Infant Mortality Rates in the United States," *Pediatrics* 118, no. 4 (October 2006): 1566–1573.

10. "It's Always a Party with Triplets": The Advent of High-Order Multiples

221 So inured was the public: Katharine Q. Seelye, "First Black Sextuplets Belatedly Win Public Notice," *New York Times,* January 8, 1998, p. A12.

221 Each December the children: In the December 2005 issue of *Ladies' Home Journal,* for example, the children were all on the cover, dressed in green and red, and the cover line read: "The Septuplets at Eight: All-New Photos."

223 In 2003: Melissa Middleton, interview, February 14, 2005.

226 "throw a bunch of spaghetti": Sher is quoted in Elizabeth Weil, "Breeder Reaction," *Mother Jones,* p. 36.

226 "I heard of one practice": Lucinda Veeck Gosden, interview, December 8, 2004.

226 "He transferred eight": Janis Elspas, interview, July 8, 2005.

227 "In the 1980s": Siva Subramanian, interview, July 14, 2005.

227 "The state of the field": Jim Trimarchi, lecture on "ART, the Clinical and Basic Science Behind the Technology," at Frontiers in Reproduction class, Marine Biological Laboratory, Woods Hole, Massachusetts, June 6, 2005.

228 Any of these systems: To bear out Trimarchi's remarks—at a federal conference on fertility treatment outcomes, on September 12 and 13, 2005, Catherine Racowsky, associate professor of obstetrics, gynecology, and reproductive biology at Harvard Medical School, lectured on embryos, culture, and the gaps in knowledge, saying, "We do not know whether one incubator brand is better than another, we don't know about the optimal gas phase . . . We also have to be concerned about which shelf in the incubator. What we have found is that although most incubators perform uniformly . . . there are a couple that have a significant increase in pregnancy rate in the middle as compared to either the top or bottom. So generally speaking

we avoid those shelves." She talked about the difficulty created by the fact that culture manufacturers are not required to reveal the concentration of certain ingredients. "We still are really bad at assessing embryos in the lab. I hate to admit this but it is the fact," she said. "We really have no idea whether anything we're doing in the lab is associated with adverse outcome . . . If we could, we could transfer a single embryo. We could completely eliminate multiple pregnancies."

229 PGD probably does damage the embryo: Dagan Wells, a scientist at Yale University medical school, in a lecture on "Preimplantation Genetic Diagnosis: Assessing Chromosomes and Viability Markers in Human Embryos," at the Frontiers in Reproduction course in Woods Hole, Massachusetts, on June 11, 2005, said, "Most people who do PGD think taking a cell does have an impact on development, but they think it's a small impact."

230 "The current problems": Trounson is quoted in Hall, *Merchants*, p. 258.

231 "That was a bad decision": Tammy LaMantia graciously sat for two interviews, on November 11 and 14, 2004.

236 "I can imagine a scenario": David Keefe, public remarks during a postgraduate seminar on "Genetic and Congenital Anomalies After IVF," at the 2004 ASRM annual meeting, October 16, 2004.

237 A Fairfax County, Virginia, fertility doctor: Teri Banas, "Fertility Doctor Found Guilty," United Press International, March 4, 1992.

237 "There is so much lying": Sam Thatcher, interview, April 8, 2005.

239 "Most people are not rational when they make decisions": Linda Gulyn, interview, April 15, 2005.

239 "It's not your problem": David Keefe, interview, January 19, 2005.

240 In 2003, police officers: A description of the Seymore household can be found in an undated Affidavit of Probable Cause filed by the Office of the District Attorney in Montgomery County, Pennsylvania, and released to the public; and in "Attorney Says Parents Charged in Quads Death Were Overwhelmed," Associated Press, July 26, 2003, and Dalondo Moultrie, "Towamencin Couple Sentenced for Death of Quadruplet Son," *Morning Call (Allentown, Pa.)*, September 30, 2004, p. B1.

240 Every year around Christmas: "Four Babies at Once? Now That's Stress," *New York Times*, November 21, 2004, p. 44.

241 "They told me there was a one in 30": Teresa Anderson, quoted in Sylvia Pagan Westphal, "Baby Boom: Multiple Births Persist as Doctors Buck Guidelines: At Some Clinics, Women Get Five or More Embryos, Raising Risks to Health; One Cause Is Patient Demand," *Wall Street Journal*, October 7, 2005, p. A1.

242 "membership in SART": David Grainger, e-mail, October 27, 2006.

242 "My biggest problem": Alfred Khoury, interview, November 30, 2004.

243 "Sometimes I feel like our subspecialties": Barbara Nies, interview, February 3, 2005.

244 Kristina Jorgensen-Harrigan, who lives: Kristina Jorgensen-Harrigan, interview, April 21, 2005.

247 "I have a foursome": Scott Coleman, interview with Scott and Tracey, October 28, 2004.

11. Deleting Fetuses: Selective Reduction, ART's Best-Kept Secret

254 According to one textbook: M. I. Evans, Y. Yaron, A. Drugan, A. Johnson, and H. R. Belkin, "Multifetal Pregnancy Reduction," in *Iatrogenic Multiple Pregnancy: Clinical Implications*, Isaac Blickstein, MD, and Louis G. Keith, MD, eds. (New York: Parthenon, 2001), pp. 147–155.

254 "This is a very sensitive": David Grainger, e-mail, October 27, 2006.

266 All of which has compelled him: To show how thinking about reducing twins has evolved: In a 1996 editorial in the *British Medical Journal* 313 (August 17, 1996): 373–374, the perinatologist Richard Berkowitz noted that "the announcement last week that a British gynaecologist had electively reduced a twin pregnancy to a singleton for social indications has caused a furor in the international press." Exploring the moral complexities, Berkowitz ventured that while it wasn't clear, at the time, that singletons did better than twins, it was true that the social and financial pressures of raising twins are greater, and that in a country like the United States where elective abortion is available, reduction of twin pregnancies should be decided on a "case by case" basis. He also said it was understandable if doctors didn't want to perform reductions of twins. In 2004, Evans and colleagues discussed the issue in light of the most recent research, showing that most twin-to-singleton reductions, which constitute less than 3 percent of all cases, are performed for women over thirty-five. They concluded that the miscarriage rate is lower in a pregnancy reduced to a singleton, and that the "likelihood of taking home a baby is higher after reduction than remaining with twins." They proposed that reductions of twins to singletons "might be considered with appropriate constraints and safeguards." M. I. Evans, M. I. Kaufman, A. J. Urban, D. W. Britt, J. C. Fletcher, "Fetal Reduction from Twins to Singleton; A Reasonable Consideration?" *Obstetrics and Gynecology* 104, no. 1 (July 2004): 102–109.

267 Among these is the nutritionist: In B. Luke "Nutrition and Multiple Gestation," *Seminars in Perinatology* 29, no. 5 (October 2005): 349–354, Luke points out that multiple pregnancy results in an "accelerated depletion of nutritional reserves," and recommends mothers follow a "diabetic regimen" that includes 20 percent of calories from protein, 40 percent from carbohydrate, and 40 percent from fat, as well as various vitamin and mineral supplements. In Luke et al., "Risk Factors for Adverse Outcomes in Spontaneous Versus Assisted Conception Pregnancies," cited above, researchers found that pregnancies that were reduced to twins tended to have a greater risk of prematurity, low birth weight, and fetal growth restriction than twins pregnancies that started out as twins, and argued that there were "residual effects of fetal reduction on the growth and subsequent birth rate of the remaining fetuses." Evans and other perinatologists argue that the increased risk of adverse outcomes is the residual effect of starting out as high-order multiples, not the residual effect of the reduction itself.

268 One study found that one-third: These conclusions—women are depressed, and gradually feel better—are explored in M. Garel, C. Stark, B. Blondel, G. Lefebvre, D. Vauthier-Brouzes, and J. R. Zorn, "Psychological Reactions After Multifetal Pregnancy Reduction: A 2-Year Follow-up Study," *Human Reproduction* 12, no. 3 (1997): 617–622. In D. W. Britt, S. T. Risinger, M. Mans, and M. I. Evans, "Anxiety Among Women Who Have Undergone Fertility Therapy and Who Are Considering Multifetal Pregnancy Reduction: Trends and Implications," *Journal of Maternal-Fetal Health and Neonatal Medicine* 13, no. 4 (April 2003): 271–278, researchers traced the fluctuating anxiety levels of women from the start of fertility treatment, when they hope just to get pregnant, to the new anxiety of what to do about multiples, through the "morally complicated resolution" of selective reduction and its aftermath. Patricia Schreiner-Engel, Virginia Walther, Janet Mindex, Lauren Lynch, and Richard Berkowitz, "First-trimester Multifetal Pregnancy Reduction: Acute and Persistent Psychological Reactions," *American Journal of Obstetrics and Gynecology* 172, no. 2 (February 1995): 541–547, surveyed 100 of the first women to undergo selective reduction, and found that "persistent depres-

sive symptoms were mild, although moderately severe levels of sadness and guilt continued for many. Nonetheless, 93 percent would make the same decision again." The quote from Isaac Blickstein is in *Iatrogenic Multiple Pregnancy*, p. 72.

12. Twins: The New Singleton

273 Tammy LaMantia is showing: Tammy LaMantia, interview, November 11 and 14, 2004.

274 Placenta previa has been shown: In 2006, Norwegian researchers reported that the risk of placenta previa in an assisted reproduction pregnancy is 16 in 1,000, compared to 3 in 1,000 in a naturally conceived pregnancy. L. B. Romundstad, P. R. Romundstad, A. Sunde, V. von During, R. Skjaerven, and L. J. Vatten, "Increased Risk of Placenta Previa in Pregnancies Following IVF/ICSI: A Comparison of ART and Non-ART Pregnancies in the Same Mother," *Human Reproduction* 21, no. 9 (2006): 2353–2358.

276 "You develop these techniques": Linda Gulyn, interview, April 15, 2005.

277 "The human uterus": Mark Evans, interview, August 24, 2005; twinning rates are from the National Center for Health Statistics.

277 In April 2006, in Forrest City: "Newborn Twins Taken from Mother as She Fights for Life," April 12, 2006, KATV website, http://www.katv.com/news/stories/0406/318752.html.

277 "I tell [twins] families": Emily Glickman is quoted in Susan Saulny, "In Baby Boomlet, Preschool Derby Is the Fiercest Yet," *New York Times*, March 3, 2006, p. 1.

278 Babies born as twins: Elizabeth Weil, "Breeder Reaction," MotherJones, July/August, p. 36.

278 "I have to admit, I'm equivocal": Sam Thatcher, interview, April 8, 2005.

278 Similarly, at Maryland-based: Robert Stillman, interview, September 1, 2005.

279 "the pregnancy rate is down": Lucinda Veeck Gosden, interview, December 8, 2004.

279 In 2005, a report was released: These figures were cited during an oral presentation by G. Adamson et al., "ICMART World Report on In Vitro Fertilization 2000: How Does the United States Compare?," delivered at the 2005 ASRM annual meeting, October 18, 2005, and published as an abstract in *Fertility and Sterility* 84, no. 1 (September 2005): suppl. no. 1, p. S86. The results were also discussed in Michael Smith, "ASRM: U.S. Leads in IVF Successes—at a Steep Price," *MedPage Today*, October 18, 2005, http://www.medpagetoday.com/ OBGYN/infertility/tb/1952.

279 Nobody knows why this is: The higher twinning figures for IVF embryos were discussed, and Judith Hall made her comment, at the NICHD fertility treatment outcomes conference, September 12 and 13, 2005. A discussion of the frequency of splitting in IVF embryos is also found in Blickstein and Keith, eds., *Iatrogenic Multiple Pregnancy*, p. 11.

279 A 2006 article in the *Wall Street Journal*: Elizabeth Bernstein, "Hearing Your Biological Clock Tick—at Age 22," *Wall Street Journal*, July 13, 2006, p. D1. Clomid results in twins in 6 to 12 percent of pregnancies, according to Blickstein and Keith, eds., *Iatrogenic Multiple Pregnancy*, p. 39.

279 "Twins are not normal": Alfred Khoury, interview, November 30, 2004.

279 "We can't stop this train": Meredith Kramer Hay, interview, August 22, 2005.

281 "We had so many consumers": Chip Whalen, interview, February 2005.

282 "I didn't leave the house": Rachel Haas, interview, April 15, 2005.

283 "I had never held a baby": Nancy Thiel, interview, August 20, 2005.

284 "Andrew is bigger, strong": Rob Thiel interview, August 20, 2005.

285 "Sarah wants to make sure": Rachel Haas, interview, April 15, 2005.

286 Nancy Segal, a fraternal twin: Nancy L. Segal, PhD, *Entwined Lives: Twins and What They Tell Us About Human Behavior* (New York: Plume, 1999), p. xiv.

286 It is true that twins have a higher rate of handicap: The infant mortality rate for twins is 5.7 percent, compared to 0.9 percent in singletons. (Infant mortality is death in the first year of life; the infant mortality rate for triplets is 16.7 percent.) The rate of handicap in twins is 12.6 percent, compared to 9 percent in singletons; the rate of severe handicap in twins is 3.4 percent, compared to 2 percent in singletons. The intrauterine death of a single fetus occurs in between 0.5 and 6.8 percent of all twin pregnancies. The fetal death of a co-twin significantly increases the risk of cerebral palsy in the surviving twin. The prevalence of CP in twins is about 7 in 1,000, compared to 1 in 1,000 for singletons. The rule of thumb for multifetal pregnancies is that the gestational age for delivery is about three weeks less, per additional fetus; since most singleton pregnancies deliver at 40 weeks, the average delivery age for twins is 37. These figures can be found in Blickstein and Keith, ed., *Iatrogenic Multiple Pregnancy*, pp. 26–32. The comparative risks of long-term sequelae are on p. 154.

286 That leaves 93 percent: B. Alin Akerman, "Long-Term Health and Psychosocial Outcomes," in Blickstein and Keith, eds., *Iatrogenic Multiple Pregnancy*, pp. 199–209.

287 "People wrongly equate": Nancy Segal, quoted in "Born Together, Raised Together, So Why Not in Classroom, Too?" Ginia Bellafante, *New York Times*, February 24, 2006, p.1.

287 "It is vital that [twins]": All of these letters ran in the letters to the editor section, "Twins in the Classroom: Go Your Own Way?" *New York Times*, March 1, 2006, p. 18.

13. Souls on Ice: America's Frozen Human Embryo Glut

290 "As time goes by": Tammy LaMantia, interview, October 10, 2006.

290 In 2002, SART, the research arm: David I. Hoffman, Gail L. Zellman, C. Christine Fair, Jacob F. Mayer, Joyce G. Zeitz, William E. Gibbons, and Thomas G. Turner, "Cryopreserved Embryos in the United States and Their Availability for Research," *Fertility and Sterility* 79, no. 5 (May 2003): 1063–1069.

290 Early on in the development: Challoner, *The Baby Makers*, describes the process of freezing an embryo (the secret is to dehydrate it) and the birth of the first "ice baby" from a frozen-thawed embryo in Holland in December 1983, pp. 53–57. The birth of a child from an embryo that had been frozen, and thawed, and frozen, and thawed, is recorded in L. Keith Smith et al., "Live Birth of a Normal Healthy Baby After a Frozen Embryo Transfer with Blastocysts That Were Frozen and Thawed Twice," *Fertility and Sterility* 83, no. 1 (January 2005): 198–200.

291 "Until recently, I don't know": Robert Nachtigall, lecture, "The Frozen Embryo Disposition Decision: Why Is It So Difficult?" at 2005 ASRM annual meeting, October 16, 2005.

292 Nachtigall and colleagues decided: Robert D. Nachtigall, Gay Becker, Carrie Friese, Anneliese Butler, and Kirstin MacDougall, "Parents' Conceptualization of Their Frozen Embryos Complicates the Disposition Decision," *Fertility and Sterility* 84, no. 2 (August 2005): 431–434.

292   An earlier study, done by psychologist: Susan C. Klock, Sandra Sheinin, and Ralph R. Kazer, letter to the editor, "The Disposition of Unused Frozen Embryos," *New England Journal of Medicine* 345, no.1 (July 5, 2001): 69. In the letter, Klock and colleagues emphasize that clinics need to do a better job of staying in touch with couples to help them assess and reassess their decision about the fate of their embryos.

293   "Some saw them as biological": Robert Nachtigall, lecture, October 16, 2005.

294   As University of Wisconsin: R. Alta Charo, lecture, "Frozen Dreams," at the 2005 ASRM annual meeting, October 16, 2005. The Tennessee case in question is *Davis v. Davis,* one of the first embryo disposition cases, resolved by the Tennessee Supreme Court in 1992.

295   In the face of pressure: President George W. Bush, radio address to the nation, August 13, 2001, available on the White House official website at http://www.whitehouse.gov/news/releases/2001/08/20010811-1.html.

296   James Dobson: In Lois Uttley, MP, Ronnie Pawelko, JD, and Rabbi Dennis Ross, "Embryo Politics: Implications for Reproductive Rights and Biotechnology," a briefing paper based on a presentation at the annual meeting of the American Public Health Association, November 8, 2004, and published in June 2005 by the MergerWatch project, the authors explore the conservative effort to fold embryo rights into the burgeoning fetal rights movement. Dobson is quoted on p. 2, Doerflinger on p. 5. The commentator Deroy Murdock referred to human embryos as "Microscopic-Americans" in a syndicated column, Scripps Howard News Service, August 23, 2001. DeLay is quoted in William Saletan, "Leave No Embryo Behind: The Coming War over In Vitro Fertilization," *Slate,* Friday, June 3, 2005, http://www.slate.com/id/2120222.

296   As *Slate*'s William Saletan: Saletan, "Leave No Embryo Behind."

296   These efforts are all: Susan L. Crockin, a reproductive technology and adoption lawyer, described conservative efforts to build an embryo rights movement, and inflate the moral status of the pre-embryo, in an op-ed piece, "Embryo Wars: How Do You 'Adopt' a Frozen Egg?" *Boston Globe,* December 4, 2005, p. D12. The mistake committed by an editor writing in the headline, where "egg" is used instead of "embryo," is probably a pretty good indication of public confusion over the scientific entities whose moral standing is being so vehemently debated.

297   In Chicago, a couple sued: In 2000, Alison Miller and Todd Parrish learned that their frozen IVF embryo had been mistakenly thrown out by a clinic worker, and in response, they sued for wrongful death. Judge Jeffrey Lawrence II of Cook County, Illinois, ruled that they did have standing, relying on a state law holding that "an unborn child is a human being from the time of conception and is, therefore, a legal person." In "Embryo-Death Suit Could End Fertility Treatments," Associated Press, February 10, 2005, the couple's attorney is quoted as saying that they had no desire to influence the nation's abortion debate, but simply wanted their day in court. Still, the implications of the judge's ruling on abortion rights is explored in Lois Uttley et al., "Embryo Politics."

297   "For the moment": R. Alta Charo, lecture, October 16, 2005.

298   To dramatize his opposition: For example, the president was shown holding a Snowflakes baby in a photo that ran on the front page of the *New York Times,* on May 25, 2005, to dramatize his opposition to a House bill authorizing federal funding for embryonic stem-cell research.

299   Bush's newly minted: Snow's apology, saying he had "overstated the president's position" by saying that Bush equated embryonic stem-cell research with "mur-

der," is reported in "White House Press Secretary Retracts Statement That President Bush Believes Embryonic Stem Cell Research is 'Murder'," Kaiser Daily Women's Health Policy e-mail briefing, a daily report on reproductive issues issued by the Henry J. Kaiser Family Foundation, on July 25, 2006.

299   "For a long time I've been pro-life": California Republican Rep. Dana Rohrabacher, interview, August 13, 2004.

300   Though small, Nightlight Christian: "Abortion politics was the river running through this entire, contested domain," Stephen Hall writes in *Merchants of Immortality*, in a fascinating section showing how in 2001 Nightlight Christian Adoptions filed suit in U.S. District Court in Washington, D.C., seeking an injunction that would prevent the National Institutes of Health from funding embryonic stem-cell research, pp. 250–252, 256–259.

301   "I've done a severe 180": Lori Maze, interview, August 16, 2004.

302   At a press conference in 2005: Sean Tipton, public remarks at a press conference where infertility patients' advocacy groups expressed support for embryonic stem-cell research, May 23, 2005.

302   The Unitarian Universalist: A summary of various religious teachings on the embryo, its status and moral standing, is provided in Lois Uttley, "Embryo Politics," pp. 10–13.

304   "Nobody does it": Alan DeCherney, public comments during a session, "Legal Complications of ART," 2005 ASRM annual meeting, October 17, 2005.

304   "I have tons of embryos": Vicken Sahakian, interview, August 17, 2005.

305   "We end up being the counselors": Russell Bierbaum, interview, April 19, 2006.

305   I think it's a mistake": Robert Nachtigall, interview, May 2, 2006.

306   "I have a family": Lori Maze, interview, August 16, 2004.

14. Choice Revisited: Ethics, Feminism, and ART

307   I mean, I had often reflected: Natalie Angier, *Woman: An Intimate Geography* (New York: Anchor, 1999), p. 2.

308   "Right now, there's no way": These and subsequent comments by David Keefe are from a series of interviews during several days, from January 18 to 21, 2005.

311   "The cytoplasm is not like tea": Jim Trimarchi, interview, January 20, 2006.

311   "Have you meditated": Carl Gunderson, interview, April 27, 2005.

312   There are prominent feminists: "The risks to would-be egg donors are not worth the hypothetical benefits," Norsigian is quoted in Claudia Kalb, "Ethics, Eggs and Embryos," *Newsweek*, June 20, 2005, p. 52. In Lois Uttley, "Embryo Politics," there is a very good discussion of the schism in feminist thinking about oocyte donation, particularly in California, where a host of pro-choice groups, including local Planned Parenthood affiliates, supported Proposition 71, the measure authorizing $3 billion in state funding for stem-cell research, while other feminist women's groups, including the Pro-Choice Alliance, opposed the funding measure, saying that there would not be enough scientific and ethical control exerted over the research.

312   "We have a donor in the hospital": Robert Stillman, interview, September 1, 2005.

312   Recently the world was handed: Rick Weiss, "Mature Human Embryos Cloned," *Washington Post*, February 12, 2004, A1; David Brown, "South Korean Researcher Is Said to Admit Stem Cell Fakes: Team Will Ask Journal to Retract Its Report," *Washington Post*, December 16, 2005, p. A1; Rick Weiss, "Korean Stem

Cell Lines Fakes: Scientist Resigns After Probe Discredits 9 of 11 Colonies," *Washington Post,* December 23, 2005, p. A1; Rick Weiss, "Stem Cell Fraud Worries U.S. Scientists," *Washington Post,* December 24, 2005, p. A2; Rick Weiss, "Stem Cell Advance Is Fully Refuted," *Washington Post,* December 30, 2005, p. A1; William Saletan, "Breaking Eggs," *Slate,* January 4, 2006; Michael Lemonick, "The Rise and Fall of the Cloning King," *Time,* January 9, 2006, p. 40.

313 "I am terrified": Gerald Schatten, public lecture, "Somatic Cell Nuclear Transfer in Non-Human Primates," at Marine Biological Laboratory, Frontiers in Reproduction course, Woods Hole, Massachusetts, June 9, 2005.

313 Soon afterward: Rick Weiss, "U.S. Scientist Leaves Joint Stem Cell Project: Alleged Ethical Breaches by South Korean Cited," *Washington Post,* November 12, 2005, p. A2.

313 After the full scandal broke: Rick Weiss, "U.S. Stem Cell Researcher Rebuked; Panel Says Schatten Sought Gains but Did Not Falsify Data," *Washington Post,* February 11, 2006.

313 Kathy Hudson, a biologist: Kathy Hudson, director of the Genetics and Public Policy Center, writes about the deliberations of the International Society for Stem Cell Research, and shares her own thoughts on oocyte donation, in an "e-news" bulletin on the organization's website, at http://www.dnapolicy.org/news.enews .article.nocategory.php?action=detail&newsletter_id=13&article_id=31.

314 "Unless Planned Parenthood": Paul Root Wolpe, public comments at a Planned Parenthood conference, "Beyond Abortion: Critical Bioethical Issues in Reproductive Health for the 21st Century," July 23–25, 2003, in Snowbird, Utah.

315 Also in 2003, the President's: The drafts of recommendations for oversight of IVF technology can be found on the Web page of the President's Council on Bioethics, at http://www.bioethics.gov/background/. The final recommendations are contained in the committee's final report, "Reproduction and Responsibility: The Regulation of New Biotechnologies" (Washington, D.C.), March 2004, available at the same Web address.

316 Also in 2003: Debra Rosenberg, "The War over Fetal Rights," *Newsweek,* June 9, 2003, p. 40.

316 As Kirsten Moore: Kirsten Moore, interview, December 17, 2003.

317 "Every state is already proposing": Patti Caldwell, public comments at Planned Parenthood bioethics conference, July 23–25, 2003.

318 A recent study: The survey, published by a researcher at the University of Illinois at Chicago, showed that 41 percent of infertile women would choose sex selection if it did not cost anything; but a British study showed that 80 percent of those surveyed feel that social sex selection is wrong. These studies, and the clinical trials that are being launched by Sandra Carson and others at Baylor College of Medicine, are discussed in *BioEdge,* a weekly online newsletter of bioethical issues, November 15, 2005, Issue 182. The potential market for sex selection is also demonstrated by the fact that in January 2006, sixteen women filed a lawsuit against the makers of a home-testing sex-determination kit, claiming that the "test got the genders of their babies wrong, causing them confusion and distress," according to *BioNews,* March 6, 2006. The test, Baby Gender Monitor, was being marketed to the type of women "who can't wait to open their Christmas presents." A description of the lawsuit can also be found in "Women Sue Lab That Promised to Tell Gender of Embryo," Associated Press, March 1, 2006.

318 "Most of the time": Keith Blauer, interview, March 2005.

319 "We think it's a slightly different situation": Kirsten Moore, interview, July 23, 2006.

322   "Most of them don't want to consider": Kelly Pagidas, interview, January 20, 2005.

322   "We don't want the government touching": Pamela Madsen, interview, November 2003.

323   "I see a lot of women": Cari Elam, quoted in Bernstein, "Hearing Your Biological Clock Tick," *Wall Street Journal,* July 13, 2006, D1.

323   "Women in their twenties": Kirsten Moore, interview, December 17, 2003.

323–324   "When I was sick? And I saw this need?": Lindsay Nohr Beck, interview, July 27, 2005.

325   A Cornell doctor: Kutluk Oktay, Erkan Buyuk, Lucinda Veeck, Nikica Zaninovic, Kangpu Xu, Takumi Takeuchi, Michael Opsahl, Zev Rosenwaks, "Embryo Development After Heterotopic Transplantation of Cryopreserved Ovarian Tissue," *Lancet* 363, no. 9412 (March 13, 2004): 837–40.

326   "Now [egg freezing] has a range": Christy Jones, public remarks at a fertility-awareness breakfast in Boston, Massachusetts, February 1, 2005.

327   As noted in a 2004 article: This quote, and some details of Christy Jones's personal story, are taken from Diana Kapp, "Ice, Ice Baby," *Elle,* April 2004, p. 242.

327   The plan she and some Harvard classmates: "Winners Named in Harvard Business School's Eighth Annual Business Plan Contest," press release, Harvard Business School, April 27, 2004.

328   "Ethically, it's wrong": Vicken Sahakian, interview, August 17, 2004.

329   "Should we offer it even if it works?": Michael Feinman, August 17, 2004.

330   "I found myself at thirty-seven": Lisa Sterman, interview, March 11, 2005, and subsequent e-mails.

Epilogue: Reproductive Science and the Future of Our Families

337   "Very bloody": Mary Croughan, public comments, NICHD conference on fertility treatment outcomes, September 12 and 13, 2005.

337   "They were not received well": Richard Schultz, public comment, NICHD conference.

337   In 2002, a report: The literature on the health of singleton IVF babies is becoming considerable, even though there are major challenges to performing these studies, among them the fact that few registries are kept of IVF births, making it difficult to track long-term outcomes. A few governments do keep registries of IVF babies, though, including the state of Western Australia. In 2002, using data from these registries, Australian researchers found that infants conceived with ART—either IVF or ICSI—were twice as likely as naturally conceived infants to have "multiple major defects and to have chromosomal and musculoskeletal defects." The rates of birth defects for ART babies was found to be 9 percent, compared to 4.2 percent in babies conceived naturally. Michele Hansen, Jennifer Kurinczuk, Carol Bower, and Sandra Webb, "The Risk of Major Birth Defects After Intracytoplasmic Sperm Injection and In Vitro Fertilization," *New England Journal of Medicine* 346, no. 10 (March 7, 2002): 725–730. Other studies have shown no increased risk of malformations. In the United States, a 2004 study found that IVF singletons have a higher rate of preterm delivery, low birth weight (a weight of less than 5 pounds, 8 ounces at birth), very low birth weight (less than 3 pounds, 5 ounces), and perinatal mortality, and are more likely to be small for their gestational age. Rebecca A. Jackson, Kimberly A. Gibson, Yvonne W. Wu, and Mary S. Croughan, "Perinatal Outcomes in Singletons Following In Vitro Fertilization: A Meta-Analysis," *Obstetrics and Gynecology* 103, no. 3 (March 2004): 551–563. Another study in *Obstetrics*

and Gynecology did not find an association between fertility treatment (both IVF and super-ovulation) and fetal abnormalities, but did find that IVF was associated with an increase in preeclampsia, gestational hypertension, placental abruption, placenta previa, and risk of C-section delivery. Tracy Shevell, Fergal D. Malone, John Vidaver, T. Flint Porter, David A. Luthy, Christine H. Comstock, Gary D. Hankins, Keith Eddleman, Siobhan Dolan, Lorraine Dugoff, Sabrina Craigo, Ilan E. Timor, Stephen R. Carr, Honor M. Wolfe, Diana W. Bianchi, and Mary E. D'Alton, "Assisted Reproductive Technology and Pregnancy Outcome," *Obstetrics and Gynecology* 106, no. 5 (November 2005): 1039–1045. A review of the literature by U.S. government researchers concluded in 2004 that "there is evidence of an increase in chromosomal abnormalities among pregnancies conceived with intracytoplasmic sperm injection and low birth weight and preterm delivery among singletons conceived with all types of ART; however, there remains uncertainty about whether these risks stem from the treatment or the parental infertility." Laura A. Schieve, Sonja A. Rasmussen, Germaine M. Buck, Diana E. Schendel, Meredith A. Reynolds, and Victoria Wright, "Are Children Born After Assisted Reproduction at Increased Risk for Adverse Health Outcomes?" *Obstetrics and Gynecology* 103, no. 6 (June 2004): 1154–1163. And in 2004, a major report issued in the U.K. by the Medical Research Council, "Assisted Reproduction: A Safe, Sound Future," concluded that "Although there is widespread acceptance, based on experience, that current ART procedures are generally safe, the evidence for this, particularly in terms of long-term safety, is relatively weak."

338   The work of these scientists: The higher proportion of IVF offspring among children afflicted with Beckwith-Wiedemann syndrome is discussed in Michael R. DeBaun, Emily L. Niemitz, and Andrew P. Feinberg, "Association of In Vitro Fertilization with Beckwith-Wiedemann Syndrome and Epigenetic Alterations of LIT1 and H19," *American Journal of Human Genetics* 72 (2003): 156–160; and Emily L. Niemitz and Andrew P. Feinberg, "Epigenetics and Assisted Reproductive Technology: A Call for Investigation," *American Journal of Human Genetics* 74 (2004): 599–609.

339   But in 1999, Gearhart: Richard I. Silver, Ronald Rodriguez, Thomas S. K. Chang, and John P. Gearhart, "In Vitro Fertilization Is Associated with an Increased Risk of Hypospadias," *Journal of Urology* 161 (June 1999): 1954–1957.

340   "This is not an extreme price": John Gearhart, interview, April 5, 2005.

340   All of this evidence: On September 12 and 13, 2005, a major conference, "Infertility Treatment and Adverse Pregnancy Outcomes," was convened by the National Institute for Child Health and Human Development. "We know that multifetal pregnancies present a problem, but there is increasing evidence that singletons do, too," said Uma Reddy, medical director of the pregnancy and perinatology branch of the Center for Developmental Biology and Perinatal Medicine at the NICHD. There, many scientists made presentations. Mary Croughan, a professor of epidemiology and biostatistics at the University of California, San Francisco, discussed preliminary data from a federally funded longitudinal study of the outcomes of 50,000 women who used fertility treatment, looking at certain physical and behavioral disorders and whether their incidence may be greater in children conceived through fertility treatment. Dolores Lamb, a scientist at Baylor College of Medicine who is one of the world's experts on male genetic infertility, reviewed the evidence that ICSI is permitting the transmission from father to son not only of infertility but potentially a host of other genetic issues, and in some cases may be magnifying them. Linda Giudice, a reproductive endocrinologist who

now practices at UCSF, called for a registry of IVF and ICSI babies. Germaine Buck Louis, a federal epidemiologist, said, "If we were looking at PCB components, metals, or phthalates, and we had a collection of evidence with the odds ratios that have been reported, those environmental agents would be moved into a category of probable reproductive or developmental consequence."

341 Thirty years ago, the scientist Richard Dawkins: Richard Dawkins, *The Selfish Gene* (Oxford: Oxford University Press, 1976).

BIBLIOGRAPHY

Books

Angier, Natalie. *Woman: An Intimate Geography.* New York: Houghton Mifflin, 1999.

Becker, Gay. *How Women and Men Approach New Reproductive Technologies.* Berkeley: University of California Press, 2000.

Blickstein, Isaac, and Louis G. Keith, eds. *Iatrogenic Multiple Pregnancy: Clinical Implications.* New York: Parthenon, 2001.

Brinsden, Peter, ed. *Textbook of In Vitro Fertilization and Assisted Reproduction: The Bourn Hall Guide to Clinical and Laboratory Practice.* London: Taylor and Francis, 2005.

Challoner, Jack. *The Baby Makers.* London: Channel 4 Books, 1999.

Dawkins, Richard. *The Selfish Gene.* Oxford: Oxford University Press, 1976.

Drexler, Peggy, PhD. *Raising Boys Without Men: How Maverick Moms Are Creating the Next Generation of Exceptional Men.* Emmaus, Pennsylvania: Rodale, 2005.

Fessler, Ann. *The Girls Who Went Away: The Hidden History of Women Who Surrendered Children for Adoption in the Decades before Roe v. Wade,* London: The Penguin Press, 2006.

Friedan, Betty. *The Feminine Mystique.* New York: W. W. Norton, 1963.

Fukuyama, Francis. *Our Posthuman Future: Consequences of the Biotechnology Revolution.* New York: Farrar, Straus and Giroux, 2002.

Gilbert, Scott F., ed. *Developmental Biology.* 7th ed. Sunderland, Mass., 2003.

Gleick, James. *Faster: The Acceleration of Just About Everything.* New York: Vintage, 1999.

Goffman, Erving. *Stigma: Notes on the Management of Spoiled Identity.* New York: Simon & Schuster, 1963.

Hall, Stephen S. *Merchants of Immortality: Chasing the Dream of Human Life Extension.* Boston: Houghton Mifflin, 2003.

Henig, Robin Marantz. *Pandora's Baby: How the First Test Tube Babies Sparked the Reproductive Revolution.* Boston: Houghton Mifflin, 2004.

Hewlett, Sylvia Ann. *Creating a Life: Professional Women and the Quest for Children.* New York: Hyperion, 2002.

Hulbert, Ann. *Raising America: Experts, Parents, and a Century of Advice About Children.* New York: Alfred A. Knopf, 2003.

Inhorn, Marcia, and Frank van Balen, eds. *Infertility Around the Globe: New Thinking on Childlessness, Gender, and Reproductive Technologies.* Berkeley: University of California Press, 2002.

Marsh, Margaret, and Wanda Ronner. *The Empty Cradle: Infertility in America from Colonial Times to the Present.* Baltimore: Johns Hopkins University Press, 1996.

McGee, Glenn. *The Perfect Baby: Parenthood in the New World of Cloning and Genetics.* Lanham, Md.: Rowman and Littlefield, 2000.

Paulson, Richard, and Judith Sachs. *Rewinding Your Biological Clock: Motherhood Late in Life.* New York: W. H. Freeman, 1998.

Plotz, David. *The Genius Factory: The Curious History of the Nobel Prize Sperm Bank.* New York: Random House, 2005.

President's Council on Bioethics. *Beyond Therapy: Biotechnology and the Pursuit of Happiness.* Report of the President's Council on Bioethics. New York: Regan Books, 2003.

Rensberger, Boyce. *Life Itself: Exploring the Realm of the Living Cell.* Oxford: Oxford University Press, 1997.

Schwartz, Barry. *The Paradox of Choice: Why More Is Less.* New York: HarperCollins, 2004.

Segal, Nancy L., Ph.D., *Entwined Lives: Twins and What They Tell Us About Human Behavior.* New York: Plume, 1999.

Spar, Debora. *The Baby Business: How Money, Science, and Politics Drive the Commerce of Conception.* Boston: Harvard Business School Press, 2006.

Stock, Gregory. *Redesigning Humans: Choosing Our Genes, Changing Our Future.* Boston: Houghton Mifflin, 2003.

Tasker, Fiona, and Susan Golombok. *Growing Up in a Lesbian Family.* New York: Guilford, 1997.

Trounson, Alan O., and Roger G. Gosden, eds. *Biology and Pathology of the Oocyte: Role in Fertility and Reproductive Medicine,* Cambridge: Cambridge University Press, 2003.

Newspaper Articles

Ahuja, Anjana. "God Is Not in Charge, We Are." *The Times (London),* July 24, 2003.

Associated Press, "Attorney Says Parents Charged in Quads Death Were Overwhelmed," July 6, 2003.

———. "Embryo-Death Suit Could End Fertility Treatments," February 10, 2005.

———. "Women Sue Lab That Promised to Tell Gender of Embryo," March 1, 2006.

Banas, Teri. "Fertility Doctor Found Guilty." United Press International, March 4, 1992.

Bellafante, Ginia. "Born Together, Raised Together, so Why Not in Classroom, Too?" *New York Times,* February 24, 2006.

———"Surrogate Mothers' New Niche: Bearing Babies for Gay Couples." *New York Times,* May 27, 2005.

Bernstein, Elizabeth. "Hearing Your Biological Clock Tick—at Age 22." *Wall Street Journal,* July 13, 2006.

Boggan, Steven. "Children Born of Sperm Donors Win Right to Take On Government." *The Independent,* July 27, 2002.

Brown, David. "South Korean Researcher Is Said to Admit Stem Cell Fakes: Team Will Ask Journal to Retract Its Report." *Washington Post,* December 16, 2005.

Crockin, Susan L. "Embryo Wars: How Do You 'Adopt' a Frozen Egg?" *Boston Globe,* December 4, 2005.

Dillon, Nancy. "Their Joy Is Fourfold." *New York Daily News,* December 28, 2004.

Eban, Katherine. "Key Work on In Vitro Fertilization Crippled by Federal Oversight." *New York Sun,* June 5, 2002.

Egan, Jennifer. "Looking for Mr. Good Sperm." *New York Times Magazine,* March 19, 2006.

Fiely, Dennis. "Registry Connects Sisters Who Shared Same Donor Dad." *Columbus (Ohio) Dispatch,* April 8, 2006.

Gandy, Kim. "Campaign Goes Too Far," *USA Today,* September 6, 2002.

Goodchild, Sophie, and Jonathan Owen. "The Baby Millionaires." *The Independent,* January 8, 2006.

Harmon, Amy. "Hello, I'm Your Sister. Our Father Is Donor 150." *New York Times,* November 20, 2005.

Hartcollis, Anemona. "New York Judges Reject Any Right to Gay Marriage." *New York Times,* July 7, 2006.

Hawkes, Nigel. "Retired Lecturer, 67, Set to Be Oldest Mother, and It's Twins." *The Times (London),* December 31, 2004.

Henderson, Mark. "Donor Children Win Right to Find Biological Parents." *The Times (London),* January 22, 2004.

Leahy, Michael. "Family Vacation." *Washington Post Magazine,* June 19, 2005.

Leland, Elizabeth. "Expecting 5." *Charlotte (N.C.) Observer,* September 8, 2002.

Lister, Sam. "How a Donor Sperm Boy Traced His Father Using the Internet." *The Times (London),* November 3, 2005.

McGinn, Daniel. "Marriage by the Numbers." *Newsweek,* June 5, 2006.

McKay, Rich. "Orlando Hospital Triples Its Pleasure Three Times Over." *Bradenton (Fla.) Herald,* May 6, 2005.

Moultrie, Dalondo. "Towamencin Couple Sentenced for Death of Quadruplet Son." *Morning Call (Allentown, Pa.),* September 30, 2004.

Mundy, Liza. "The Impossible Dream: Are Solutions to Infertility Only for the Rich?" *Washington Post Magazine,* April 20, 2003.

Murdock, Deroy. "A Solution for 'Surplus' Embryos: Adopt Them." Scripps Howard News Service, August 23, 2001.

New York Times, "Four Babies at Once? Now That's Stress." *New York Times,* November 21, 2004.

Perez, Erica, and Mayrav Saar. "California Woman in Her 50s Gives Birth to Second Child." *Orange County (Calif.) Register,* February 5, 2005.

Romano, Lois. "Multiple Single Moms, One Nameless Donor." *Washington Post,* Feb. 27, 2006.

———. "Popular Donor's Family Tree May Keep On Growing." *Washington Post,* May 1, 2006.

Saulny, Susan. "In Baby Boomlet, Preschool Derby Is the Fiercest Yet." *New York Times,* March 3, 2006.

Seelye, Katharie Q. "First Black Sextuplets Belatedly Win Public Notice." *New York Times,* January 8, 1998.

Shear, Michael D, and Rob Stein, "Woman, 55, Gives Birth to Grandchildren." *Washington Post,* December 29, 2004.

Weinstein, Henry, and Lee Romney, "Court Affirms Gay Couples' Parental Status." *Los Angeles Times,* August 23, 2005.

Weiss, Rick. "Korean Stem Cell Lines Fakes: Scientist Resigns After Probe Discredits 9 of 11 Colonies." *Washington Post,* December 23, 2005.

———. "Mature Human Embryos Cloned." *Washington Post,* February 12, 2004.

———. "Pioneering Fertility Technique Resulted in Abnormal Fetuses." *Washington Post,* May 18, 2001.

———. "Stem Cell Advance Is Fully Refuted." *Washington Post,* December 30, 2005.

———. "Stem Cell Fraud Worries U.S. Scientists." *Washington Post,* December 24, 2005.

———. "U.S. Scientist Leaves Joint Stem Cell Project: Alleged Ethical Breaches by South Korean Cited." *Washington Post,* November 12, 2005.

———. "U.S. Stem Cell Researcher Rebuked; Panel Says Schatten Sought Gains but Did Not Falsify Data," *Washington Post,* February 11, 2006.

Westphal, Sylvia Pagan. "Baby Boom: Multiple Births Persist as Doctors Buck Guide-lines: At Some Clinics, Women Get Five or More Embryos, Raising Risks to Health; One Cause is Patient Demand," *Wall Street Journal,* October 7, 2005.

Selected Magazine Articles

Brounstein, Laura. "I Get Up Every Morning and Smile." *Ladies' Home Journal,* May 2004.
Economist. "Infertility and Inheritance." U.S. Edition, June 10, 2000.
Gottlieb, Lori. "The XY Files." *Atlantic Monthly,* September 2005.
Hass, Nancy. "Whose Life Is It, Anyway?" *Elle,* September 2005.
Kalb, Claudia. "Ethics, Eggs and Embryos." *Newsweek,* June 20, 2005.
———. "Should You Have Your Baby Now?" *Newsweek,* August 13, 2001.
Kapp, Diana. "Ice, Ice Baby." *Elle,* April 2004.
Lemonick, Michael. "The Rise and Fall of the Cloning King." *Time,* January 9, 2006.
Mundy, Liza. "Souls on Ice: America's Human Embryo Glut and the Unbearable Light-ness of Almost Being." *Mother Jones,* July/August 2006.
Newman, Judith. "Triplet Epidemic: Why New Jersey Leads the Nation in Multiple Births." *National Geographic,* October 2005.
Pollitt, Katha. "The Strange Case of Baby M," *The Nation,* May 23, 1987. Available on the Web at http://www.thenation.com/doc/19870523/19870523pollitt.
Rosenberg, Debra. "The War over Fetal Rights." *Newsweek,* June 9, 2003.
Schindehette, Susan. "My Life as a Sperm Donor," *People,* June 5, 2006.
Weil, Elizabeth. "Breeder Reaction: Does Everybody Have the Right to a Baby? And Who Should Pay When Nature Alone Doesn't Work?" *Mother Jones,* July/August 2006.

Selected Online Articles, E-mail Alerts, and Websites

BioEdge, a weekly online newsletter about bioethics, "Creating Kids for Sex Selection Research," November 15, 2005, Issue 82.
BioNews, a weekly online newsletter about bioethics, "Women Sue Makers of Embryo Gender Testing Kit," March 6, 2006.
Dreaper, Jane. "IVF Donor Sperm Shortage Revealed," BBC News, September 13, 2006. http://news.bbc.co.uk/go/pr/fr/-/2/hi/health/5341982.stm.
European Society of Human Reproduction and Embryology, World Report on ART fact sheet, http://www.eshre.com/emc.asp?pageId=807.
Henderson, Mark. "'No Father Needed' Under Shake-up of UK Fertility Rules," *Times Online,* July 12, 2006. http://www.timesonline.co.uk/article/0,,200-2267114,00.html.
Horsey, Kirsty. "Cash and Sperm Shortages Threaten UK Fertility Services," commen-tary published in *BioNews,* May 30, 2006.
Hudson, Kathy, e-news bulletin on the deliberations of the International Society for Stem Cell Research about oocyte donation, http://www.dnapolicy.org/news.enews.article.nocategory.php?action=detail&newsletter_id=13&article_id=31.
Kaiser Daily Women's Health Policy e-mail briefing, a daily report on reproductive issues issued by the Henry J. Kaiser Family Foundation, "White House Press Secre-tary Retracts Statement That President Bush Believes Embryonic Stem Cell Research Is 'Murder,'" July 25, 2006.
Langlands, Eva. "Clinic Closes as IVF Crisis Hits Home," *Times Online,* June 4, 2006, www.timesonline.c.uk./article/0,,2090-2210281,00.html.

President's Council on Bioethics, drafts of recommendations for oversight of IVF technology, http://www.bioethics.gov/background/.

Redfearn, Suz. "Did Elizabeth Edwards Use Donor Eggs? All Signs Point to Yes," *Slate,* October 29, 2004. http://www.slate.com/id/2108863/.

Saletan, William. "Leave No Embryo Behind: The Coming War over In Vitro Fertilization," *Slate,* June 3, 2005. http://www.slate.com/id/2120222.

———. "Breaking Eggs," *Slate,* January 4, 2006.

Walsh, Joan. "The Baby Panic." *Salon,* April 23, 2002. http://archive.salon.com/mwt/feature/2002/04/23/hewlett_book/print.html.

Scientific Papers

Adamson, G., P. Lancaster, J. De Mouzon, K. Nygren, E. Sullivan, and F. Zegers-Hochschild. "ICMART World Report on In Vitro Fertilization 2000: How Does the United States Compare?," delivered at the 2005 ASRM annual meeting, October 18, 2005, and published as an abstract in *Fertility and Sterility* 84, no. 1 (September 2005), suppl. no. 1, p. S86.

American College of Obstetricians and Gynecologists, "Perinatal Risks Associated with Assisted Reproductive Technology," ACOG Committee Opinion No. 324, *Obstetrics and Gynecology* 106 (2005): 1143–1146.

Bachrach, Christine A., Kathy Shepherd Stolly, and Kathryn London. "Relinquishment of Premarital Births: Evidence from National Survey Data," *Family Planning Perspectives* 24, no. 1 (January–February 1992): 27–32.

Becker, Gay, Annaliese Butler, and Robert D. Nachtigall. "Resemblance Talk: A Challenge for Parents Whose Children Were Conceived with Donor Gametes in the U.S.," *Social Science and Medicine* 61 (2005): 1300–1309.

Behrman, Richard E. and Adrienne Stith Butler, eds. Committee on Understanding Premature Birth and Assuring Healthy Outcomes, Board on Health Sciences Policy, Institute of Medicine of the National Academies. "Preterm Birth: Causes, Consequence, and Prevention." Washington, D.C.: National Academies Press, 2006.

Bellis, Mark, Karen Hughes, Sara Hughes, and John R. Ashton. "Measuring Paternal Discrepancy and Its Public Health Consequences," *Journal of Epidemiology and Community Health* 59 (2005): 749–754.

Berkowitz, Richard. "From Twin to Singleton," *British Medical Journal* 313 (August 17, 1996): 373–374.

Bitler, Marianne, and Lucie Schmidt. Written summary of a talk delivered at the Health Disparities in Infertility conference convened by the National Institutes of Child Health and Human Development, Washington, D.C., March 10–11, 2005.

Bos, H. M., F. van Balen, and D. C. van den Boom. "Planned Lesbian Families: Their Desire and Motivation to Have Children," *Human Reproduction* 18, no. 10 (October 2003): 2216–2224.

Brewaeys, A., J. K. de Bruyn, F. M. Helmerhorst. "Anonymous or Identity-Registered Sperm Donors? A Study of Dutch Recipients' Choices," *Human Reproduction* 20, no. 3 (March 2005): 820–824.

Brewaeys, A., I. Ponjaert, E. V. Van Hall, S. Golombok. "Donor Insemination: Child Development and Family Functioning in Lesbian Mother Families," *Human Reproduction* 12, no. 6 (June 1997): 1349–1359.

Britt, D. W., S. T. Risinger, M. Mans, and M. I. Evans. "Anxiety Among Women Who Have Undergone Fertility Therapy and Who Are Considering Multifetal Pregnancy

Reduction: Trends and Implications," *J Fetal Neonatal Medicine* 13, no. 4 (April 2003): 271–278.

Callaghan, William, Marian MacDorman, Sonja Rasmussen, Cheng Qin, and Eve Lackritz. "The Contribution of Preterm Birth to Infant Mortality Rates in the United States," *Pediatrics* 118, no. 4 (October 2006): 1566–1573.

Carlsen, E., A. Giwercman, N. Keiding, and N. Skakkebaek. "Evidence for Decreasing Quality of Semen During Past 50 Years," *Br Med J* 305 (1992): 609–613.

Chandra, A., G. M. Martinez, W. D. Mosher, J. C. Abma, and J. Jones. "Fertility, Family Planning, and Reproductive Health of U.S. Women: Data from the 2002 National Survey of Family Growth," National Center for Health Statistics, *Vital Health Statistics* 23, no. 25.

Cook, R., S. Golombok, A. Bish, and C. Murray. "Disclosure of Donor Insemination: Parental Attitudes," *American Journal of Orthopsychiatry* 65, no. 4 (October 1995): 549–559.

Cook, R., I. Vatev, Z. Michova, and S. Golombok. "The European Study of Assisted Reproduction Families: A Comparison of Family Functioning and Child Development Between Eastern and Western Europe," *J. Psychosom. Obstet. Gynecol.* 18 (1997): 203–212.

DeBaun, Michael R., Emily L. Niemitz, and Andrew P. Feinberg. "Association of In Vitro Fertilization with Beckwith-Wiedemann Syndrome and Epigenetic Alterations of LIT1 and H19," *American Journal of Human Genetics* 72 (2003): 156–160.

Evans, M. I., M. I. Kaufman, A. J. Urban, D. W. Britt, and J. C. Fletcher. "Fetal Reduction from Twins to a Singleton: A Reasonable Consideration?" *Obstetrics and Gynecology* 104, no. 1 (July 2004): 102–109.

Faddy, Malcolm J., Sherman J. Silber, and Roger G. Gosden. "Intra-cytoplasmic Sperm Injection and Infertility," *Nature Genetics* 29 (October 2001): 131.

Fretts, Ruth C., Julie Schmittdiel, Frances McLean, Robert H. Usher, and Marlene Goldman. "Increased Maternal Age and the Risk of Fetal Death," *New England Journal of Medicine* 333, no. 15 (October 12, 1995): 953–957.

Garel, M., C. Stark, B. Blondel, G. Lefebvre, D. Vauthier-Brouzes, and J. R. Zorn, "Psychological Reactions After Multifetal Pregnancy Reduction: A 2-year Follow-up Study," *Human Reproduction* 12, no. 3 (1997): 617–622.

Golombok, S., A. Brewaeys, M. T. Giavazzi, D. Guerra, F. MacCallum, and J. Rust, "The European Study of Assisted Reproduction Families: The Transition to Adolescence," *Human Reproduction* 17, no. 3 (2002): 830–840.

———. "Families Created by the New Reproductive Technologies: Quality of Parenting and Social and Emotional Development," *Child Development* 66 (1995): 285–298.

Golombok, Susan, Clare Murray, Peter Brinsden, and Hossam Abdalla, "Social Versus Biological Parenting: Family Functioning and the Socioemotional Development of Children Conceived by Egg or Sperm Donation," *Journal Child Psychol. Psychiat.* 40, no. 4 (1999): 519–527.

Golombok, Susan, and Fiona Tasker. "Do Parents Influence the Sexual Orientation of Their Children? Findings from a Longitudinal Study of Lesbian Families," *Developmental Psychology* 32 (1996): 3–11.

Gonsalves, Joanna, Fei Sun, Peter N. Schlegel, Paul J. Turek, Carin V. Hopps, Calvin Greene, Renee H. Martin, and Renee A. Reijo Pera. "Defective Recombination in Infertile Men," *Human Molecular Genetics* 13, no. 22 (2004): 2875–2883.

Grifo, Jamie, David Hoffman, and Philip McNamee. "We Are Due for a Correction . . . and We Are Working to Achieve One," commentary in *Fertility and Sterility* 75, no. 1 (January 2002): 14.

Gurmankin, A., A. Caplan, and A. Braverman. "Screening Practices and Beliefs of Assisted Reproductive Technology Programs," *Fertility and Sterility* 83, no. 1 (January 2005): 61–67.

Guzick, David, and Shanna Swan. "The Decline of Infertility: Apparent or Real?" *Fertility and Sterility* 86, no. 3 (September 2006): 524–526.

Hansen, Michele, Jennifer Kurinczuk, Carol Bower, and Sandra Webb. "The Risk of Major Birth Defects After Intracytoplasmic Sperm Injection and In Vitro Fertilization," *New England Journal of Medicine* 346, no. 10 (March 7, 2002): 725–730.

Heffner, Linda J. "Advanced Maternal Age—How Old Is Too Old?" *New England Journal of Medicine* 351, no. 19 (November 4, 2004).

Hoffman, David I., Gail L. Zellman, C. Christine Fair, Jacob F. Mayer, Joyce G. Zeitz, William E. Gibbons, and Thomas G. Turner. "Cryopreserved Embryos in the United States and Their Availability for Research," *Fertility and Sterility* 79, no. 5 (May 2003): 1063–1069.

Jackson, Rebecca A., Kimberly A. Gibson, Yvonne W. Wu, and Mary S. Croughan, "Perinatal Outcomes in Singletons Following In Vitro Fertilization: A Meta-analysis," *Obstetrics and Gynecology* 103, no. 3 (March 2004): 551–563.

Jensen, Tina Kold, Elisabeth Carlsen, Neils Jorgensen, Jorgen G. Berthelsen, Niels Keiding, Kaare Christensen, Jorgen Holm Petersen, Lisbeth B. Knudsen, and Niels E. Skakkebaek. "Poor Semen Quality May Contribute to Recent Decline in Fertility Rates," *Human Reproduction* 17, no. 6 (2002): 1437–1440.

Klipstein, Sigal, Meredith Regan, David A. Ryley, Marlene B. Goldman, Michael M. Alper, and Richard Reindollar. "One Last Chance for Pregnancy: A Review of 2,705 In Vitro Fertilization Cycles in Women Age 40 Years and Above," *Fertility and Sterility* 84, no. 2 (August 2005): 435–445.

Klock, S. C., and D. A. Greenfeld, "Parents' Knowledge About the Donors and Their Attitudes Toward Disclosure in Oocyte Donation," *Human Reproduction* 19, no. 7 (2004): 1575–1579.

Klock, S. C., Sandra Sheinin, and Ralph R. Kazer. Letter to the Editor, "The Disposition of Unused Frozen Embryos," *New England Journal of Medicine* 345, no. 1 (July 5, 2001): 69.

Kremer, Jan A. M., Joep H. A. M. Tuerlings, George Borm, Lies H. Hoefsloot, Eric J. H. Meuleman, Didi D. M. Braat, Han G. Brunner, and Hans M. W. M. Merkus. "Does Intracytoplasmic Sperm Injection Lead to a Rise in the Frequency of Microdeletions in the *AZFc* Region of the Y Chromosome in Future Generations?" *Human Reproduction* 13, no. 10 (1998): 2808–2811.

Leridon, Henri. "Can Assisted Reproduction Technology Compensate for the Natural Decline in Fertility with Age? A Model Assessment," *Human Reproduction* 19, no. 7 (June 2004): 1553.

Levine, R. J., R. M. Mathew, C. B. Chenault, M. H. Brown, M. E. Hurtt, K. S. Bentley, K. L. Mohr, and P. K. Working. "Differences in the Quality of Semen in Outdoor Workers During Summer and Winter," *New England Journal of Medicine* 323, no. 1 (July 5, 1990): 12–16.

Lobo, Rogerio A. "Potential Options for Fertility Preservation in Women," *New England Journal of Medicine* 353, no. 1 (July 7, 2005): 64–73.

Lockwood, Gillian. "The Diagnostic Value of Inhibin in Infertility Evaluation," *Semin Reprod Med* 22, no. 3 (2004): 195–208.

Luke, B. "Nutrition and Multiple Gestation," *Semin Perinatol* 29, no. 5 (October 2005): 349–354.

Luke, B., M. B. Brown, C. Nugent, V. H. Gonzalez-Quintero, F. R. Witter, and R. B. New-man, "Risk Factors for Adverse Outcomes in Spontaneous Versus Assisted Conception Twin Pregnancies," *Fertility and Sterility* 81, no. 2 (February 2004): 315–319.

Lycett, E., K. Daniels, R. Curson, and S. Golombok. "School-aged Children of Donor Insemination: A Study of Parents' Disclosure Patterns," *Human Reproduction* 20, no. 3 (March 2005): 810–819.

MacKay, Andrea, Cynthia Berg, Jeffrey King, Catherine Duran, and Jeani Chang, "Pregnancy-Related Mortality Among Women with Multifetal Pregnancies," *Obstetrics and Gynecology* 107 (2006): 563–568.

Marlow, Neil, Dieter Wolke, Melanie Bracewell, and Muthanna Samara, "Neurologic and Developmental Disability at Six Years of Age After Extremely Preterm Birth," *New England Journal of Medicine* 352, no. 1 (January 6, 2005): 9–19.

Murray, C., and S. Golombok "Going It Alone: Solo Mothers and Their Infants Conceived by Donor Insemination," *American Journal of Orthopsychiatry* 75, no. 2 (April 2005): 242–253.

———. "To Tell or Not to Tell: The Decision-Making Process of Egg-Donation Parents," *Human Fertility* 6 (2003): 83–89.

Nachtigall, Robert, written summary of presentation on "International Comparisons of Access to Infertility Services," delivered at the Health Disparities in Infertility conference convened by the National Institute of Child Health and Human Development, Washington, D.C., March 10–11, 2005.

Nachtigall, Robert D., Gay Becker, Carrie Friese, Anneliese Butler, and Kirstin MacDougall, "Parents' Conceptualization of Their Frozen Embryos Complicates the Disposition Decision," *Fertility and Sterility* 84, no. 2 (August 2005): 431–434.

Nachtigall, Robert D., Gay Becker, and Mark Wozny, "The Effects of Gender-Specific Diagnosis on Men's and Women's Response to Infertility," *Fertility and Sterility* 57, no. 1 (January 1992): 113–121.

Niemitz, Emily L., and Andrew P. Feinberg, "Epigenetics and Assisted Reproductive Technology: A Call for Investigation," *American Journal of Human Genetics* 74 (2004): 599–609.

Oktay, Kutluk, Erkan Buyuk, Lucinda Veeck, Nikica Zaninovik, Kangpu Xu, Takumi Takeuchi, Michael Opsahl, Zev Rosenwaks. "Embryo Development After Heterotopic Transplantation of Cryopreserved Ovarian Tissue," *Lancet* 363, no. 9412 (March 13, 2004): 837–40.

Page, David C., Sherman Silber, and Laura G. Brown., "Men with Infertility Caused by AZFc Deletion Can Produce Sons by Intracytoplasmic Sperm Injection, but Are Likely to Transmit the Deletion and Infertility," *Human Reproduction* 14, no. 7 (1999): 1722–1726.

Palermo, G., H. Joris, P. Devroey, and A. Van Steirteghem. "Pregnancies After Intracytoplasmic Injection of Single Spermatozoon into an Oocyte," *Lancet* 340 (1992): 17–8.

Patterson, C. J. "Children of Lesbian and Gay Parents," *Child Development* 63, no. 5 (October 1992): 1025–1042.

Paulson, Richard J., Melvin H. Thornton, Mary M. Francis, and Herminia S. Salvador. "Successful Pregnancy in a 63-Year-Old Woman," *Fertility and Sterility* 67, no. 5 (May 1997): 949–951.

Reynolds, Meredith. Laura Schieve, Joyce Martin, Gary Jeng, and Maurizio Macaluso, "Trends in Multiple Births Conceived Using Assisted Reproductive Technology, United States, 1997–2000," *Pediatrics* 111 (2003): 1159–1162.

Romundstad, L. B., P. R. Romundstad, A. Sunde, V. von During, R. Skjaerven, and L. J. Vatten. "Increased Risk of Placenta Previa in Pregnancies Following IVF/ICSI: A

Comparison of ART and Non-ART Pregnancies in the Same Mother," *Human Reproduction* 21, no. 9 (2006): 2353–2358.

Ryan, G. L., R. A. Maassen, A. Dokras, C. H. Syrop, and B. J. Van Voorhis. "Infertile Women Are Twice as Likely to Desire Multiples as Fertile Controls," oral presentation at 2005 annual convention of the American Society for Reproductive Medicine, October 17, 2005, abstract published in *Fertility and Sterility* 84 (September 2005), Suppl. no. 1, p. S22.

Sauer, M. V., and R. J. Paulson. "Quadruplet Pregnancy in a 51-Year-Old Menopausal Woman Following Oocyte Donation," *Human Reproduction* 12 (December 8, 1993): 2243–2244.

Sauer, M. V., R. J. Paulson, and R. A. Lobo. "Pregnancy After Age 50: Application of Oocyte Donation to Women After Natural Menopause," *Lancet* 341, no. 8841 (February 1993): 321–323.

———. "A Preliminary Report on Oocyte Donation Extending Reproductive Potential to Women over 40," *New England Journal of Medicine* 323, no. 17 (October 25, 1990): 1157–1160.

Scheib, J. E., M. Riordan, and S. Rubin. "Adolescents with Open-Identity Sperm Donors: Reports from 12–17 Year Olds," *Human Reproduction* 20, no. 1 (2005): 239–252.

Schieve, Laura A., Sonja A. Rasmussen, Germaine M. Buck, Diana E. Schendel, Meredith A. Reynolds, and Victoria Wright. "Are Children Born After Assisted Reproduction at Increased Risk for Adverse Health Outcomes?" *Obstetrics and Gynecology* 103, no. 6 (June 2004): 1154–1163.

Schreiner-Engel, Patricia, Virginia Walther, Janet Mindex, Lauren Lynch, and Richard Berkowitz. "First-Trimester Multifetal Pregnancy Reduction: Acute and Persistent Psychological Reactions," *American Journal of Obstetrics and Gynecology* 172, no. 2 (February 1995): 541–547.

Shevell, Tracy, Fergal D. Malone, John Vidaver, T. Flint Porter, David A. Luthy, Christine H. Comstock, Gary D. Hankins, Keith Eddleman, Siobhan Dolan, Lorraine Dugoff, Sabrina Craigo, Ilan E. Timor, Stephen R. Carr, Honor M. Wolfe, Diana W. Bianchi, and Mary E. D'Alton. "Assisted Reproductive Technology and Pregnancy Outcome," *Obstetrics and Gynecology* 106, no. 5 (November 2005): 1039–1045.

Sheynkin, Yefim, Michael Jung, Peter Yoo, David Schulsinger, and Eugene Komaroff. "Increase in Scrotal Temperature in Laptop Computer Users," *Human Reproduction* 20, no. 2 (August 2005): 452–455.

Silber, Sherman J., and Sjoerd Repping. "Transmission of Male Infertility to Future Generations: Lessons from the Y Chromosome," *Human Reproduction Update* 8, no. 3 (2002): 217–229.

Silver, Richard I., Ronald Rodriguez, Thomas S. K. Chang, and John P. Gearhart. "In Vitro Fertilization Is Associated with an Increased Risk of Hypospadias," *Journal of Urology* 161 (June 1999): 1954–1957.

Skakkebaek, Neils E., Niels Jorgensen, Katharina M. Main, Ewa Rajpert-De Meyts, Henrik Leffers, Anna-Maria Andersson, Anders Juul, Elisabeth Carlsen, Gerda Krog Mortensen, Tina Kold Jensen, and Jorma Toppari. "Is Human Fecundity Declining?" *International Journal of Andrology* 29 (2006): 2–11.

Smith, L. Keith, Ellen H. Roots, and M. Janelle Odom Dorsett. "Live Birth of a Normal Healthy Baby After a Frozen Embryo Transfer with Blastocysts That Were Frozen and Thawed Twice," *Fertility and Sterility*, 83, no. 1 (January 2005): 198–200.

Spandorfer, S. D., K. Bendickson, K. Dragesic, G. Schattman, O. K. Davis, and Z. Rosenwaks. "IVF Outcome in Women 45 Years and Older Utilizing Autologous Oocytes: Success Is Limited to Women at 45 years of Age with a Good Response," oral presentation at the ASRM meeting, October 19, 2005, abstract published in *Fertility and Sterility* 84 (September 2005), suppl. no. 11; S139.

Stacey, J., and T. Biblarz. "(How) Does the Sexual Orientation of Parents Matter?" *American Sociological Review* 66, no. 2 (April 2001): 159–183.

Stephen, Elizabeth Hervey, and Anjani Chandra, "Use of Infertility Services in the United States: 1995," *Family Planning Perspectives* 32, no. 3 (May/June 2000): 132–137.

———. "Declining Estimates of Infertility in the United States: 1982–2002, *Fertility and Sterility* 86, no. 3 (September 2006): 516–523.

Swan, Shanna, and Irva Hertz-Picciotto. "Reasons for Infecundity," *Family Planning Perspectives* 31, no. 3 (May–June 1999): 156–157.

Tasker, F., and S. Golombok, "Adults Raised as Children in Lesbian Families," *Am J Orthopsychiatry* 65, no. 2 (April 1995): 203–215.

Turek, Paul J., and Renee A. Reijo Pera, "Current and Future Genetic Screening for Male Infertility," *Urologic Clinics of North America* 29 (2002): 767–792.

Uttley, Lois, Ronnie Pawelko, and Rabbi Dennis Ross, "Embryo Politics: Implications for Reproductive Rights and Biotechnology," a briefing paper based on a presentation at the annual meeting of the American Public Health Association, November 8, 2004, and published in June 2005 by the MergerWatch project.

Wainright, J. L., S. T. Russell, and C. J. Patterson. "Psychosocial Adjustment, School Outcomes, and Romantic Relationships of Adolescents with Same-Sex Parents," *Child Development* 75, no. 6 (November–December 2004): 1886–1898.

Selected Government and Other Reports

Marquardt, Elizabeth. "The Revolution in Parenthood: The Emerging Global Clash Between Adult Rights and Children's Needs." Commission on Parenthood's Future. New York: Institute for American Values, 2006.

Martin, J. A., B. E. Hamilton, P. D. Sutton, et al., "Births: Final Data for 2003," *National Vital Statistics Reports* 54, no.2, National Center for Health Statistics, September 8, 2005.

Medical Research Council, United Kingdom. "Assisted Reproduction: A Safe, Sound Future." 2004. Available at www.mrc.ac.uk.

President's Council on Bioethics. *Reproduction and Responsibility: The Regulation of New Technologies.* Report of the President's Council on Bioethics. Washington, D.C., March 2004. Available at www.bioethics.gov.

U.S. Department of Health and Human Services, Centers for Disease Control and Prevention. 2004 *Assisted Reproductive Technology Success Rates: National Summary and Fertility Clinic Reports,* December 2006.

Women's Health @ Stanford and the Collaborative on Health and the Environment. "Challenged Conceptions: Environmental Chemicals and Fertility," October 2005. Available at www.healthandenvironment.org.

Selected Lectures and Presentations

Becker, Gay, "Resemblance Talk: A Challenge for Parents Whose Children Were Conceived with Donor Gametes," lecture delivered during the session on "Counseling Couples About Collaborative Reproduction: The Ethical, Cultural and Psychological Dimensions Following ART," at the 2005 annual meeting of the American Society for Reproductive Medicine, October 15–16, 2005, in Montreal, Canada.

Benward, Jean, "Adoption and Gamete Donation: Similarities and Differences," lecture at collaborative reproduction session at the 2005 ASRM annual meeting, October 15–16, 2005.

Broderick, Pia. "Lessons from Down Under: Can Disclosure Really Be Legislated?," lecture at collaborative reproduction session at 2005 ASRM annual meeting, October 15–16, 2005.

Charo, R. Alta. "Frozen Dreams," lecture on frozen embryos at the collaborative reproduction session of the 2005 ASRM annual meeting, October 15–16, 2005.

Croughan, Mary. "Childhood Outcomes Following Infertility and Infertility Treatment," lecture at a conference on "Infertility Treatment and Adverse Outcomes" convened by the National Institute of Child Health and Human Development, Washington, D.C., September 12 and 13, 2005.

Dobrinski, Ina, "Germ Cell Transplantation," lecture delivered at the Frontiers in Reproduction 2005 course at the Marine Biological Laboratory, Woods Hole, Mass., June 3, 2005.

Golombok, Susan, keynote address given at "Real Families, Real Facts," a conference organized by Family Pride, an organization dedicated to gay and lesbian family-building, on May 22, 2006, in Philadelphia, Pa.

Hunt, Patricia A. "Control of Mammalian Meiosis," lecture at Frontiers in Reproduction 2005, June 2, 2006.

Keefe, David, lecture on "Genetic and Congenital Anomalies After IVF," at the 2004 ASRM annual meeting, October 16, 2004.

Lamb, Dolores, "Male Infertility Treatment and Pregnancy Outcomes," lecture at NICHD conference on infertility treatment and adverse outcomes, September 12 and 13, 2005.

Louis, Germaine Buck, "Data Sources for Assessing Fertility Treatment and Human Development," lecture at NICHD conference on infertility treatment and adverse outcomes, September 12 and 13, 2005.

Nachtigall, Robert D., lecture, "The Frozen Embryo Disposition: Why Is It So Difficult?," delivered during the collaborative reproduction session at the 2005 ASRM annual meeting, October 15–16, 2005.

Racowsky, Catherine, "Embryo Culture Variables That May Influence Pregnancy Outcome," lecture at NICDH conference on infertility treatment and adverse outcomes, September 12–13, 2005.

Schatten, Gerald, "Somatic Cell Nuclear Transfer in Non-Human Primates," lecture at Frontiers in Reproduction 2005, June 9, 2005.

Stacey, Judith, "How Does the Gender of Parents Matter?" lecture at Family Pride Conference, on May 22, 2003, in Philadelphia, Pa.

Stevens, Barry, lecture delivered in session on "Competing Rights in Third-Party Reproduction: Past, Present, and Future," at 2005 ASRM annual meeting.

Trimarchi, Jim, "ART: The Clinical and Basic Science Behind the Technology," lecture at Frontiers in Reproduction 2005, June 6, 2005.

Wells, Dagan, "Preimplantation Genetic Diagnosis: Assessing Chromosomes and Viability Markers in Human Embryos," lecture at Frontiers in Reproduction 2005, June 11, 2005.

Wolpe, Paul Root, lecture at Planned Parenthood conference, "Beyond Abortion: Critical Bioethical Issues in Reproductive Health for the 21st Century," July 23–25, 2003, in Snowbird, Utah.

INDEX